Lecture Notes Electrical Engineering

Volume 18

Stanisław Rosłoniec

Fundamental Numerical Methods for Electrical Engineering

 Springer

Prof. Dr. Hab. Ing. Stanisław Rosłoniec
Institute of Radioelectronics
Warsaw University of Technology
Nowowiejska 15/19
00-665 Warsaw
Poland
s.rosloniec@ire.pw.edu.pl

ISBN: 978-3-540-79518-6   e-ISBN: 978-3-540-79519-3

Library of Congress Control Number: 2008927874

Cover design: eStudio Calamar S.L.

Printed on acid-free paper

9 8 7 6 5 4 3 2 1

springer.com

Contents

About the Author

Stanisław Rosłoniec received his M.Sc. degree in electronic engineering from the Warsaw University of Technology, Warsaw, in 1972. After graduation he joined the Department of Electronics, (Institute of Radioelectronics), Warsaw University of Technology where in 1976 he was granted with distinction his doctor's degree (Ph.D). The thesis has been devoted to nonlinear phenomena occurring in microwave oscillators with avalanche and Gunn diodes. In 1991, he received Doctorate in Science degree in electronic engineering from the Warsaw University of Technology for a habilitation thesis on new methods of designing linear microwave circuits. Finally, he received in 2001 the degree of professor of technical science. In 1996, he was appointed as associate professor in the Warsaw University of Technology, where he lectured on *"Fundamentals of radar and radionavigation techniques"*, *"UHF and microwave antennas"*, *"Numerical methods"* and *"Methods for analysis*

and synthesis of microwave circuits". His main research interest is computer-aided design of different microwave circuits, and especially planar multi-element array antennas. He is the author of more than 80 scientific papers, 30 technical reports and 6 books, viz. *"Algorithms for design of selected linear microwave circuits"* (in Polish), WkŁ, Warsaw 1987, *"Mathematical methods for designing electronic circuits with distributed parameters"* (in Polish), WNT, Warsaw 1988, *"Algorithms for computer-aided design of linear microwave circuits"*, Artech House, Inc. Boston–London 1990, *"Linear microwave circuits – methods for analysis and synthesis"* (in Polish), WKŁ, Warsaw 1999 and *"Fundamentals of the antenna technique"* (in Polish), Publishing House of the Warsaw University of Technology, Warsaw 2006. The last of them is the present book *"Fundamental Numerical Methods for Electrical Engineering"*. Since 1992, Prof. Rosloniec has been tightly cooperating with the Telecommunications Research Institute (PIT) in Warsaw. The main subject of his professional activity in PIT is designing the planar, in-phase array antennas intended for operation in long-range three-dimensional (3D) surveillance radar stations. A few of two-dimensional (planar) array antennas designed by him operate in radars of type TRD-12, RST-12M, CAR 1100 and TRS-15. These modern radar stations have been fabricated by PIT for the Polish Army and foreign contractors.

Introduction

Stormy development of electronic computation techniques (computer systems and software), observed during the last decades, has made possible automation of data processing in many important human activity areas, such as science, technology, economics and labor organization. In a broadly understood technology area, this development led to separation of specialized forms of using computers for the design and manufacturing processes, that is:

– computer-aided design (CAD)
– computer-aided manufacture (CAM)

In order to show the role of computer in the first of the two applications mentioned above, let us consider basic stages of the design process for a standard piece of electronic system, or equipment:

– formulation of requirements concerning user properties (characteristics, parameters) of the designed equipment,
– elaboration of the initial, possibly general electric structure,
– determination of mathematical model of the system on the basis of the adopted electric structure,
– determination of basic responses (frequency- or time-domain) of the system, on the base of previously established mathematical model,
– repeated modification of the adopted diagram (changing its structure or element values) in case, when it does not satisfy the adopted requirements,
– preparation of design and technological documentation,
– manufacturing of model (prototype) series, according to the prepared documentation,
– testing the prototype under the aspect of its electric properties, mechanical durability and sensitivity to environment conditions,
– modification of prototype documentation, if necessary, and handing over the documentation to series production.

The most important stages of the process under discussion are illustrated in Fig. I.1.

Fig. I.1

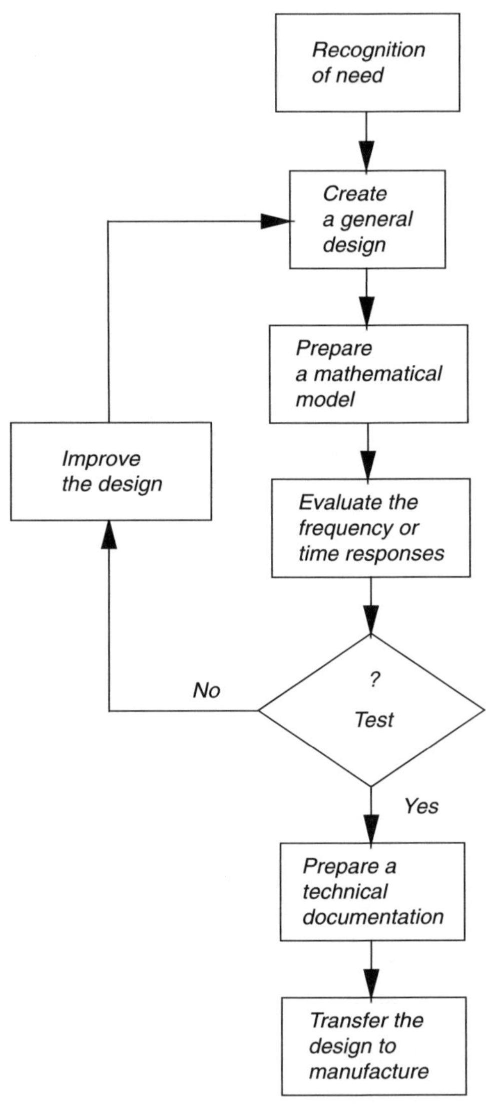

According to the diagram presented above, the design process begins with the
formulation of user requirements, which should be satisfied by the designed system
in presence of the given construction and technological limitations. Next, among
various possible solutions (electrical structures represented by corresponding structures), the ones, which best satisfy the requirements adopted at the start are chosen.
During this stage, experience (knowledge and intuition) of the designer has decisive
influence on the design process. For general solution chosen in this manner (values
of system elements can be changed), mathematical model, in the form of transfer
function, insertion losses function or state equations, is next determined. On the

base of the adopted mathematical model, frequency- or time-domain responses of the designed system are then calculated. These characteristics are analyzed during the next design stage. In case when the system fully satisfies the requirements taken at the start, it is accepted and its electric structure elaborated in this manner can be considered as the base for preparation of the construction and technological documentation. In the opposite case, the whole design cycle is repeated for changed values of elements of the adopted electrical structure. When modification of the designed system is performed with participation of the designer (manual control), the process organized in this way is called interactive design. It is also possible to modify automatically the parameters of the designed system, according to appropriate improvement criterions (goal function), which should take usually minimal or maximal values. Design process is then called optimization. During the stage of constructing mathematical model of the designed system, as well as during the stage of analysis, there is a constant need for repeated performing of basic mathematical procedures, such as:

– solving systems of linear algebraic equations,
– solving systems of nonlinear algebraic equations,
– approximation or interpolation of one or many variable functions,
– integration of one or many variable functions,
– integration of ordinary differential equations,
– integration of partial differential equations,
– solving optimization problems, the minimax problem included.

The second process mentioned above, namely the CAM, can be considered in a similar way. The author is convinced that efficient use of computer in both processes considered, requires extensive knowledge of mathematical methods for solving the problems mentioned above, known commonly under the name of numerical methods. This is, among other things the reason, why numerical methods became one of the basic courses, held in technical universities and other various kinds of schools with technical profile Considerable cognitive virtues and specific beauty of this modern area of mathematics is the fact, which should also be emphasized here.

This book was worked out as education aid for the course "Numerical Methods in Radio Electronics" lead by the author on the Faculty of Electronics and Information Technology of Warsaw University of Technology. During its elaboration, considerable emphasis was placed on the transparency and completeness of discussed issues, and presented contents constitute sufficient base for writing calculation programs in arbitrary programming language, as for example in Turbo Pascal. Each time, when it was justified for editorial reasons, vector notation of the equation systems and vector operations were deliberately abandoned, the fact that facilitates undoubtedly the understanding of methods and numerical algorithms explained in this book. Numerous examples of engineering problems taken from electronics and high-frequency technology area serve for the same purpose.

Chapter 1
Methods for Numerical Solution
of Linear Equations

As already mentioned in the Introduction, in many engineering problems there is a constant need for solving systems of linear equations. It could be said with full responsibility that solving of such equations constitutes one of the most common and important problems of the numerical mathematics [1–5]. The system of n linear equations can be written in the following expanded form:

$$\begin{aligned}
a_{11}x_1 + a_{12}x_2 + \cdots + a_{1n}x_n &= b_1 \\
a_{21}x_1 + a_{22}x_2 + \cdots + a_{2n}x_n &= b_2 \\
&\vdots \\
a_{n1}x_1 + a_{n2}x_2 + \cdots + a_{nn}x_n &= b_n
\end{aligned} \tag{1.1}$$

Using the definitions (notions) of the square matrix and the column matrix (vector), the system (1.1) can be represented by the following equivalent matrix equation:

$$\mathbf{A} \cdot \mathbf{X} = \mathbf{B} \tag{1.2}$$

where

$$\mathbf{A} = \begin{bmatrix} a_{11} & a_{12} & \cdots & a_{1n} \\ a_{21} & a_{22} & \cdots & a_{2n} \\ \cdots & \cdots & \cdots & \cdots \\ a_{n1} & a_{n2} & \cdots & a_{nn} \end{bmatrix} \quad \text{is the square matrix of coefficients}$$

$\mathbf{B} = [b_1, b_2, \ldots, b_n]^{\mathrm{T}}$ is the vector of free terms
$\mathbf{X} = [x_1, x_2, \ldots, x_n]^{\mathrm{T}}$ is the vector of variables

The transposition symbol "T" is used for the vectors $\mathbf{B} = [b_1, b_2, \ldots, b_n]^{\mathrm{T}}$ and $\mathbf{X} = [x_1, x_2, \ldots, x_n]^{\mathrm{T}}$, which are in fact column matrices. Solution of the equation system (1.1) consists in finding such values for every component of the vector of unknowns $\mathbf{X}$ that all equations of the system (1.1) are simultaneously satisfied. This assertion is legitimate only when it is assumed that such solution exists. In

S. Rosłoniec, *Fundamental Numerical Methods for Electrical Engineering*,
© Springer-Verlag Berlin Heidelberg 2008

the opposite case, the whole effort, undertaken in order to determine such solution, would be in vain. In order to avoid such undesirable conditions, we should investigate in advance the existence of a unique nontrivial solution – the task for which the analysis of the square coefficient matrix $\mathbf{A}$ and calculation of its determinant can help. The fundamental forms of square matrices and the formula used for calculating their determinants are given below for the particular case of the third-order square matrix ($n = 3$).

Symmetric matrix

$$\mathbf{A} = \begin{bmatrix} 2 & 1 & -1 \\ 1 & 3 & 2 \\ -1 & 2 & 4 \end{bmatrix}, \quad (a_{ij} = a_{ji})$$

Upper triangular matrix

$$\mathbf{U} = \begin{bmatrix} 1 & 2 & 3 \\ 0 & -1 & 1 \\ 0 & 0 & 2 \end{bmatrix}$$

Lower triangular matrix

$$\mathbf{L} = \begin{bmatrix} 1 & 0 & 0 \\ 2 & -1 & 0 \\ 4 & 0 & 2 \end{bmatrix}$$

Diagonal unitary matrix

$$\mathbf{E} = \begin{bmatrix} 1 & 0 & 0 \\ 0 & 1 & 0 \\ 0 & 0 & 1 \end{bmatrix}$$

Zero matrix

$$\mathbf{0} = \begin{bmatrix} 0 & 0 & 0 \\ 0 & 0 & 0 \\ 0 & 0 & 0 \end{bmatrix}$$

The variable $D(\det \mathbf{A})$ defined by Eq. (1.3) is called the determinant of the square matrix $\mathbf{A}$ of order n:

$$D = \begin{vmatrix} a_{11} & a_{12} & \dots & a_{1n} \\ a_{21} & a_{22} & \dots & a_{2n} \\ \dots & \dots & \dots & \dots \\ a_{n1} & a_{n2} & \dots & a_{nn} \end{vmatrix} = \sum (-1)^k a_{1\alpha} a_{2\beta} \dots a_{n\omega} \tag{1.3}$$

where the indexes $\alpha, \beta, \ldots, \omega$ denote all among the $n!$ possible permutations of the numbers $1, 2, 3, \ldots, n$, and k is the number of inversions in a particular permutation.

According to this definition, the determinant of the second-order matrix $(n = 2)$ is

$$\det \mathbf{A} = a_{11}a_{22} - a_{12}a_{21} \tag{1.4}$$

In case of the third-order matrix $(n = 3)$, we have

$$\det \mathbf{A} = a_{11}a_{22}a_{33} + a_{12}a_{23}a_{31} + a_{21}a_{32}a_{13} - a_{13}a_{22}a_{31} - a_{11}a_{23}a_{32} - a_{33}a_{12}a_{21} \tag{1.5}$$

In the general case $(n > 3)$, the calculation of the determinant of the square matrix is a cumbersome task. Therefore, as a rule, we use an indirect method based on the properties of the triangular matrix $\mathbf{T}$ (upper or lower), having the determinant equal to

$$\det \mathbf{T} = t_{11} \cdot t_{22} \cdot t_{33} \cdot \ldots \cdot t_{kk} \cdot \ldots \cdot t_{nn} \tag{1.6}$$

where t_{kk} is the kth element of the main diagonal of this matrix. Another property that can also be used for this purpose is the equality of determinants of a square matrix $\mathbf{A}$ and the equivalent triangular matrix $\mathbf{T}$, if only the rows and columns are not permuted in the process of the transformation of the matrix $\mathbf{A}$ into the matrix $\mathbf{T}$ [3, 6]. This transformation can be made by eliminating the unknowns, i.e., in the same way as in case of the elimination process described step by step in Sect. 1.1.1. According to the formula (1.6), we have $\det \mathbf{E} = 1$ and $\det \mathbf{0} = 0$. The necessary and sufficient condition for the existence of a solution of the equation system (1.1) is that the determinant D of the coefficient matrix $\mathbf{A}$ is distinct from zero. The matrix for which this condition is satisfied is called nonsingular. When $D = 0$, the equation system under consideration can have either no solution or an infinite number of solutions. This property has the following simple geometrical interpretation in the case $n = 3$. Each equation of the system (1.1) describes a plane in the three-dimensional space, as proved in Appendix A. The intersection of the two planes P_1 and P_2, shown in Fig. 1.1, represents the straight line that intersects the third plane P_3 at point S. The coordinates (x_{1s}, x_{2s}, x_{3s}) of this point represent the solution being sought.

In case when $D = 0$, some of these planes are parallel or identical. The methods used for numerical solution of the systems of linear equations can be classified as follows:

- direct (simple) methods
- iteration methods

In case of the direct methods, explicit recursive formulas are used to determine the components of the vector $\mathbf{X}$ constituting the solution, and it is not necessary to know the initial approximate solution (starting point). A classical example of such

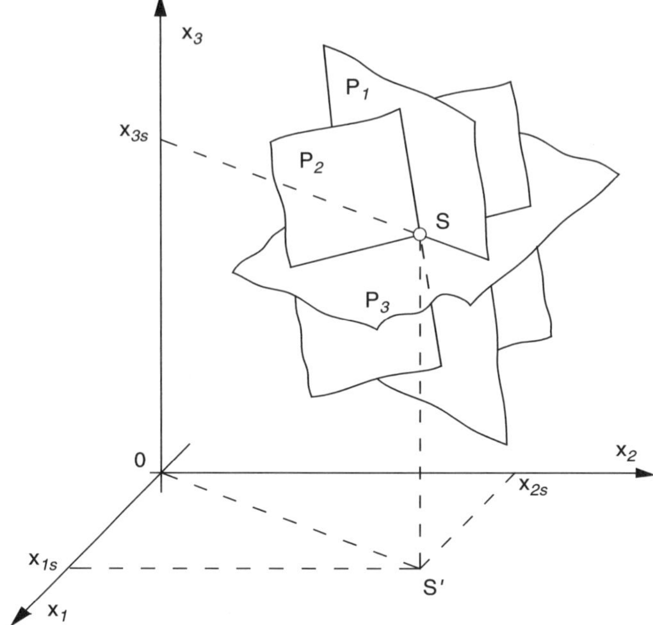

Fig. 1.1

direct method is the Cramer's rule explained below for solving a system of two equations.

$$a_{11}x_1 + a_{12}x_2 = b_1$$
$$a_{21}x_1 + a_{22}x_2 = b_2 \tag{1.7}$$

$$D = \begin{vmatrix} a_{11} & a_{12} \\ a_{21} & a_{22} \end{vmatrix} = a_{11}a_{22} - a_{12}a_{21}, \quad D_1 = \begin{vmatrix} b_1 & a_{12} \\ b_2 & a_{22} \end{vmatrix} = b_1 a_{22} - b_2 a_{21}$$

$$D_2 = \begin{vmatrix} a_{11} & b_1 \\ a_{21} & b_2 \end{vmatrix} = a_{11}b_2 - b_1 a_{21}, \quad x_1 = D_1/D, \quad x_2 = D_2/D \text{ when } D \neq 0.$$

In case of larger equation systems ($n > 2$), this rule is numerically ineffective and hence is of little practical use. The main advantages of direct methods are their simplicity and universality. The most important disadvantages are the necessity to store (in the computer memory) the whole coefficient matrix **A** during the computing process and the effect of computing error accumulation, which is specially inconvenient in case of very large equation systems, such as for $n > 100$. The effect of computing error accumulation is absent when we use the iteration methods often called the consecutive iterations methods. They are mainly used for solving the large equation systems. Unfortunately, the knowledge of an approximate solution, ensuring convergence of the computation process, is necessary to start the solving procedure. The basic methods belonging to both groups will be discussed in this chapter, see Fig. 1.2.

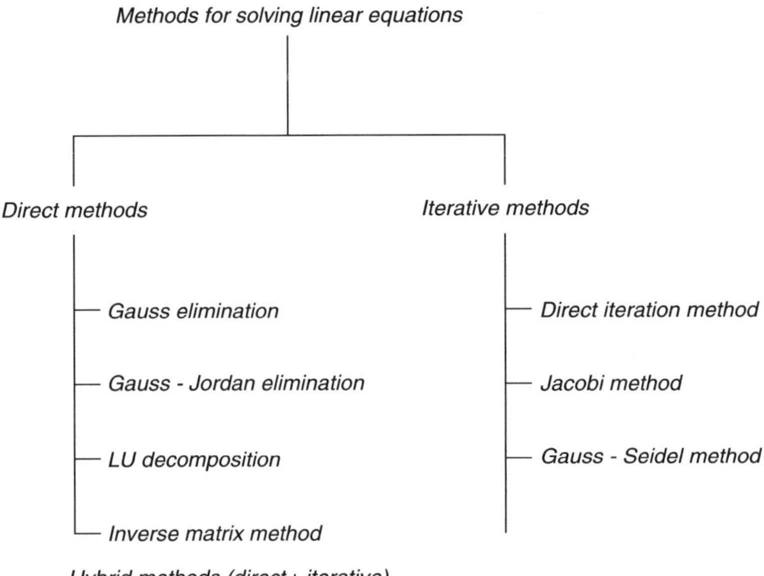

Fig. 1.2

1.1 Direct Methods

1.1.1 The Gauss Elimination Method

In order to explain the main concept of the Gauss elimination method, let us consider the following system of three equations in three unknowns:

$$
\begin{aligned}
a_{11}x_1 + a_{12}x_2 + a_{13}x_3 &= b_1 \\
a_{21}x_1 + a_{22}x_2 + a_{23}x_3 &= b_2 \\
a_{31}x_1 + a_{32}x_2 + a_{33}x_3 &= b_3
\end{aligned}
\tag{1.8}
$$

In order to eliminate the unknown variable x_1 from the second equation, we add the first equation to the second equation multiplied by $-a_{21}/a_{11}$. In a similar way, we multiply the first equation by $-a_{31}/a_{11}$ and we add it to the third equation. Now we have the following equation system:

$$
\begin{aligned}
a_{11}x_1 + a_{12}x_2 + a_{13}x_3 &= b_1 \\
a'_{22}x_2 + a'_{23}x_3 &= b'_2 \\
a'_{32}x_2 + a'_{33}x_3 &= b'_3
\end{aligned}
\tag{1.9}
$$

where $a'_{22} = a_{22} - a_{12}a_{21}/a_{11}$, $a'_{23} = a_{23} - a_{13}a_{21}/a_{11}$, $b'_2 = b_2 - b_1a_{21}/a_{11}$, $a'_{32} = a_{32} - a_{12}a_{31}/a_{11}$, $a'_{33} = a_{33} - a_{13}a_{31}/a_{11}$, and $b'_3 = b_3 - b_1a_{31}/a_{11}$

Subsequently, we can eliminate the variable x_2 from the third equation of the system (1.9). For this end, we multiply the second equation by the term $-a'_{32}/a'_{22}$ and add it to the third equation of the system. Finally, we obtain the system:

$$\begin{aligned}
a_{11}x_1 + a_{12}x_2 + a_{13}x_3 &= b_1 \\
a'_{22}x_2 + a'_{23}x_3 &= b'_2 \\
a''_{33}x_3 &= b''_3
\end{aligned} \tag{1.10}$$

where $a''_{33} = a'_{33} - a'_{23}a'_{32}/a'_{22}$ and $b''_3 = b'_3 - b'_2 a'_{32}/a'_{22}$.

This transformed equation system (1.10) is mathematically completely equivalent to the system (1.8). The coefficient matrix of this system has the triangular form, which means that the first stage of the procedure, called the elimination stage or, more colloquially, the upward movement, is completed. According to the formula (1.6), the determinant of this matrix is equal to $D = a_{11} \cdot a'_{22} \cdot a''_{33}$ and has the same value as the determinant of the coefficient matrix $\mathbf{A}$ of the equation system (1.8). In case this determinant is different from zero, the transition to the second stage of the procedure is legitimate. This second stage is called the substitution stage, or the backward movement, and begins with the determination of x_3 from the third equation and substitution of the value of x_3 obtained in this way in the second equation. After this substitution, the second equation of the system (1.10) contains only one unknown variable, namely x_2, which can be found in the elementary way. The values of x_3 and x_2 calculated in this way are then substituted in the first equation of the system (1.10), which reduces to the one-variable equation in x_1, which can be solved in an equally simple way. The second and the third equation of the system (1.9) can be interchanged in order to obtain the condition in which the coefficients in the main diagonal have the maximum absolute values. In this manner, the numerical error of the method is reduced. Additionally, the interruption of the count after the incidental occurrence of the division by zero becomes impossible. The method improved in this way is called the Gauss elimination method with the choice of the main (pivotal) element, called the pivoting strategy. For the arbitrary $n \geq 3$, the variable elimination process used in the Gauss method (stage 1) can be considered as determination of the matrix series: $\mathbf{A}^{(1)} = \mathbf{A}, \mathbf{A}^{(2)}, \mathbf{A}^{(3)}, \ldots, \mathbf{A}^{(i)}, \ldots, \mathbf{A}^{(n)}; \mathbf{B}^{(1)} = \mathbf{B}, \mathbf{B}^{(2)}, \mathbf{B}^{(3)}, \ldots, \mathbf{B}^{(i)}, \ldots, \mathbf{B}^{(n)}$, where the matrices $\mathbf{A}^{(i)}$ and $\mathbf{B}^{(i)}$ have the form

$$\mathbf{A}^{(i)} = \begin{bmatrix}
a_{11}^{(1)} & a_{12}^{(1)} & \cdots & a_{1i}^{(1)} & \cdots & a_{1n}^{(1)} \\
0 & a_{22}^{(2)} & \cdots & a_{2i}^{(2)} & \cdots & a_{2n}^{(2)} \\
\vdots & \vdots & \cdots & \vdots & \cdots & \vdots \\
\cdot & \cdot & \cdots & \cdot & \cdots & \cdot \\
\cdot & \cdot & \cdots & \cdot & \cdots & \cdot \\
0 & 0 & \cdots & a_{ii}^{(i)} & \cdots & a_{in}^{(i)} \\
\cdot & \cdot & \cdots & \cdot & \cdots & \cdot \\
\cdot & \cdot & \cdots & \cdot & \cdots & \cdot \\
\cdot & \cdot & \cdots & \cdot & \cdots & \cdot \\
0 & 0 & \cdots & a_{ni}^{(i)} & \cdots & a_{nn}^{(i)}
\end{bmatrix}
\quad
\mathbf{B}^{(i)} = \begin{bmatrix}
b_1^{(1)} \\
b_2^{(2)} \\
\cdot \\
\cdot \\
b_i^{(i)} \\
\cdot \\
\cdot \\
b_n^{(i)}
\end{bmatrix} \tag{1.11}$$

According to the procedure of elimination of the consecutive unknown variables described above, the elements of these matrices are determined by using the following expressions:

$$a_{jk}^{(i+1)} = a_{jk}^{(i)} - \frac{a_{ji}^{(i)}}{a_{ii}^{(i)}} \cdot a_{ik}^{(i)}, \quad b_j^{(i+1)} = b_j^{(i)} - \frac{a_{ji}^{(i)}}{a_{ii}^{(i)}} \cdot b_i^{(i)} \qquad (1.12)$$

where $i = 1, 2, 3, \ldots, n$; $j = i + 1, i + 2, i + 3, \ldots, n$; and $k = i + 1, i + 2, i + 3, \ldots, n$.

In a case when $a_{ii}^{(i)} \approx 0$ dividing by zero may accour, see formula (1.12). Such computational menace can be eliminated by an appropriate choice of the pivotal element, also called pivot. This protection consists in the choice, among elements $a_{ii}^{(i)}, a_{i+1.i}^{(i)}, a_{i+2.i}^{(i)} \cdots, a_{n,i}^{(i)}$ belonging to the i – column, of the element having the largest nonzero absolute value. Let the row k – of the matrix $\mathbf{A}^{(i)}$ be the row, for which

$$\left| a_{ki}^{(i)} \right| = \max \left| a_{ji}^{(i)} \right|, \quad i \le j \le n \qquad (1.13)$$

Then, the row k of the matrix $\mathbf{A}^{(i)}$ should be interchanged with the ith row. At the same time, the elements $b_i^{(i)}$ and $b_k^{(i)}$ of the column matrix (vector) $\mathbf{B}^{(i)}$ should be interchanged. In a similar way, the process of calculating the matrices $\mathbf{A}^{(i)}$ and $\mathbf{B}^{(i)}$ should be continued up to the position $i = n$. The resulting equation system, equivalent to the original system $\mathbf{A} \cdot \mathbf{X} = \mathbf{B}$, is

$$\mathbf{A}^{(n)} \cdot \mathbf{X} = \mathbf{B}^{(n)} \qquad (1.14)$$

in which the coefficient matrix $\mathbf{A}^{(n)}$ is the upper triangular matrix. In the process of finding the pivot in the ith iteration, if we obtain

$$\left| a_{ki}^{(i)} \right| = \max \left| a_{ji}^{(i)} \right| \le \varepsilon \qquad (1.15)$$

where ε is a given, sufficiently small positive number (e.g., $\varepsilon = 10^{-16}$), the whole process should be interrupted, because it means that the determinant D of the matrix of coefficients $(\mathbf{A}, \mathbf{A}^{(n)})$ is equal to zero. In the opposite case ($D \ne 0$), we should pass to the next substitution stage. The essential meaning of this uncomplicated stage was explained above using an example of the three equations in three variables. When $n \ge 3$, the particular terms of the desired solution can be found by using the following recursive formula:

$$x_i = \frac{1}{a_{ii}^{(n)}} \left[b_i^{(n)} - \sum_{j=i+1}^{n} a_{ij}^{(n)} \cdot x_j \right] \qquad (1.16)$$

where $i = n, n - 1, n - 2, \ldots, 1$.

Example 1.1 Let us solve the system of three linear equations by using the Gauss elimination method:

$$10x_1 - 7x_2 + 0x_3 = 7$$
$$-3x_1 + 2x_2 + 6x_3 = 4$$
$$5x_1 - 1x_2 + 5x_3 = 6$$

In the first step, we eliminate the variable x_1 from the second and the third equation. For this purpose, we multiply both sides of the first equation by the number 0.3 and add it to the second equation. In a similar way, we multiply the first equation by -0.5 and add it to the third equation. As a result of these operations, we obtain

$$10x_1 - 7x_2 + 0x_3 = 7$$
$$-0.1x_2 + 6x_3 = 6.1$$
$$2.5x_2 + 5x_3 = 2.5$$

In order to reduce the rounding error, the second and the third equation should be interchanged.

$$10x_1 - 7x_2 = 7$$
$$2.5x_2 + 5x_3 = 2.5$$
$$-0.1x_2 + 6x_3 = 6.1$$

Our next step is the elimination of the variable x_2 from third equation. For this end, the second equation should be multiplied by $1/25$ and then added to the third equation. After this operation, we obtain the system

$$10x_1 - 7x_2 = 7$$
$$2.5x_2 + 5x_3 = 2.5$$
$$6.2x_3 = 6.2$$

The coefficient matrix of this transformed equation system has the form of a triangular matrix. The determinant of this matrix, calculated using the formula (1.6), equals $D = 10 \cdot 2.5 \cdot 6.2 = 155$. The nonzero value of this determinant shows that the equation system under consideration has one nontrivial solution. It follows directly from the third equation that $x_3 = 1$. After substituting the value $x_3 = 1$ in the second equation, we obtain $2.5x_2 + 5 \cdot 1 = 2.5$. This equation is satisfied by $x_2 = -2.5/2.5 = -1$. Substituting $x_3 = 1$ and $x_2 = -1$ in the first equation, we obtain $x_1 = 0$. Finally, our complete solution is as follows: $x_1 = 0$, $x_2 = -1$, and $x_3 = 1$.

The formulas and the procedure for choosing the pivoting element, explained in this section, served as a base for the design of the computer program GAUSS. This program was subsequently used to solve the following system of six linear equations ($n = 6$):

$$\begin{bmatrix} 6 & -3 & 2 & 1 & -1 & 1 \\ -3 & -7 & 0 & 4 & -2 & 1 \\ 4 & -3 & 6 & -1 & 2 & 1 \\ 2 & 4 & 5 & -7 & -3 & 2 \\ -1 & 5 & -4 & 0 & 8 & -2 \\ 3 & 0 & 4 & -2 & 5 & -6 \end{bmatrix} \cdot \begin{bmatrix} x_1 \\ x_2 \\ x_3 \\ x_4 \\ x_5 \\ x_6 \end{bmatrix} = \begin{bmatrix} 11 \\ -5 \\ 28 \\ -6 \\ 25 \\ -4 \end{bmatrix}$$

During the elimination process, accompanied by an operation of choosing the pivots, this equation system is being transformed to the following equivalent form:

$$\begin{bmatrix} 6 & -3 & 2 & 1 & -1 & 1 \\ 0 & -8.5 & 1 & 4.5 & -2.5 & 1.5 \\ 0 & 0 & 4.921569 & -4.686275 & -4.137255 & 2.549020 \\ 0 & 0 & 0 & 2.135458 & 6.784861 & -2.199203 \\ 0 & 0 & 0 & 0 & 5.264925 & 0.1343285 \\ 0 & 0 & 0 & 0 & 0 & -6.612686 \end{bmatrix} \cdot \begin{bmatrix} x_1 \\ x_2 \\ x_3 \\ x_4 \\ x_5 \\ x_6 \end{bmatrix}$$

$$= \begin{bmatrix} 11 \\ 0.5 \\ -9.372549 \\ 29.270920 \\ 27.1306 \\ -39.67612 \end{bmatrix}$$

Solution of this equation system, determined during the second, substitution stage, is the vector $\mathbf{X} \equiv (x_1 = 1, x_2 = 2, x_3 = 3, x_4 = 4, x_5 = 5, x_6 = 6)$.

1.1.2 The Gauss–Jordan Elimination Method

In order to explain the essence of the Gauss–Jordan method, let us reconsider the system of three linear equations in three unknowns $\mathbf{I}_1$, $\mathbf{I}_2$, and $\mathbf{I}_3$, interpreted here as the complex amplitudes of the currents in a three-mesh electrical circuit.

$$\begin{aligned} R_{11}\mathbf{I}_1 + R_{12}\mathbf{I}_2 + R_{13}\mathbf{I}_3 &= \mathbf{V}_1 \\ R_{21}\mathbf{I}_1 + R_{22}\mathbf{I}_2 + R_{23}\mathbf{I}_3 &= \mathbf{V}_2 \\ R_{31}\mathbf{I}_1 + R_{32}\mathbf{I}_2 + R_{33}\mathbf{I}_3 &= \mathbf{V}_3 \end{aligned} \tag{1.17}$$

It is obvious that multiplying both sides of each of the equations (1.17) by a constant and summing them up does not change the values of the unknown currents $\mathbf{I}_1$, $\mathbf{I}_2$, and $\mathbf{I}_3$. Hence, by using this operation repeatedly, in order to eliminate some of the unknowns, it is possible to transform the equation system (1.17) to the following form:

$$1 \cdot \mathbf{I}_1 + 0 \cdot \mathbf{I}_2 + 0 \cdot \mathbf{I}_3 = C_1$$
$$0 \cdot \mathbf{I}_1 + 1 \cdot \mathbf{I}_2 + 0 \cdot \mathbf{I}_3 = C_2 \qquad (1.18)$$
$$0 \cdot \mathbf{I}_1 + 0 \cdot \mathbf{I}_2 + 1 \cdot \mathbf{I}_3 = C_3$$

in which the transformed matrix of coefficients $\mathbf{A}$ is the diagonal unitary matrix. It follows directly from the equations (1.18) that $\mathbf{I}_1 = C_1$, $\mathbf{I}_2 = C_2$, and $\mathbf{I}_3 = C_3$. One of the procedures serving to eliminate some unknowns from the particular equations was demonstrated using Example 1.2.

Example 1.2 Assume that the equation system (1.17) has the same coefficients, as the system analyzed in Example 1.1, i.e.,

$$10\mathbf{I}_1 - 7\mathbf{I}_2 = 7$$
$$-3\mathbf{I}_1 + 2\mathbf{I}_2 + 6\mathbf{I}_3 = 4$$
$$5\mathbf{I}_1 - \mathbf{I}_2 + 5\mathbf{I}_3 = 6$$

Using the transformations shown in Example 1.1, this equation system can be written in the following equivalent triangular form:

$$10\mathbf{I}_1 - 7\mathbf{I}_2 = 7$$
$$2.5\mathbf{I}_2 + 5\mathbf{I}_3 = 2.5$$
$$6.2\mathbf{I}_3 = 6.2$$

Dividing all three equations by their diagonal coefficients, we obtain

$$\mathbf{I}_1 - 0.7\mathbf{I}_2 = 0.7$$
$$\mathbf{I}_2 + 2\mathbf{I}_3 = 1$$
$$\mathbf{I}_3 = 1$$

Now we shall eliminate the variable $\mathbf{I}_2$ from the first equation. To do this, we may add to it the second equation multiplied by 0.7. Resulting equation system has the form

$$\mathbf{I}_1 + 0 \cdot \mathbf{I}_2 + 1.4\mathbf{I}_3 = 1.4$$
$$\mathbf{I}_2 + 2\mathbf{I}_3 = 1$$
$$\mathbf{I}_3 = 1$$

Next we shall eliminate the variable $\mathbf{I}_3$ from the first and the second equation. It can be done by multiplying the third equation by the constant -1.4 and adding it to the first equation. Similarly, the third equation should be multiplied by -2 and added to the second equation. Finally, we obtain the following system of equations:

$$\mathbf{I}_1 + 0 \cdot \mathbf{I}_2 + 0 \cdot \mathbf{I}_3 = 0$$
$$\mathbf{I}_2 + 0 \cdot \mathbf{I}_3 = -1$$
$$\mathbf{I}_3 = 1$$

for which the matrix of coefficients $\mathbf{A}$ is diagonal and unitary and has the solution $\mathbf{I}_1 = 0, \mathbf{I}_2 = -1$, and $\mathbf{I}_3 = 1$. During the transformation of the equation system (1.17) to the form (1.18), the vector of currents $\mathbf{I}$ remains unchanged and the operation is made with respect to the matrix of coefficients and vector of voltages. Creation of the so-called augmented matrix of order $n \times (n+1)$ therefore proves very useful. In case of the system of three equations ($n = 3$) discussed in this section, the augmented matrix has the form:

$$\mathbf{R} = \begin{bmatrix} R \vdots V \end{bmatrix} = \begin{bmatrix} R_{11} & R_{12} & R_{13} & V_1 \\ R_{21} & R_{22} & R_{23} & V_2 \\ R_{31} & R_{23} & R_{33} & V_3 \end{bmatrix} \qquad (1.19)$$

Matrix $\mathbf{R}$ may be transformed through the full elimination process, after which it takes the form

$$\begin{bmatrix} 1 & 0 & 0 & I_1 \\ 0 & 1 & 0 & I_2 \\ 0 & 0 & 1 & I_3 \end{bmatrix} \qquad (1.20)$$

For this purpose, the computation algorithm given in the literature can be used [8, 9].

1.1.3 The LU Matrix Decomposition Method

Let us now consider the task of solving repeatedly the system of linear equations

$$\mathbf{A} \cdot \mathbf{X} = \mathbf{B} \qquad (1.21)$$

each time for the same matrix of coefficients $\mathbf{A}$, but for different excitation vectors $\mathbf{B}$. The Gauss and Gauss–Jordan elimination methods are not effective for solving this particular problem, because the repeated transformation of the matrix $\mathbf{A}$ and vector $\mathbf{B}$ is needed even though the matrix $\mathbf{A}$ remains the same always. In such case, one of the LU decomposition methods, as for example the Crout method [7, 8], may prove to be more convenient. In this last method, decomposition of the nonsingular matrix $\mathbf{A}$ of the order n into the product $\mathbf{A} = \mathbf{L} \cdot \mathbf{U}$ of the two triangular matrices (lower $\mathbf{L}$ and upper $\mathbf{U}$) is used. Structures of the two matrices $\mathbf{L}$ and $\mathbf{U}$ of this product are described by the following general relation:

$$
\begin{bmatrix}
a_{11} & a_{12} & a_{13} & \cdots & a_{1n} \\
a_{21} & a_{22} & a_{23} & \cdots & a_{2n} \\
a_{31} & a_{32} & a_{33} & \cdots & a_{3n} \\
\cdot & \cdot & \cdot & \cdots & \cdot \\
a_{n-1,1} & a_{n-1,2} & a_{n-1,3} & \cdots & a_{n-1,n} \\
a_{n1} & a_{n2} & a_{n3} & \cdots & a_{nn}
\end{bmatrix}
$$

$$
=
\begin{bmatrix}
l_{11} & 0 & 0 & \cdots & 0 \\
l_{21} & l_{22} & 0 & \cdots & 0 \\
l_{31} & l_{32} & l_{33} & \cdots & 0 \\
\cdot & \cdot & \cdot & \cdots & \cdot \\
l_{n-1,1} & l_{n-1,2} & l_{n-1,3} & \cdots & 0 \\
l_{n1} & l_{n2} & l_{n3} & \cdots & l_{nn}
\end{bmatrix}
\cdot
\begin{bmatrix}
1 & u_{12} & u_{13} & \cdots & u_{1n} \\
0 & 1 & u_{23} & \cdots & u_{2n} \\
0 & 0 & 1 & \cdots & u_{3n} \\
\cdot & \cdot & \cdot & \cdots & \cdot \\
0 & 0 & 0 & \cdots & u_{n-1,n} \\
0 & 0 & 0 & \cdots & 1
\end{bmatrix}
$$

The equations of the system (1.21), which we want to solve, should be written in such an order that the diagonal elements a_{ii} of the coefficient matrix $\mathbf{A}$ are different from zero and possibly have the greatest absolute values. Then the diagonal elements l_{ii} of the lower triangular matrix $\mathbf{L}$ will also be different from zero. Substituting the relation $\mathbf{A} = \mathbf{L} \cdot \mathbf{U}$ in Eq. (1.21), we obtain

$$\mathbf{L} \cdot \mathbf{U} \cdot \mathbf{X} = \mathbf{B} \tag{1.22}$$

Assume initially that the triangular matrices $\mathbf{L}$ and $\mathbf{U}$ are known. In consequence, solving the equation system (1.22) with respect to the column vector $\mathbf{X}$ can be performed in the two simple stages. In the first stage, from the equation

$$\mathbf{L} \cdot \mathbf{D} = \mathbf{B} \tag{1.23}$$

we determine the vector $\mathbf{D}$. According to Eq. (1.22), this vector satisfies the equation

$$\mathbf{U} \cdot \mathbf{X} = \mathbf{D} \tag{1.24}$$

involving also the desired solution. The second stage therefore consists in determining the vector $\mathbf{X}$ from Eq. (1.24). Both stages of the solution process mentioned above can be performed in a relatively simple way, thanks to the triangular form of the $\mathbf{L}$ and $\mathbf{U}$ matrices. For example, in case of three equations, the system (1.23) takes the form

$$
\begin{aligned}
l_{11}d_1 &= b_1 \\
l_{21}d_1 + l_{22}d_2 &= b_2 \\
l_{31}d_1 + l_{32}d_2 + l_{33}d_3 &= b_3
\end{aligned}
\tag{1.25}
$$

and its solution with respect to the vector $\mathbf{D} \equiv [d_1, d_2, d_3]^{\mathrm{T}}$ may be obtained without serious difficulties. In the general case ($n > 3$), the components d_i of the auxiliary vector $\mathbf{D}$ can be found by using the following recursive formula

$$d_1 = \frac{b_1}{l_{11}}$$

$$d_k = \frac{1}{l_{kk}} \cdot \left[b_k - \sum_{i=1}^{k-1} l_{ki} \cdot d_i \right], \quad k = 2, 3, \ldots, n \tag{1.26}$$

When the column vector $\mathbf{D}$ is known we can solve the matrix equation (1.24) which for $n = 3$ takes the following form.

$$\begin{aligned}
1x_1 + u_{12}x_2 + u_{13}x_3 &= d_1 \\
1x_2 + u_{23}x_3 &= d_2 \\
1x_3 &= d_3
\end{aligned} \tag{1.27}$$

The solution $\mathbf{X} \equiv [x_1, x_2, x_3]^{\mathrm{T}}$ of these equations can be found in a similarly uncomplicated way, i.e., using the method of consecutive substitutions.

For an arbitrary $n > 3$, the matrix equation (1.24) has the form

$$\begin{bmatrix}
1 & u_{12} & u_{13} & \cdots & u_{1n} \\
0 & 1 & u_{23} & \cdots & u_{2n} \\
0 & 0 & 1 & \cdots & u_{3n} \\
\cdot & \cdot & \cdot & \cdots & \cdot \\
0 & 0 & 0 & \cdots & u_{n-1,n} \\
0 & 0 & 0 & \cdots & 1
\end{bmatrix} \cdot \begin{bmatrix}
x_1 \\ x_2 \\ x_3 \\ \cdot \\ x_{n-1} \\ x_n
\end{bmatrix} = \begin{bmatrix}
d_1 \\ d_2 \\ d_3 \\ \cdot \\ d_{n-1} \\ d_n
\end{bmatrix} \tag{1.28}$$

In order to find the solution vector $\mathbf{X} \equiv [x_1, x_2, x_3, \ldots, x_n]^{\mathrm{T}}$, the method of consecutive substitutions should be applied. It is defined this time by the following recursive computation formula

$$x_j = d_j - \sum_{k=j+1}^{n} u_{jk} \cdot x_k \tag{1.29}$$

where $j = n, n - 1, n - 2, \ldots, 1$.

According to Eqs. (1.23) and (1.24), after substituting the new vector $\mathbf{B}$, we need to determine only the new vector $\mathbf{D}$ and next we must calculate the vector $\mathbf{X}$, which is the desired solution to our problem. The matrices $\mathbf{L}$ and $\mathbf{U}$ need not be reprocessed and this fact diminishes essentially the amount of calculations. It is due to the fact that these matrices were assumed to be known. In the general case, they can be determined using the following recursive relations:

$$l_{i1} = a_{i1}$$

$$l_{ij} = a_{ij} - \sum_{k=1}^{j-1} l_{ik} \cdot u_{kj} \qquad \text{for } i \geq j > 1$$

$$u_{1j} = \frac{a_{1j}}{l_{11}}$$

$$u_{ij} = \frac{1}{l_{ii}} \left(a_{ij} - \sum_{k=1}^{i-1} l_{ik} \cdot u_{kj} \right) \qquad \text{for } 1 < i < j$$

$$(1.30)$$

which are in the literature often referred to as the Doolittle formulas [9, 10]. The term a_{ij}, where $1 \leq i \leq n$, $1 \leq j \leq n$, appearing in these relations, is the element of a given nonsingular matrix of coefficients **A**. These relations are developed in Appendix D, using the standard rule of multiplying two matrices of the same order.

Example 1.3 Consider the following system of equations

$$\begin{bmatrix} 1 & 2 & 3 & 4 \\ 2 & 11 & 20 & 29 \\ 3 & 8 & 16 & 24 \\ 4 & 14 & 25 & 40 \end{bmatrix} \cdot \begin{bmatrix} x_1 \\ x_2 \\ x_3 \\ x_4 \end{bmatrix} = \begin{bmatrix} 4.0 \\ 20.6 \\ 17.4 \\ 27.8 \end{bmatrix}$$

The triangular matrices **L** and **U** determined using the relations (1.30) are equal to

$$\mathbf{L} = \begin{bmatrix} 1 & 0 & 0 & 0 \\ 2 & 7 & 0 & 0 \\ 3 & 2 & 3 & 0 \\ 4 & 6 & 1 & 4 \end{bmatrix} \qquad \mathbf{U} = \begin{bmatrix} 1 & 2 & 3 & 4 \\ 0 & 1 & 2 & 3 \\ 0 & 0 & 1 & 2 \\ 0 & 0 & 0 & 1 \end{bmatrix}$$

The determinant of the matrix of coefficients satisfies the equation $\det \mathbf{A} = \det \mathbf{L} \cdot \det \mathbf{U} = 84 \cdot 1 = 84$. The solution obtained by using the LU decomposition method is $x_1 = 1.0$, $x_2 = 0.7$, $x_3 = 0.4$, and $x_4 = 0.1$.

1.1.4 The Method of Inverse Matrix

The method of inverse matrix also finds an application for the task of repeatedly solving the system of linear equations

$$\mathbf{A} \cdot \mathbf{X} = \mathbf{B} \qquad (1.31)$$

for which the matrix of coefficients $\mathbf{A}$ remains unchanged. In other words, the equation system is being solved for different values of the free terms forming the vector $\mathbf{B}$.

As we know from the extensive literature on the subject, application of the above method is legitimate, if the number of solution processes applied to the equation system (1.31) is greater than $2n$, where n is the rank of the matrix $\mathbf{A}$, equal to the number of equations in the system. The inverse $\mathbf{A}^{-1}$ of a nonsingular square matrix $\mathbf{A}$ (having the determinant D different from zero) is also the nonsingular square matrix of the same rank. Product of these matrices, i.e.,

$$\mathbf{A}^{-1} \cdot \mathbf{A} = \mathbf{A} \cdot \mathbf{A}^{-1} = \mathbf{E} \tag{1.32}$$

is equal to the unitary matrix $\mathbf{E}$, having also the same rank. The equation system (1.31) will remain unchanged after multiplication of both sides by an inverse matrix, i.e.,

$$\mathbf{A}^{-1} \cdot \mathbf{A} \cdot \mathbf{X} = \mathbf{A}^{-1} \cdot \mathbf{B} \tag{1.33}$$

Substituting relation (1.32) in the expression (1.33), we obtain

$$\mathbf{E} \cdot \mathbf{X} = \mathbf{A}^{-1} \cdot \mathbf{B} \tag{1.34}$$

The product of the unitary matrix $\mathbf{E}$ of the rank n by a column vector $\mathbf{X}$ with n elements is identical to the vector $\mathbf{X}$. Due to this property, Eq. (1.34) can be written as

$$\mathbf{X} = \mathbf{A}^{-1} \cdot \mathbf{B} \tag{1.35}$$

expressing the essence of the method under consideration. It follows from the above equation that the solution vector $\mathbf{X}$ can be found by simple multiplication of the inverse matrix $\mathbf{A}^{-1}$ by the vector of free terms $\mathbf{B}$. Determination of the inverse matrix $\mathbf{A}^{-1}$ therefore constitutes an essential and most difficult problem, which must be solved in the first stage. Different algorithms available in the literature on linear algebra can be used for this purpose. In case of the matrix of a small rank ($n \leq 3$), the relations given in Appendix B may prove to be useful. One of the most popular algorithms used for calculating the inverse matrix is presented below. Assume that a square nonsingular matrix $\mathbf{A}$ is given. Denote the elements of this matrix by a_{ij}, where $1 \leq i \leq n, 1 \leq j \leq n$. Elements (terms) of the inverse matrix $\mathbf{A}^{-1}$ are denoted by x_{ij}, where $1 \leq i \leq n, 1 \leq j \leq n$. Product of this two matrices, i.e.,

$$\mathbf{A} \cdot \mathbf{A}^{-1} = \mathbf{E} \tag{1.36}$$

can be presented in the following equivalent form:

$$\sum_{k=1}^{n} a_{ik}x_{kj} = \delta_{ij} \tag{1.37}$$

where δ_{ij} is the Kronecker symbol taking the value 1 for $i = j$ and the value 0 for $i \neq j$. It follows from Eq. (1.37) that, if we want to determine elements of the column j of the matrix $\mathbf{A}^{-1}$, the following system of equations should be solved:

$$
\begin{aligned}
a_{11}x_{1j} + a_{12}x_{2j} + \ldots + a_{1n}x_{nj} &= 0 \\
a_{21}x_{1j} + a_{22}x_{2j} + \ldots + a_{2n}x_{nj} &= 0 \\
\cdots\cdots\cdots\cdots\cdots\cdots\cdots\cdots\cdots\cdots\cdots\cdots \\
a_{j1}x_{1j} + a_{j2}x_{2j} + \ldots + a_{jn}x_{nj} &= 1 \\
\cdots\cdots\cdots\cdots\cdots\cdots\cdots\cdots\cdots\cdots\cdots\cdots \\
a_{n1}x_{1j} + a_{n2}x_{2j} + \ldots + a_{nn}x_{nj} &= 0
\end{aligned}
\tag{1.38}
$$

In order to find all elements of the matrix $\mathbf{A}^{-1}$, the equation system (1.38) should be solved n times, namely for $j = 1, 2, 3, \ldots, n$. The matrix of coefficients $\mathbf{A}$ of this system remains unchanged, and therefore, it can be effectively solved by using the LU decomposition method described in the previous section. The product (1.32) can be used to evaluate precision obtained for the calculated inverse matrix $\mathbf{A}^{-1}$. In case this precision is not satisfactory, the main equation system can be solved once more, this time by using the relation (1.35).

Example 1.4 Solve the following system of equations using the method of inverse matrix

$$
\begin{bmatrix} 1 & -2 & 3 \\ -1 & 1 & 2 \\ 2 & -1 & -1 \end{bmatrix} \cdot \begin{bmatrix} x_1 \\ x_2 \\ x_3 \end{bmatrix} = \begin{bmatrix} 12 \\ 8 \\ 4 \end{bmatrix}
$$

The inverse $\mathbf{A}^{-1}$ of the coefficients matrix $\mathbf{A}$ of the system given below is equal to (see Appendix B)

$$
\mathbf{A}^{-1} = \frac{1}{8} \begin{bmatrix} -1 & 5 & 7 \\ -3 & 7 & 5 \\ 1 & 3 & 1 \end{bmatrix}
$$

According to the relation (1.35), we have

$$
\begin{bmatrix} x_1 \\ x_2 \\ x_3 \end{bmatrix} = \frac{1}{8} \begin{bmatrix} -1 & 5 & 7 \\ -3 & 7 & 5 \\ 1 & 3 & 1 \end{bmatrix} \cdot \begin{bmatrix} 12 \\ 8 \\ 4 \end{bmatrix} = \begin{bmatrix} 7 \\ 5 \\ 5 \end{bmatrix}
$$

Finally, we find our solution: $x_1 = 7$, $x_2 = 5$, and $x_3 = 5$.

1.2 Indirect or Iterative Methods

1.2.1 The Direct Iteration Method

In this section, we consider the direct iteration method, the first one belonging to the class of iterative methods. Let us consider the system of n linear equations:

$$
\begin{aligned}
a_{11}x_1 + a_{12}x_2 + \ldots + a_{1n}x_n &= b_1 \\
a_{21}x_1 + a_{22}x_2 + \ldots + a_{2n}x_n &= b_2 \\
&\cdots\cdots\cdots \\
a_{n1}x_1 + a_{n2}x_2 + \ldots + a_{nn}x_n &= b_n
\end{aligned}
\tag{1.39}
$$

Assume that the approximate solution $[x_1^{(0)}, x_2^{(0)}, x_3^{(0)}, \ldots, x_n^{(0)}]$ was previously found by using one of the direct methods described earlier in this chapter. After substituting this approximate solution in the equation system (1.39), we obtain

$$
\begin{aligned}
a_{11}x_1^{(0)} + a_{12}x_2^{(0)} + \cdots + a_{1n}x_n^{(0)} &= b_1^{(0)} \\
a_{21}x_1^{(0)} + a_{22}x_2^{(0)} + \cdots + a_{2n}x_n^{(0)} &= b_2^{(0)} \\
&\cdots\cdots\cdots \\
a_{n1}x_1^{(0)} + a_{n2}x_2^{(0)} + \cdots + a_{nn}x_n^{(0)} &= b_n^{(0)}
\end{aligned}
\tag{1.40}
$$

Let us now introduce the corrections determined with respect to the final solution

$\mathbf{X} = [x_1, x_2, x_3, \ldots, x_n]^{\mathrm{T}}$ and to the vector $\mathbf{B} = [b_1, b_2, b_3, \ldots, b_n]^{\mathrm{T}}$, i.e.,
$\varepsilon_i^{(0)} = x_i - x_i^{(0)}$ for $i = 1, 2, 3, \ldots, n$
$r_i^{(0)} = b_i - b_i^{(0)}$ for $i = 1, 2, 3, \ldots, n$

By subtracting the equation system (1.40) from (1.39), we obtain the system of n linear equations in which the unknowns form the appropriate correction vector:

$$
\begin{aligned}
a_{11}\varepsilon_1^{(0)} + a_{12}\varepsilon_2^{(0)} + \ldots + a_{1n}\varepsilon_n^{(0)} &= r_1^{(0)} \\
a_{21}\varepsilon_1^{(0)} + a_{22}\varepsilon_2^{(0)} + \ldots + a_{2n}\varepsilon_n^{(0)} &= r_2^{(0)} \\
&\cdots\cdots\cdots \\
a_{n1}\varepsilon_1^{(0)} + a_{n2}\varepsilon_2^{(0)} + \ldots + a_{nn}\varepsilon_n^{(0)} &= r_n^{(0)}
\end{aligned}
\tag{1.41}
$$

Solving the system (1.41) with respect to the corrections $\varepsilon_1^{(0)}, \varepsilon_2^{(0)}, \ldots, \varepsilon_n^{(0)}$, we obtain the second, more accurate approximation of the desired solution, i.e.,

$$
\begin{aligned}
x_1^{(1)} &= x_1^{(0)} + \varepsilon_1^{(0)} \\
x_2^{(1)} &= x_2^{(0)} + \varepsilon_2^{(0)} \\
&\cdots\cdots \\
x_n^{(1)} &= x_n^{(0)} + \varepsilon_n^{(0)}
\end{aligned}
\tag{1.42}
$$

Repeating the process described above several times, one can obtain such accuracy that two solutions obtained in the two consecutive iterations will differ negligibly. It means that the vector $[r_1^{(k)}, r_2^{(k)}, r_3^{(k)}, \ldots, r_n^{(k)}]$ will approach the zero vector $[0, 0, 0, \ldots, 0]$. We shall underline here the fact that in the process of solving the equation systems, similar to the one described by Eq. (1.41), with respect to the consecutive corrections, the matrix of coefficients $\mathbf{A}$ remains unchanged and only the column vector $[r_1, r_2, \ldots, r_n]$ varies from one consecutive solution to the next. Hence, application of the LU decomposition method appears to be useful in case of such equation systems.

1.2.2 Jacobi and Gauss–Seidel Methods

Let us consider the following system of n linear equations in n unknowns, for which the coefficient matrix $\mathbf{A}$ is nonsingular.

$$\begin{aligned}
1 \cdot x_1 + a_{12}x_2 + a_{13}x_3 + \ldots + a_{1n}x_n &= b_1 \\
a_{21}x_1 + 1 \cdot x_2 + a_{23}x_3 + \ldots + a_{2n}x_n &= b_2 \\
\cdots\cdots\cdots\cdots\cdots\cdots\cdots\cdots\cdots\cdots\cdots \\
a_{n1}x_1 + a_{n2}x_2 + a_{n3}x_3 + \ldots + 1 \cdot x_n &= b_n
\end{aligned} \tag{1.43}$$

Now assume that the initial approximation of the desired solution $[x_1^{(0)}, x_2^{(0)}, x_3^{(0)}, \ldots, x_n^{(0)}]$ is known. The majority of linear equation systems, formulated in connection with various engineering problems, can be transformed into the canonical form (1.43) simply by interchanging individual equations and dividing each of them by its respective diagonal coefficient. The equations should be arranged in such a way that the nonzero coefficients having the largest modules (absolute values) occupy the main diagonal.

1.2.2.1 The Jacobi Method

The matrix of coefficients $\mathbf{A}$ of the equation system (1.43) can be expressed as the sum of three square matrices of the same rank, i.e.,

$$\mathbf{A} = \mathbf{L} + \mathbf{E} + \mathbf{U} \tag{1.44}$$

where $\mathbf{L}$ is the lower diagonal matrix, $\mathbf{E}$ the diagonal unitary matrix, and $\mathbf{U}$ an upper diagonal matrix. After substituting relation (1.44) in Eq. (1.43), we obtain

$$\mathbf{A} \cdot \mathbf{X} = (\mathbf{L} + \mathbf{E} + \mathbf{U}) \cdot \mathbf{X} = \mathbf{B} \tag{1.45}$$

Eq. (1.45) can be used to obtain directly iterative Jacobi formula.

$$\mathbf{E} \cdot \mathbf{X} = -(\mathbf{L} + \mathbf{U}) \cdot \mathbf{X} + \mathbf{B} \tag{1.46}$$

which after some elementary matrix transformations can be written as

$$\mathbf{X} = -(\mathbf{L} + \mathbf{U}) \cdot \mathbf{X} + \mathbf{B}$$

This formula can often be found in the literature in the different, but equivalent form:

$$\mathbf{X} = \mathbf{C} \cdot \mathbf{X} + \mathbf{B} \tag{1.47}$$

where $\mathbf{C} = -(\mathbf{L} + \mathbf{U}) = \mathbf{E} - \mathbf{A}$ is the Jacobi matrix. According to this formula, elements c_{ij} of the matrix $\mathbf{C}$ are equal to

$$c_{ij} = \begin{cases} -a_{ij} & \text{for } i \neq j, \quad i = 1, 2, 3, \ldots, n, \quad j = 1, 2, 3, \ldots, n \\ 1 - a_{ij} & \text{for } i = j \end{cases}$$

Consecutive $(k + 1)$ approximation of an exact solution can be calculated from the formula (1.47), based on the approximation obtained from the previous iteration, k:

$$\mathbf{X}^{(k+1)} = \mathbf{C} \cdot \mathbf{X}^{(k)} + \mathbf{B} \tag{1.48}$$

where $k = 0, 1, 2, 3, \ldots$. The series of consecutive iterations obtained in this way is convergent to an exact solution, if the coefficient matrix $\mathbf{A}$ is strictly diagonally dominant or strict column diagonally dominant [3, 7].

1.2.2.2 Supplement

The square nonsingular matrix $\mathbf{A}$ is called diagonally dominant if the sum of the absolute values of its elements on the main diagonal is greater than or equal to the sum of the absolute values of the remaining elements of the analyzed row of the matrix, i.e.,

$$|a_{ii}| \geq \sum_{\substack{k=1 \\ k \neq i}}^{n} |a_{ik}|$$

where $i = 1, 2, 3, \ldots, n$. The matrix $\mathbf{A}$ is strictly diagonally dominant if all above inequalities are strict. The square matrix $\mathbf{A}$ is column diagonally dominant if its transpose $\mathbf{A}^{\mathrm{T}}$ is diagonally dominant.

Table 1.1

↓ Iteration	x_1	x_2	x_3	x_4
0	1.000000000	1.000000000	1.000000000	1.000000000
1	2.500000000	1.050000000	2.200000000	3.050000000
2	1.585000000	−0.495000000	1.070000000	1.992500000
3	2.059750000	0.449250000	1.985000000	3.063750000
4	1.592475000	−0.246750000	1.272700000	2.322187500
5	1.954811250	0.316201250	1.801980000	2.922920000
…	…	…	…	…
10	1.767227058	−0.080484505	1.492187444	2.595988039
…	…	…	…	…
50	1.810208660	0.060535363	1.558382308	2.668661047

Example 1.5 Solve the following equation system using the Jacobi method

$$
\begin{bmatrix}
5 & -1 & 3 & 0.5 \\
0.6 & 0.3 & 1 & 2 \\
0.6 & 0.6 & 3 & 1.2 \\
2 & 4 & 0.6 & 1.2
\end{bmatrix}
\cdot
\begin{bmatrix}
x_1 \\ x_2 \\ x_3 \\ x_4
\end{bmatrix}
=
\begin{bmatrix}
15 \\ 8 \\ 9 \\ 8
\end{bmatrix}
$$

taking the initial approximation $[x_1^{(0)} = 1, \ x_2^{(0)} = 1, x_3^{(0)} = 1, x_4^{(0)} = 1]$. Proceeding according to the algorithm described above, in the first stage the given equation system is transformed to the canonical form

$$
\begin{bmatrix}
1 & -0.2 & 0.6 & 0.1 \\
0.5 & 1 & 0.15 & 0.3 \\
0.2 & 0.2 & 1 & 0.4 \\
0.3 & 0.15 & 0.5 & 1
\end{bmatrix}
\cdot
\begin{bmatrix}
x_1 \\ x_2 \\ x_3 \\ x_4
\end{bmatrix}
=
\begin{bmatrix}
3 \\ 2 \\ 3 \\ 4
\end{bmatrix}
$$

The Jacobi iteration formula obtained based on this system of equations gives

$$
\begin{bmatrix}
x_1^{(k+1)} \\ x_2^{(k+1)} \\ x_3^{(k+1)} \\ x_4^{(k+1)}
\end{bmatrix}
=
\begin{bmatrix}
0 & 0.2 & -0.6 & -0.1 \\
-0.5 & 0 & -0.15 & -0.3 \\
-0.2 & -0.2 & 0 & -0.4 \\
-0.3 & -0.15 & -0.5 & 0
\end{bmatrix}
\cdot
\begin{bmatrix}
x_1^{(k)} \\ x_2^{(k)} \\ x_3^{(k)} \\ x_4^{(k)}
\end{bmatrix}
+
\begin{bmatrix}
3 \\ 2 \\ 3 \\ 4
\end{bmatrix}
$$

Some consecutive approximations of the exact solution obtained from this formula are given in Table 1.1.

1.2.2.3 The Gauss–Seidel Method

Consider the equally simple algorithm of Gauss–Seidel iteration method, presented below for the case of the system of three linear equations:

$$a_{11}x_1 + a_{12}x_2 + a_{13}x_3 = b_1$$
$$a_{21}x_1 + a_{22}x_2 + a_{23}x_3 = b_2 \tag{1.49}$$
$$a_{31}x_1 + a_{32}x_2 + a_{33}x_3 = b_3$$

Assume that the elements a_{ii} of the main diagonal are different from zero. In the opposite case, the equations should be rearranged. We determine the unknowns x_1, x_2, and x_3 from the first, second, and third equation, respectively:

$$x_1 = \frac{1}{a_{11}}\left(b_1 - a_{12}x_2 - a_{13}x_3\right), \quad a_{11} \neq 0$$

$$x_2 = \frac{1}{a_{22}}\left(b_2 - a_{21}x_1 - a_{23}x_3\right), \quad a_{22} \neq 0 \tag{1.50}$$

$$x_3 = \frac{1}{a_{33}}\left(b_3 - a_{31}x_1 - a_{32}x_2\right), \quad a_{33} \neq 0$$

We assume a certain nonzero approximate solution: $x_1 = x_1^{(0)}$, $x_2 = x_2^{(0)}$, and $x_3 = x_3^{(0)}$. Substituting these values in the first equation of the system (1.50) yields

$$x_1^{(1)} = \frac{1}{a_{11}}\left(b_1 - a_{12}x_2^{(0)} - a_{13}x_3^{(0)}\right)$$

Using $x_1^{(1)}$ and $x_3^{(0)}$ we determine $x_2^{(1)}$ from the second equation of system (1.50), i.e.

$$x_2^{(1)} = \frac{1}{a_{22}}\left(b_2 - a_{21}x_1^{(1)} - a_{23}x_3^{(0)}\right),$$

Substituting $x_1^{(1)}$ and $x_2^{(1)}$ in the third equation, we calculate

$$x_3^{(1)} = \frac{1}{a_{33}}\left(b_3 - a_{31}x_1^{(1)} - a_{32}x_2^{(1)}\right)$$

In this way the first iteration of the solving process has been completed. Of course the whole process should be repeated many times until the solution $\mathbf{X} \equiv [x_1, x_2, x_3]$ similar to the solution resulting from the previous iteration is obtained. The process described above is convergent if

$$|a_{ii}| \geq \sum_{i \neq j} |a_{i,j}| \qquad \text{for } i = 1, 2, 3, \ldots, n \tag{1.51}$$

with a condition that at least one of the above inequalities should be strict. Condition (1.51) is sufficient but not necessary, and for some equation systems, the computation process may prove to be convergent even when this condition is not satisfied. According to this conclusion, the equations should be rearranged in such a way that the elements of the main diagonal have largest absolute values. It follows from the comparison of the algorithms of both methods presented above that they have much in common. In the Jacobi method, a consecutive, $(k + 1)$ approximation of the desired exact solution is determined exclusively based on the approximation obtained in previous k iteration. In case of the Gauss–Seidel method, individual components of each consecutive approximation $x_j^{(k+1)}$, where $j = 1, 2, 3, \ldots, n$, are determined based on an approximation obtained in the previous k iteration and of newly calculated components $x_i^{(k+1)}$, where $i < j$. This property guarantees faster convergence and numerical stability of the Gauss–Seidel method. During the analysis of many technical and economical problems, we meet the necessity of solving large systems of equations (e.g., for $n \geq 100$). To solve very large equation systems, iterative methods are chiefly used, as for example the Gauss–Seidel method described above. At the beginning, we meet here the problem of initial approximation of the desired solution, which can be obtained solving the system in question by the Gauss elimination method. The initial solution obtained in this way should guarantee convergence of the calculation process.

Example 1.6 In order to illustrate an application of the Gauss–Seidel method, we will find some approximate solutions to the following equation system:

$$\begin{bmatrix} 4 & -1 & 1 & 0 & 0 & 0 \\ 2 & 6 & -1 & 0 & 0 & 2 \\ 1 & 2 & -5 & 1 & 0 & 1 \\ 1 & -1 & 1 & 4 & 0 & 0.5 \\ 1 & 0 & 0 & 1 & 5 & 2 \\ 0 & 0 & 1 & -1 & 2 & 7 \end{bmatrix} \cdot \begin{bmatrix} x_1 \\ x_2 \\ x_3 \\ x_4 \\ x_5 \\ x_6 \end{bmatrix} = \begin{bmatrix} 4 \\ 6.4 \\ 0.3 \\ 2.9 \\ 2.6 \\ -1.1 \end{bmatrix}$$

Calculation process begins with the following initial approximation: $x_1^{(0)} = 1.0, x_2^{(0)} = 0.8,\quad x_3^{(0)} = 0.6, x_4^{(0)} = 0.4, x_5^{(0)} = 0.2,$ and $x_6^{(0)} = 0.0$. Diagonal elements of the matrix of coefficients (elements on the main diagonal) satisfy condition (1.51), which is the sufficient condition for convergence of the calculation process. Numerical values of some approximations of the exact solution ($x_1 = 1.077511712, x_2 = 0.883357158, x_3 = 0.573309477, x_4 = 0.563250046, x_5 = 0.288218927, x_6 = -0.240928187$) are given in Table 1.2

Table 1.2

X	Number of iterations		
	1	5	10
x_1	1.049999999	1.077607126	1.077511712
x_2	0.816666687	0.883261359	0.883357158
x_3	0.556666673	0.573347946	0.573309477
x_4	0.527500028	0.563184419	0.563250046
x_5	0.204499975	0.288186709	0.288218927
x_6	-0.219738086	-0.240933952	-0.240928187

In case of rearranging an arbitrary equation pair of the analyzed system (as, for example, Eqs. (1.5) and (1.6)), the solution obtained for the vector $\mathbf{X}$ would remain unchanged. Unfortunately, condition (1.51) will no longer be satisfied and the calculation process may become divergent.

1.3 Examples of Applications in Electrical Engineering

Example 1.7 The electric diagram of the six-element ladder circuit driven by a voltage source $e(t) = E_g \cos(\omega t + 0)$ with an internal impedance $\mathbf{Z}_g$ is shown in Fig. 1.3.

The voltage waveform $u_l(t)$ across the loading impedance $\mathbf{Z}_l$ can be found by using the mesh current method formulating balance equations for complex amplitudes of the voltage in the independent closed loops [11, 12]. This method is based on Kirchhoff's law, stating that the sum of voltages in each closed loop of an electric circuit is equal to zero. According to this law, we can formulate the following equations for the meshes chosen as in Fig. 1.3:

$$
\begin{aligned}
\mathbf{Z}_1 \mathbf{I}_1 + \mathbf{Z}_2(\mathbf{I}_1 - \mathbf{I}_2) + \mathbf{Z}_g \mathbf{I}_1 &= \mathbf{E}_g \\
\mathbf{Z}_3 \mathbf{I}_2 + \mathbf{Z}_4(\mathbf{I}_2 - \mathbf{I}_3) + \mathbf{Z}_2(\mathbf{I}_2 - \mathbf{I}_1) &= 0 \\
\mathbf{Z}_5 \mathbf{I}_3 + \mathbf{Z}_6 \mathbf{Z}_l/(\mathbf{Z}_6 + \mathbf{Z}_l)\mathbf{I}_3 + \mathbf{Z}_4(\mathbf{I}_3 - \mathbf{I}_2) &= 0
\end{aligned}
\tag{1.52}
$$

where $\mathbf{E_g} = E_g \exp(j0), \mathbf{I}_1, \mathbf{I}_2, \mathbf{I}_3$ are the complex amplitudes of the control voltage and currents flowing in the meshes having the same orientations and $\mathbf{Z}_1, \mathbf{Z}_2, \mathbf{Z}_3, \mathbf{Z}_4, \mathbf{Z}_5, \mathbf{Z}_6,$ and $\mathbf{Z}_l$ are the impedances determined according to the rules of the symbolic calculus [4]. After rearranging, the system of equations (1.52) can be written in the following matrix form:

$$
\begin{bmatrix}
\mathbf{Z}_1 + \mathbf{Z}_2 + \mathbf{Z}_g & -\mathbf{Z}_2 & 0 \\
-\mathbf{Z}_2 & \mathbf{Z}_2 + \mathbf{Z}_3 + \mathbf{Z}_4 & -\mathbf{Z}_4 \\
0 & -\mathbf{Z}_4 & \mathbf{Z}_4 + \mathbf{Z}_5 + \dfrac{\mathbf{Z}_6 \mathbf{Z}_l}{\mathbf{Z}_6 + \mathbf{Z}_l}
\end{bmatrix}
\times
\begin{bmatrix}
\mathbf{I}_1 \\ \mathbf{I}_2 \\ \mathbf{I}_3
\end{bmatrix}
=
\begin{bmatrix}
\mathbf{E}_g \\ 0 \\ 0
\end{bmatrix}
\tag{1.53}
$$

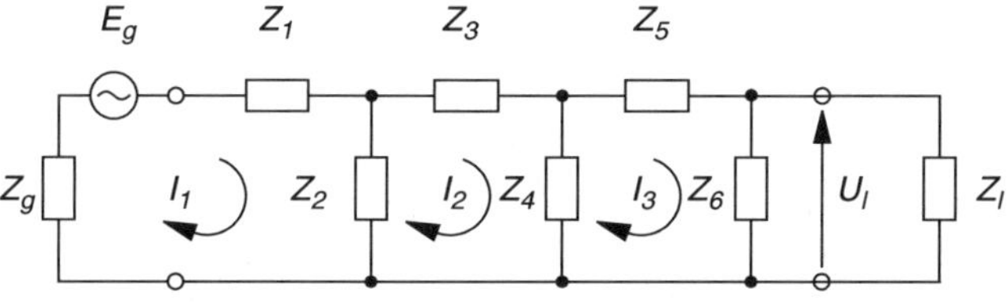

Fig. 1.3

Elements of the coefficient matrix of the equation system (1.53) are the complex numbers, as in case of the complex amplitudes of currents that we calculate. This equation system can be solved by using the Gauss elimination method described in Sect. 1.1. In this case, the operations of addition, subtraction, multiplication, and division should be performed according to the rules concerning complex numbers. Prior to solving the equation system, individual equations should be placed in such an order that the complex coefficients of the main diagonal have possibly largest absolute values. Solving the system (1.53), we obtain complex amplitudes of the mesh currents, including the amplitude $\mathbf{I}_3 = I_3 \exp(j\varphi_3)$. Finally, the evaluated voltage waveform $u_l(t)$ is described by

$$u_l(t) = \mathrm{Re}\left[\mathbf{I}_3 \frac{\mathbf{Z}_6 \mathbf{Z}_l}{\mathbf{Z}_6 + \mathbf{Z}_l} \exp(j\omega t) \right] = \left| \mathbf{I}_3 \frac{\mathbf{Z}_6 \mathbf{Z}_l}{\mathbf{Z}_6 + \mathbf{Z}_l} \right| \cos(\omega t + \varphi_3 + \psi) \qquad (1.54)$$

where the symbol Re [ ] denotes the real part of the expression given in the square brackets, and ψ is the phase angle of the complex number $\mathbf{Z}_6 \mathbf{Z}_l / (\mathbf{Z}_6 + \mathbf{Z}_l)$. In order to perform the calculations, we take the values $\mathbf{Z}_g = 1\Omega$, $\mathbf{Z}_l = 1\mathrm{M}\Omega$, $\mathbf{Z}_1 = \mathbf{Z}_3 = \mathbf{Z}_5 = 1/(j\omega C)$, $\mathbf{Z}_2 = \mathbf{Z}_4 = \mathbf{Z}_6 = 10\mathrm{k}\Omega$, where $C = 470\mathrm{pF}$, $E_g = 10\mathrm{V}$, and $\omega = 2\pi f = 2\pi(10^4)$ rad/s. The equation system (1.53) can be formulated in the following matrix form:

$$\begin{bmatrix} 10001 - j33862.75 & -10000 & 0 \\ -10000 & 20000 - j33862.75 & -10000 \\ 0 & -10000 & 19900.99 - j33862.75 \end{bmatrix} \times \begin{bmatrix} \mathbf{I}_1 \\ \mathbf{I}_2 \\ \mathbf{I}_3 \end{bmatrix}$$
$$= \begin{bmatrix} 10 \\ 0 \\ 0 \end{bmatrix}$$

The solution for these equations are the following current amplitudes: $\mathbf{I}_1 = 0.0643 + j0.2616\mathrm{mA}$, $\mathbf{I}_2 = -0.0499 + j0.0437\mathrm{mA}$, and $\mathbf{I}_3 = -0.0160 - j0.0053\mathrm{mA}$. The output voltage, determined according to the expression (1.54), is $u_l(t) = 0.3354 \cos(\omega t + 3.4614)\mathrm{V}$.

Example 1.8 An another method largely used in the electrical engineering is the analysis of nodal potentials [11, 12]. In order to explain its essence, let us evaluate the voltage $u_2(t)$ across the output terminals of the circuit "double T" shown in Fig. 1.4.

For the analysis we assume that the system is supplied by the current source $i_g(t) = I_g \cos(\omega t + 0)$ with an internal admittance $\mathbf{Y}_g$. The nodal voltage method is based on Kirchhoff's theorem, stating that the sum of currents entering and that leaving a node of the electric circuit is always equal to zero. It is not too difficult to prove that this law applies also for the case of complex amplitudes of these currents. In consequence, we can formulate the following equations for the analyzed circuit:

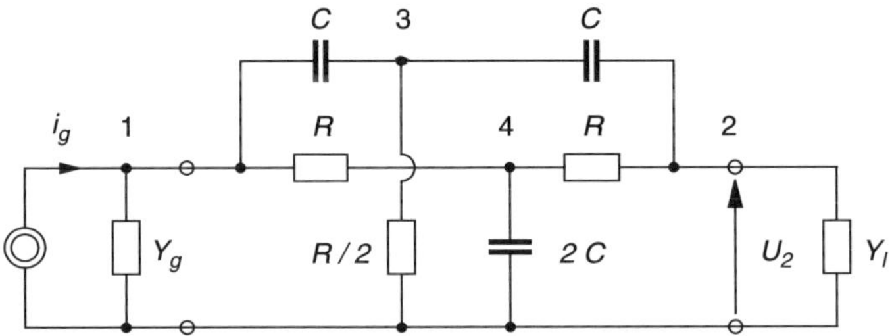

Fig. 1.4

$$\mathbf{I}_g - \mathbf{Y}_g \mathbf{U}_1 - G(\mathbf{U}_1 - \mathbf{U}_4) - pC(\mathbf{U}_1 - \mathbf{U}_3) = 0$$
$$G(\mathbf{U}_4 - \mathbf{U}_2) + pC(\mathbf{U}_3 - \mathbf{U}_2) - \mathbf{Y}_l \mathbf{U}_2 = 0 \qquad (1.55)$$
$$pC(\mathbf{U}_1 - \mathbf{U}_3) - 2G\mathbf{U}_3 - pC(\mathbf{U}_3 - \mathbf{U}_2) = 0$$
$$G(\mathbf{U}_1 - \mathbf{U}_4) - G(\mathbf{U}_4 - \mathbf{U}_2) - 2pC\mathbf{U}_4 = 0$$

where $\mathbf{I}_g = I_g \exp(j0)$, $\mathbf{U}_1$, $\mathbf{U}_2$, $\mathbf{U}_3$, $\mathbf{U}_4$ are respectively the complex amplitude of the control current and voltage complex amplitudes determined at the corresponding nodes 1, 2, 3, and 4, see Fig. 1.4, $G = 1/R$ and $\mathbf{p} = j\omega = j2\pi f$ is the operator of the applied symbolic calculus. After arranging, the equation system (1.55) takes the form

$$\begin{bmatrix} \mathbf{Y}_g + G + pC & 0 & -pC & -G \\ 0 & G + pC + \mathbf{Y}_l & -pC & -G \\ -pC & -pC & 2G + 2pC & 0 \\ -G & -G & 0 & 2G + 2pC \end{bmatrix} \times \begin{bmatrix} \mathbf{U}_1 \\ \mathbf{U}_2 \\ \mathbf{U}_3 \\ \mathbf{U}_4 \end{bmatrix} = \begin{bmatrix} \mathbf{I}_g \\ 0 \\ 0 \\ 0 \end{bmatrix}$$

$$(1.56)$$

As in case of the previous example, the equation system (1.56) can be solved by using the Gauss elimination method, and we obtain the vector of nodal voltages complex amplitudes, including $\mathbf{U}_2 = U_2 \exp(j\varphi_2)$. According to the principles of the symbolic calculus, the desired voltage waveform $u_2(t)$ is

$$u_2(t) = \mathrm{Re}[\mathbf{U}_2 \exp(j\omega t)] = U_2 \cos(\omega t + \varphi_2) \qquad (1.57)$$

Complex amplitudes of $\mathbf{U}_1$, $\mathbf{U}_2$, $\mathbf{U}_3$, and $\mathbf{U}_4$ calculated for $R = 10\,\mathrm{k\Omega}$, $C = 1\mathrm{nF}$, $Y_g = Y_l = 1\mu\mathrm{S}$, $I_g = 1\mathrm{mA}$, and for several values of the angular frequency ω are given in Table 1.3.

It follows from the analysis of amplitudes of the voltage $\mathbf{U}_2$ given in the third column of the Table 1.3 that it attains the minimum for the angular frequency $\omega_0 = 2\pi f_0 = 1/(RC) = 10^5 \mathrm{rad/s}$. In other words, the analyzed two-pole filter has a selective frequency response $\beta(\omega) = U_2(\omega)/U_1(\omega)$, Fig. 1.5(a).

Due to the above property, this two-port filter (Fig. 1.4) can be applied in the low frequency harmonic oscillator as it is shown in Fig. 1.5(b).

Table 1.3

ω (rad/s)	U_1(V)	U_2(V)	U_3(V)	U_4(V)
0.98×10^5	5.0502	−0.0495	2.4875	0.0129
	−j4.9998	−j0.0515	+j0.0126	−j2.5383
0.99×10^5	5.0249	−0.0246	2.4875	0.0126
	−j4.9749	−j0.0254	+j0.0125	−j2.5126
1.00×10^5	4.9998	0.0000	2.4875	0.0124
	−j4.9502	+j0.0000	+j0.0124	−j2.4875
1.01×10^5	4.9749	0.0244	2.4875	0.0121
	−j4.9259	+j0.0246	+j0.0123	−j2.4629
1.02×10^5	4.9503	0.0485	2.4875	0.0119
	−j4.9017	+j0.0485	+j0.0121	−j2.4387

(a)

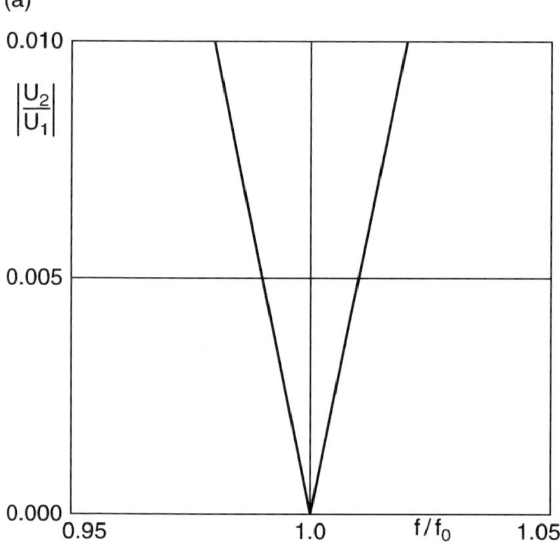

(b)

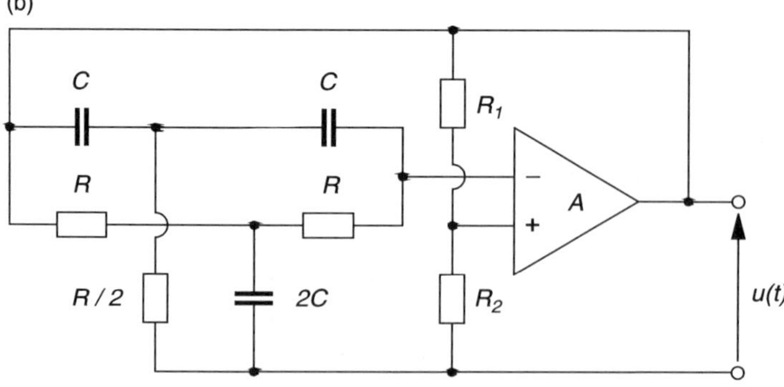

Fig. 1.5

In this circuit, the "double T" two-port network was placed in the negative feedback loop, and this feedback attains the minimum absolute value for angular frequency $\omega_0 = 1/(RC)$. The positive feedback needed for ensuring the generation condition is implemented by using a properly designed frequency-nonselective resistance divider, (R_1, R_2).

References

1. Akai T.J., Applied numerical methods for engineers. John Wiley and Sons, New York, 1994
2. Atkinson K.E., An introduction to numerical analysis (2nd edition). John Wiley and Sons, New York, 1988
3. Berezin L.S. and N.P. Zhidkov, Computing methods. Pergamon Press, Elmsford, NY, 1965
4. Faddeev D.K. and V.N. Fadeeva, Computational methods of linear algebra. W.H. Freeman and Co., San Francisco, CA, 1963
5. Forsythe G.E. and C.B. Moler, Computer solution of linear algebraic systems. Prentice-Hall, Inc., Englewood Cliffs, NJ, 1967
6. Dahlquist G. and A. Bjorck, Numerical methods. Prentice-Hall, Englewood Cliffs, NY, 1974
7. Mathews J.H., Numerical methods for mathematics, science and engineering. Prentice-Hall International, Inc., Englewood Cliffs, NJ, 1992
8. Shoup T.E., Applied numerical methods for the microcomputer. Prentice-Hall, Inc., Englewood Cliffs, NJ, 1984
9. Forsythe G.E., M.A. Malcolm and C.B. Moler, Computer methods for mathematical computations. Prentice-Hall, Englewood Cliffs, NJ, 1977
10. Mathews J.H., Numerical methods for mathematics, science and engineering. Prentice-Hall, Inc., Englewood Cliffs, NJ, 1987
11. Chua L.O. and P.M. Lin, Computer aided analysis of electronics circuits. Algorithms and computational techniques. Prentice-Hall, Englewood Cliffs, NJ, 1975
12. Irwin J.D. and C.-H. Wu, Basic engineering circuit analysis (6th edition). John Wiley and Sons, Inc., New York, 1999

Chapter 2
Methods for Numerical Solving the Single Nonlinear Equations

Numerous scientific and technical problems can be described by means of single equations with one variable or systems of n equations with n variables. According to the character of functions appearing in these equations, they can be linear or nonlinear. The corresponding classification of algebraic equations is given in the diagram of Fig. 2.1.

In this diagram, the class of single nonlinear equations is divided into polynomial and transcendent equations. They will be discussed in this order in the following sections of the present chapter. For presentation of the Bairstow's method, some formulas were used, which are developed later in Sect. 3.3. These expressions represent in fact a special case of Newton's method, used for solving systems of n nonlinear equations with n variables, [2] [7] [8].

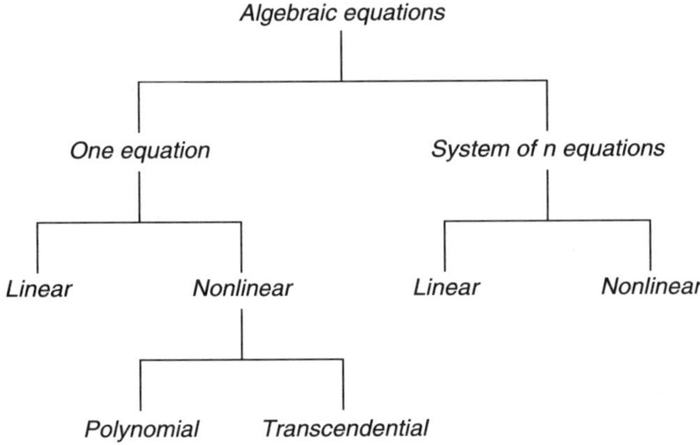

Fig. 2.1

S. Rosłoniec, *Fundamental Numerical Methods for Electrical Engineering*,
© Springer-Verlag Berlin Heidelberg 2008

2.1 Determination of the Complex Roots of Polynomial Equations by Using the Lin's and Bairstow's Methods

The polynomial equations are expressed by sums of finite number of terms containing powers of the variable x. Thus, equations of this kind can be presented in the following general form:

$$F(x) \equiv W_n(x) = x^n + a_1 x^{n-1} + a_2 x^{n-2}$$
$$+a_3 x^{n-3} + \cdots + a_{n-1} x + a_n = 0 \qquad (2.1)$$

The class of polynomial equations includes also equations, in which the function $F(x)$ has the form of a rational function, i.e., the quotient of two nonzero polynomials. In any case, equations described by means of a rational function can be transformed to the canonical form (2.1). The well-known quadratic equation:

$$ax^2 + bx + c = 0, \quad a \neq 0 \qquad (2.2)$$

is the simplest case of a nonlinear polynomial equation. Real or complex roots of this equation can be calculated by using the following simple formulas:

$$x_1 = \frac{-b + \sqrt{\Delta}}{2a}, \quad x_2 = \frac{-b - \sqrt{\Delta}}{2a} \qquad (2.3)$$

where the discriminant is equal to $\Delta = b^2 - 4ac$. The similar explicit formulas found by Nicollo Fontana (1500–1557) for cubic equations are much more complicated. In the mathematical bibliography they are often unjustly called as Cardan's formulas. In case of the equations of higher orders ($n \geq 4$) finding roots by means of analytical methods is extremely difficult. Therefore, numerical methods are used for this purpose, among which the methods of Lin's and Bairstow's [1–3] are most commonly used.

2.1.1 Lin's Method

The Lin's and Bairstow's methods are based on the fact that the equation (2.1) can be written as:

$$(x^2 + px + q)(x^{n-2} + b_1 x^{n-3} + b_2 x^{n-4} + \cdots + b_{n-3} x + b_{n-2})$$
$$+Rx + S = 0 \qquad (2.4)$$

In this expression, $Rx + S$ is a linear remainder term that we desire to be zero. A zero remainder would mean that the original polynomial (2.1) is exactly divisible by the quadratic factor $(x^2 + px + q)$. Coefficients $p, q, b_1, b_2, b_3, \ldots, b_{n-2}, R$ and S of the Eq. (2.4) are related to the coefficients $1, a_1, a_2, a_3, a_4, \ldots, a_n$ of Eq. (2.1) as follows:

$$b_1 = a_1 - p$$
$$b_2 = a_2 - pb_1 - q$$
$$b_3 = a_3 - pb_2 - qb_1$$
$$\vdots$$
$$b_i = a_i - pb_{i-1} - qb_{i-2} \qquad (2.5)$$
$$\vdots$$
$$b_{n-2} = a_{n-2} - pb_{n-3} - qb_{n-4}$$
$$R = a_{n-1} - pb_{n-2} - qb_{n-3}$$
$$S = a_n - qb_{n-2}$$

Expressions (2.5) can be obtained by comparison of coefficients of the corresponding terms of Eqs. (2.1) and (2.4) containing the same power of the unknown variable x. If the term $(Rx + S)$ appearing in the Eq. (2.4) is equal to zero $(R = 0, \ S = 0)$, then the roots of the quadratic factor $(x^2 + px + q)$ are also the roots of Eq. (2.1). Assuming that the coefficients p and q of the factor $(x^2 + px + q)$ are real, the roots $x_1 = c + jd$ and $x_2 = c - jd$ form the pair of the complex conjugate numbers. The real part c and the imaginary part d of this roots are related to the coefficients p and q in the following way:

$$p = -2c, \quad q = c^2 + d^2 \qquad (2.6)$$

The mathematical basis of the Lin's and Bairstow's methods consists in finding such values of the coefficients $p, q, b_1, b_2, b_3, \ldots, b_{n-2}$, for which $R = 0$ and $S = 0$. In other words, the polynomial (2.1) should be divisible (without remainder) by the factor $(x^2 + px + q)$. This condition can be expressed by the equations:

$$R = a_{n-1} - pb_{n-2} - qb_{n-3} = 0$$
$$S = a_n - qb_{n-2} = 0 \qquad (2.7)$$

from which consecutive more accurate values of the coefficients can be found.

$$q' = \frac{a_n}{b_{n-2}}, \quad b_{n-2} \neq 0$$
$$p' = \frac{a_{n-1} - q'b_{n-3}}{b_{n-2}} \qquad (2.8)$$

The process of determining the coefficients p and q by the Lin's method is performed in the following manner. In the first iteration, for given (initial) values of the coefficients p and q, the coefficients $b_1, b_2, b_3, \ldots, b_{n-2}, R$ and S are found using formulas (2.5). If the coefficients R and S are different from zero then, according to (2.8), the successive, more accurate values of p and q are determined. In the next iteration, they are used to calculate new values of coefficients $b_1, b_2, b_3, \ldots, b_{n-2}, R$ and S. This process is continued until the coefficients R and S become close to zero within the predetermined accuracy. It follows from the character of the expressions (2.8) that Eqs. (2.7) are solved using the Gauss–Seidel method, but the coefficients b_{n-2} and b_{n-3} of these equations have different values for each iteration. In other words, these coefficients are functions of p and q, see formulas (2.5). When conditions $R = 0$ and $S = 0$ are satisfied, Eq. (2.4) takes the form:

$$(x^2 + px + q)(x^{n-2} + b_1 x^{n-3} + b_2 x^{n-4} + \cdots + b_{n-3}x + b_{n-2})$$
$$= (x^2 + px + q) \cdot W_{n-2}(x) = 0 \tag{2.9}$$

First roots of this equation are the roots $x_1 = c + jd$ and $x_2 = x_1^*$ of the quadratic factor $(x^2 + px + q)$ determined according to (2.3) or (2.6). The consecutive roots of the Eq. (2.9), and at the same time of the Eq. (2.1), are determined by solving the equation:

$$W_{n-2}(x) = 0 \tag{2.10}$$

If the order of reduced polynomial $W_{n-2}(x)$, that is $(n - 2)$, becomes greater than 2, then these roots can be determined in much the same way as described above. The remaining roots of the Eq. (2.1) can be of course found by means of the same procedure. During the final phase of the computation, the order of the successive reduced polynomial $W_{n-2k}(x)$, where $k = 1, 2, 3, \ldots, n/2$, is not greater than 2, and determining its roots terminates the process of solving Eq. (2.1). It is known from the numerous cases described in the literature that the Lin's method just presented can be characterized by slow convergence of the process of calculating the coefficients p and q, until $R = 0$ and $S = 0$. Moreover, by unfortunate choice of the initial values of these coefficients, the whole process can be divergent. These shortcomings were eliminated in the Bairstow's method, in which the system of equations (2.7) is being solved using the Newton's method described in Sect. 3.3.

2.1.2 Bairstow's Method

By determining the roots of polynomial equation (2.1) using the Bairstow's method, the system (2.7) is being iteratively solved by means of the Newton's method. According to the Eqs. (3.16) and (3.17), the successive, $(n + 1)$ approximations of the coefficients p and q are calculated from the following formulas:

$$p^{(n+1)} = p^{(n)} - \frac{1}{J}\left(R \cdot \frac{\partial S}{\partial q} - S \cdot \frac{\partial R}{\partial q}\right)$$
$$\tag{2.11}$$
$$q^{(n+1)} = q^{(n)} + \frac{1}{J}\left(R \cdot \frac{\partial S}{\partial p} - S \cdot \frac{\partial R}{\partial p}\right)$$

where

$$J = \frac{\partial R}{\partial p} \cdot \frac{\partial S}{\partial q} - \frac{\partial S}{\partial p} \cdot \frac{\partial R}{\partial q} \neq 0$$

During the calculation of the partial derivatives appearing in the above formulas, we shall be aware of the fact that the coefficients R and S are functions of the coefficients $b_1, b_2, b_3, \ldots, b_{n-2}$. The later are in turn dependent on p and q, see formulas (2.5). It is therefore necessary to determine the sequence of partial derivatives of the coefficients $b_1, b_2, b_3, \ldots, b_{n-2}$, R and S with respect to p and q. These sequences can be described in the following way:

$$\frac{\partial b_1}{\partial p} = c_1 = -1$$

$$\frac{\partial b_2}{\partial p} = c_2 = -b_1 - p\left(\frac{\partial b_1}{\partial p}\right) = -b_1 - pc_1 = -b_1 + p$$

$$\frac{\partial b_3}{\partial p} = c_3 = -b_2 - p\left(\frac{\partial b_2}{\partial p}\right) - q\left(\frac{\partial b_1}{\partial p}\right) = -b_2 - pc_2 - qc_1$$

$$\vdots$$

$$\frac{\partial b_i}{\partial p} = c_i = -b_{i-1} - p\left(\frac{\partial b_{i-1}}{\partial p}\right) - q\left(\frac{\partial b_{i-2}}{\partial p}\right) = -b_{i-1} - pc_{i-1} - qc_{i-2}$$

$$\vdots$$

$$\frac{\partial b_{n-2}}{\partial p} = c_{n-2} = -b_{n-3} - p\left(\frac{\partial b_{n-3}}{\partial p}\right) - q\left(\frac{\partial b_{n-4}}{\partial p}\right) = -b_{n-3} - pc_{n-3} - qc_{n-4}$$

$$\frac{\partial R}{\partial p} = -b_{n-2} - p\left(\frac{\partial b_{n-2}}{\partial p}\right) - q\left(\frac{\partial b_{n-3}}{\partial p}\right) = -b_{n-2} - pc_{n-2} - qc_{n-3}$$

$$\frac{\partial S}{\partial p} = -q\left(\frac{\partial b_{n-2}}{\partial p}\right) = -qc_{n-2}$$

(2.12), (2.13)

$$\frac{\partial b_1}{\partial q} = d_1 = 0$$

$$\frac{\partial b_2}{\partial q} = d_2 = -p\left(\frac{\partial b_1}{\partial q}\right) - 1 = -pd_1 - 1 = -1$$

$$\frac{\partial b_3}{\partial q} = d_3 = -p\left(\frac{\partial b_2}{\partial q}\right) - b_1 - q\left(\frac{\partial b_1}{\partial q}\right) = -b_1 - pd_2 - qd_1$$

$$\vdots$$

$$\frac{\partial b_i}{\partial q} = d_i = -p\left(\frac{\partial b_{i-1}}{\partial q}\right) - b_{i-2} - q\left(\frac{\partial b_{i-2}}{\partial q}\right) = -b_{i-2} - pd_{i-1} - qd_{i-2}$$

$$\vdots$$

(2.14)

$$\frac{\partial b_{n-2}}{\partial q} = d_{n-2} = -p\left(\frac{\partial b_{n-3}}{\partial q}\right) - b_{n-4} - q\left(\frac{\partial b_{n-4}}{\partial q}\right) = -b_{n-4} - pd_{n-3} - qd_{n-4}$$

$$\frac{\partial R}{\partial q} = -p\left(\frac{\partial b_{n-2}}{\partial q}\right) - b_{n-3} - q\left(\frac{\partial b_{n-3}}{\partial q}\right) = -b_{n-3} - pd_{n-2} - qd_{n-3} \qquad (2.15)$$

$$\frac{\partial S}{\partial q} = -b_{n-2} - q\left(\frac{\partial b_{n-2}}{\partial q}\right) = -b_{n-2} - qd_{n-2}$$

The coefficients $R(p, q)$, $S(p, q)$ used in expressions (2.11) and their partial derivatives described by (2.13) and (2.15) are computed for $p = p^{(n)}$ and $q = q^{(n)}$. After calculating $p^{(n+1)}$ and $q^{(n+1)}$ they are used for computing the new values of the coefficients b_i, R and S of the Eq. (2.4). The coefficients b_i, R and S, where $i = 1, 2, 3, \ldots, n - 2$, can be calculated from formulas (2.5). Next, we use expressions (2.12), (2.13), (2.14) and (2.15) to find new values of partial derivatives of the coefficients R and S with respect to p and q correspondingly. Values of this derivatives, in accordance with formulas (2.11), make possible finding of the $p^{(n+2)}$ and $q^{(n+2)}$ approximations of the coefficients, which we are due to find. The successive approximations of the coefficients p and q can be found iteratively in similar way as described previously. These computations are continued until coefficients R and S become close to zero with prescribed accuracy. Assume now that, as the result of computations described above such values $p = p^*$ and $q = q^*$ were determined, for which $R(p^*, q^*) = 0$ and $S(p^*, q^*) = 0$. Then roots $x_1 = c + jd$ and $x_2 = c - jd$ of the equation to be solved are identical to the roots of quadratic equation $(x^2 + p^*x + q^*) = 0$. The real part c and the imaginary part d of these roots are:

$$c = -\frac{p^*}{2}, \quad d = \sqrt{q^* - c^2} \qquad (2.16)$$

Similarly, as for all iterative methods, computations according to the Bairstow's method begin from the (initial) approximation $p = p^{(0)}$ and $q = q^{(0)}$, which should be chosen in such a way that the computing process just described is convergent. The relations presented above form the theoretical basis to the elaboration of the computer program P2.1, in which it was taken $p^{(0)} = 0$, $q^{(0)} = 0$.

Example 2.1 The computer program mentioned above has been used to solve the equation:

$$W_7(x) = x^7 - 4x^6 + 25x^5 + 30x^4 - 185x^3 + 428x^2 - 257x - 870 = 0 \quad (2.17)$$

having the following roots: $x_1 = 2$, $x_2 = -1$, $x_3 = 1 + j2$, $x_4 = 1 - j2$, $x_5 = 2 + j5$, $x_6 = 2 - j5$, $x_7 = -3$, where $j = \sqrt{-1}$. After finding the roots x_1 and x_2, the Eq. (2.17) can be written in the following product form:

$$W_7(x) = (x-x_1)(x-x_2)W_5(x) = (x^2-x-2)(x^5-3x^4+24x^3+48x^2-89x+435) = 0$$

The successive roots of the Eq. (2.17) we solve are $x_3 = 1 + j2$ and $x_4 = 1 - j2$, being the roots of the reduced equation $W_5(x) = 0$, which can be written as:

$$W_5(x) = (x - x_3)(x - x_4)W_3(x) = (x^2 - 2x + 5)(x^3 - x^2 + 17x + 87) = 0$$

Solving the equation $W_3(x) = 0$ we obtain the roots $x_4 = 2 + j5$ and $x_5 = 2 - j5$. Hence, the equation $W_3(x) = 0$ can be written in the form of the following product:

$$W_3(x) = (x - x_4)(x - x_5)W_1(x) = (x^2 - 4x + 29)(x + 3) = 0$$

which gives the value of the last root we wanted to determine, i.e., $x_7 = -3$.

By changing the initial approximation $(p^{(0)}, q^{(0)})$ we can change also the order of finding successive roots of the equation. This order does not influence obviously the values of the roots but can indirectly decide on the convergence of the performed calculations. This conclusion justifies the fact that by changing the order of determining the roots, the coefficients of reduced polynomials $W_{n-2}(x)$, $W_{n-4}(x)$ change also their values. For example, when $p^{(0)} = 10$ and $q^{(0)} = 0$, the roots of the Eq. (2.17) calculated by means of the program P2.1 are equal to: $x_1 = -1$, $x_2 = -3$, $x_3 = 1 + j2$, $x_4 = 1 - j2$, $x_5 = 2 + j5$, $x_6 = 2 - j5$, $x_7 = 2$. In this case, the Eq. (2.17) is transformed in the following way:

$$\begin{aligned} W_7(x) &= (x - x_1)(x - x_2)W_5(x) \\ &= (x^2 + 4x + 3)(x^5 - 8x^4 + 54x^3 - 162x^2 + 301x - 290) = 0 \end{aligned}$$

where

$$W_5(x) = (x - x_3)(x - x_4)W_3(x) = (x^2 - 2x + 5)(x^3 - 6x^2 + 37x - 58)$$
$$W_3(x) = (x - x_4)(x - x_5)W_1(x) = (x^2 - 4x + 29)(x - 2)$$
$$W_1(x) = x - 2$$

The results obtained in this example are good illustration of the general rule saying that the nontrivial polynomial equation of the nth order has n roots which can be uniquely determined. If all coefficients of this equations are real numbers, then its complex roots form pairs of the complex conjugate numbers.

2.1.3 Laguerre Method

It is concluded in the monography [3] that the Laguerre method for finding the roots of polynomial equations is the most effective method and is sufficiently convergent. In order to illustrate its algorithm, let us assume that the kth approximation of any root x_k of the equation $W_n(x) = 0$ is known. Next, i.e., $(k + 1)$th approximation of this root is calculated by using the following formula:

$$x_{k+1} = x_k - \frac{n W_n(x_k)}{\dfrac{\mathrm{d}W_n(x_k)}{\mathrm{d}x} \pm \sqrt{H(x_k)}} \tag{2.18}$$

where

$$H(x) = (n-1)\left\{ (n-1)\left[\frac{\mathrm{d}W_n(x)}{\mathrm{d}x}\right]^2 - n W_n(x)\frac{\mathrm{d}^2 W_n(x)}{\mathrm{d}x^2} \right\}$$

Sign $\pm$ in the denominator of expression (2.18) should be chosen so as to obtain the smallest value of the difference $|x_{k+1} - x_k|$. Values of polynomial $W_n(x_k)$ and its derivatives needed for each iteration can be effectively calculated using the following formulas:

$$W_n(x_k) = b_n; \quad \frac{\mathrm{d}W_n(x)}{\mathrm{d}x} = c_{n-1}; \quad \frac{\mathrm{d}^2 W_n(x)}{\mathrm{d}x^2} = 2d_{n-2} \tag{2.19}$$

where

$$b_0 = 1, \quad b_i = x b_{i-1} + a_i, \quad i = 1, 2, 3, \ldots, n$$
$$c_0 = 1, \quad c_i = x c_{i-1} + b_i, \quad i = 1, 2, 3, \ldots, n-1$$
$$d_0 = 1, \quad d_i = x d_{i-1} + c_i, \quad i = 1, 2, 3, \ldots, n-2$$

After calculating the first root x_1, the equation $W_n(x) = 0$ can be written in the following product form: $W_n(x) = (x - x_1)W_{n-1}(x) = 0$. Consecutive roots of the equation $W_n(x) = 0$ are evaluated from formulas (2.18) and (2.19) with respect to the reduced equation $W_{n-1}(x) = 0$. The process is continued until all the roots of the equation $W_n(x) = 0$ are not evaluated. If all the coefficients of the polynomial $W_n(x)$ are real and the root x_i is complex, we can assume that the next root $x_{i+1} = (x_i)^*$. In this case, the product $(x - x_i)(x - x_{i+1})$ is a quadratic factor $(x^2 + px + q)$ with real coefficients p and q. Extracting this factor from the equation being solved we obtain the new equation with reduced second order.

2.2 Iterative Methods Used for Solving Transcendental Equations

According to the type of the function $F(x)$ describing the equation $F(x) = 0$, this equation belongs to the class of the algebraic or transcendental equations. If the function $F(x)$ has the form of a polynomial or of a rational function (quotient of the two polynomials), then this equation belongs to the class of the algebraic equations. Classical example of this class is the polynomial equation considered in previous sections. In case when the function $F(x)$ has the form of the exponential, logarithmic, trigonometric function, or their combination, such equation belongs undoubtedly to the class of transcendental equations. Only some simple transcendental equations can be solved by means of direct methods allowing to express the solution (roots) by means of a closed-form mathematical formulas. Hence, majority of the transcendental equations describing diverse scientific and technical problems cannot be solved by means of direct methods. In such cases, the iterative numerical methods, consisting usually of two stages [4–6] should be used. In the first stage, an initial approximation of the solution is determined, and in the next (in the second stage) the more accurate solutions are obtained, and the order of accuracy determines the prescribed admissible error. In the frame of this chapter, the following iterative methods will be described:

– bisection (dividing by two) method
– secant method
– tangents (Newton–Raphson) method
– indirect method based on transformation of the given problem into an equivalent optimization problem, which can be subsequently solved by means of the golden or Fibonacci cut methods.

If we want to perform computations using the methods mentioned above we must know the closed interval $[a, b]$, in which only one solution exists. This interval is usually determined using the uniform search method, as shown in Fig. 2.2.

Testing the function $F(x)$ is performed with a constant small step Δx. All over assumed interval $[a, b]$ we search a such small subinterval $\Delta x_i = [x_{i-1}, x_i]$ for which $F(x_{i-1}) \cdot F(x_i) < 0$. It means that the desired solution x^*, for which $F(x^*) = 0$, belongs to this interval. In the next stage of the solution method the subinterval Δx_i is treated as an initial one and often also denoted by $[a, b]$.

2.2.1 Bisection Method of Bolzano

Let us have the small closed interval $[a, b]$ including a single solution (real root) of the nonlinear equation $F(x) = 0$, see Fig. 2.3.

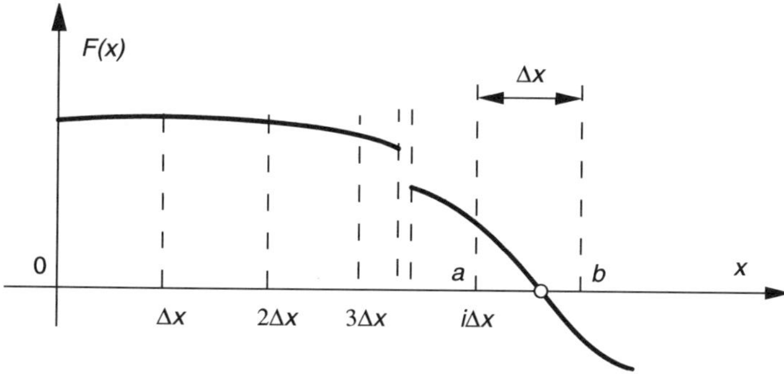

Fig. 2.2

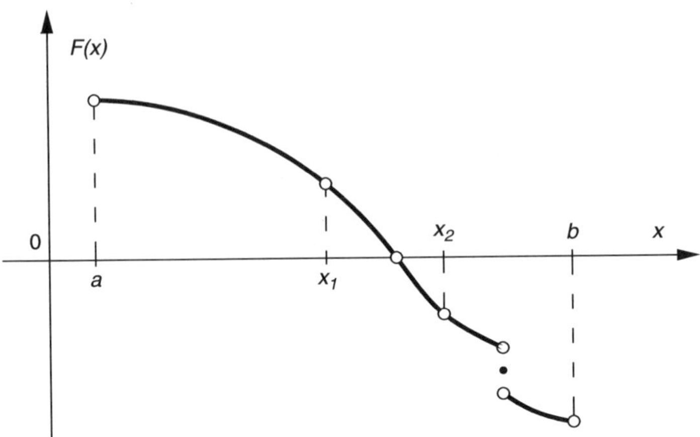

Fig. 2.3

Further search for solution can be performed by means of the bisection method of Bolzano, called briefly bisection method [3, 7]. Using this method, in the first iteration we determine $x_1 = (a+b)/2$ and calculate the value of the function $F(x_1)$. If the values $F(a)$ and $F(x_1)$ have the same signs, then the searched interval decreases, when we assume $a = x_1$. Similarly, if the values $F(x_1)$ and $F(b)$ have equal signs, then we decrease the searched interval by taking $b = x_1$. Using the same procedure, in the second iteration we calculate $x_2 = (a + b)/2$ and the value of the function $F(x_2)$. When the values $F(x_2)$ and $F(b)$ have identical signs, the searched interval becomes smaller by exclusion of the section $[x_2, b]$, that is by taking $b = x_2$. In the opposite case, i.e., when $F(a) \cdot F(x_2) > 0$ reduction of the search interval is obtained by taking $a = x_2$. This process is continued iteratively up to the moment when the length of this interval attains the value which is less than expected, or until the absolute value of the function at the ends of the interval becomes smaller than the prescribed accuracy. The criterion of ending the calculation, formulated in this way, is described by the inequality $|F(x)| \leq \varepsilon$, where ε is an arbitrarily small positive number. The bisection method is used mostly in case of solving equations for which the function $F(x)$ satisfies the Dirichlet's conditions, that it has finite number of discontinuities of the first order in the initial interval $[a, b]$, and it is bilaterally bounded inside this interval. Having in mind that we do not use the values of the derivative of the function $F(x)$ in the calculation process, this method is very effective in the sense that always leads to the solution, but in many cases is ineffective because of the amount of calculations.

2.2.2 The Secant Method

Next, we shall consider the secant method (known also as the false position method), used for the purpose of plotting the function $F(x)$ shown in Fig. 2.4.

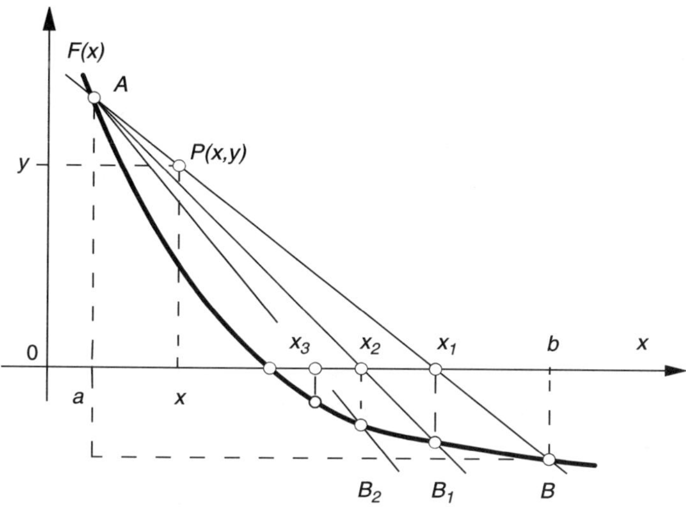

Fig. 2.4

To fix our attention, we assume that the function $F(x)$ takes on the limits of the interval $[a, b]$ the values having different signs, that is $F(b) < 0$. As the first approximation of the desired solution (root), we take the value of x being the coordinate of the intersection point of the secant, passing by the points A and B with the x-axis, i.e., x_1. In order to determine the equation of this secant, it is necessary to find the coordinates at its arbitrary point $P(x, y)$, using for this purpose the similarity of the triangles marked in Fig. 2.4.

$$\frac{F(a) - y}{F(a) - F(b)} = \frac{x - a}{b - a} \tag{2.20}$$

According to Fig. 2.4, the coordinates of the intersection point of the secant and the x-axis are equal to $x = x_1$ and $y = 0$. After introducing these values of the intersection point to the Eq. (2.20), we obtain:

$$x_1 = a - \frac{b - a}{F(b) - F(a)} F(a) \tag{2.21}$$

Comparing the signs of the values $F(a)$ and $F(x_1)$ we come to the conclusion that the desired solution belongs to the interval $[a, x_1]$, because $F(a) \cdot F(x_1) < 0$. According to this conclusion, the search interval can be narrowed by removing the section $[x_1, b]$. In the second iteration, we pose the secant through the points A and $B_1 \equiv [x_1, y = F(x_1)]$, see Fig. 2.4. This secant intersects the x-axis at the point having the coordinates $x = x_2$ and $y = 0$. Proceeding in the similar way, we narrow the searched interval unilaterally, obtaining a convergent series of approximations $\{x_1, x_2, x_3, \ldots, x_n\}$ of the solution we search for. The iteration process just presented should be repeated until the value of the function $|F(x_n)|$ is smaller than an arbitrarily small positive number ε. The method of secants, similar to the bisection method, is effective (always leads to the solution), and in comparison with bisection method is more effective in the numerical sense. This effectiveness should be interpreted as the amount of calculation necessary to obtain the solution. Similarly, as in case of the bisection method we assume that the function $F(x)$ satisfies the Dirichlet conditions in the initial interval $[a, b]$.

2.2.3 Method of Tangents (Newton–Raphson)

In order to solve the nonlinear equations with one unknown, the method of tangents is often used, which is the particular case of the Newton method [7].

In this case, it is necessary that the function $F(x)$ be bounded and differentiable at the given interval $[a, b]$, in which one single solution exists. As an illustration of the basic idea of this method, let us consider the problem of solving the equation $F(x) = 0$; geometric interpretation of which is shown in Fig. 2.5.

We start choosing an arbitrary interior point x_0 of the interval $[a, b]$, for which the function $F(x)$ takes the value $y_0 = F(x_0)$. For this value we calculate the value of derivative $F'(x_0) = F'(x = x_0)$, necessary for determining the equation of the tangent to the curve $y = F(x)$ at $P(x_0, y_0)$. The coordinates x and y at an arbitrary point $P(x, y)$ lying on this tangent satisfy the equation:

$$y - F(x_0) = F'(x_0) \cdot (x - x_0) \tag{2.22}$$

According to Fig. 2.5 the tangent intersects the x-axis at the point having the coordinates $x = x_1$ and $y = 0$. Therefore, after introducing these values into Eq. (2.22)

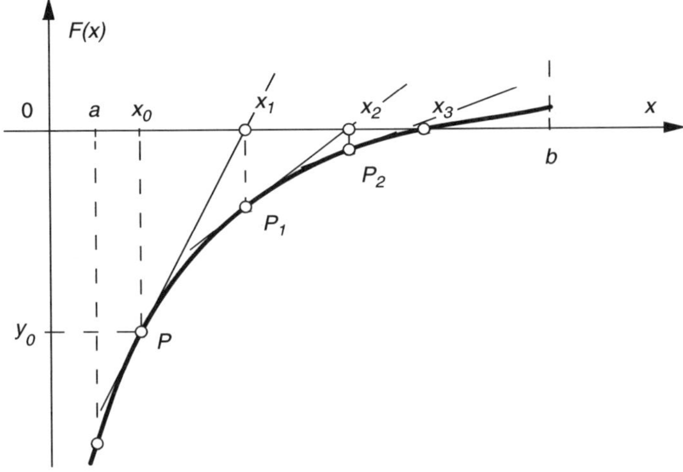

Fig. 2.5

we obtain:

$$x_1 = x_0 - \frac{F(x_0)}{F'(x_0)}, \quad F'(x_0) \neq 0 \qquad (2.23)$$

It is easy to show that $(n + 1)$th approximation of the solution is related to the previous, nth solution, by the following formula:

$$x_{n+1} = x_n - \frac{F(x_n)}{F'(x_n)}, \quad F'(x_n) \neq 0 \qquad (2.24)$$

The criterion at the end of calculations for this method has often been the form of the following inequality:

$$|F(x_n)| \leq \varepsilon \qquad (2.25)$$

where ε is an optionally small positive number. On the basis of the literature on the subject, it is possible to formulate the conclusion that the amount of calculations in each iteration of the tangent method is greater than the corresponding amount of calculations performed by using bisection or secant method. It is mainly due to the necessity of calculation of the derivative of the function $F(x)$ during each iteration. Nevertheless, the number of iterations necessary to determine the solution is much lower, which makes this method more convergent. It was proved in the literature that the tangent method has very good convergence in the (near) neighborhood of the desired solution. This method is therefore willingly used in the final stage of the mixed method and used in case when an extremely high accuracy is required. At the preliminary stages the methods of uniform search and bisection are usually applied.

One disadvantage of the method under discussion is that it requires to evaluate the derivative of function $F(x_n)$ where $n = 0, 1, 2, 3, \ldots$. Unfortunately, for many real functions it is inconvenient to find their derivatives analytically. In such situations a corresponding approximation (difference formula) should be used. If the derivative $F'(x_n)$ in the Newton method formula (2.24) is replaced by means of two successive functional approximations in the formula

$$F'(x_n) \approx F_r^{(1)}(x_n) = \frac{F(x_n) - F(x_{n-1})}{x_n - x_{n-1}}, \quad x_n \neq x_{n-1} \qquad (2.26)$$

the new iteration formula becomes

$$x_{n+1} = x_n - \frac{F(x_n)}{F_r^{(1)}(x_n)} = x_n - \frac{x_n - x_{n-1}}{1 - F(x_{n-1})/F(x_n)}, \quad F(x_n) \neq F(x_{n-1}) \neq 0 \quad (2.27)$$

The Newton–Raphson method modified in this manner is known as the secant method [7]. As with the Newton method, the search of the root by the secant technique may be terminated when consecutive values of x agree to be within some acceptable error or when the function value, $F(x_n)$, is acceptably close to zero, see condition (2.25). The secant method has the same convergence difficulties at a multiple root as does the method of Newton iteration discussed earlier.

2.3 Optimization Methods

The problem of solving the nonlinear equation with one unknown can be transformed into the corresponding, one-dimensional optimization problem. Solution of the equation $F(x) = 0$ is obviously equivalent to the problem of finding the global minimum (equal to zero) of the unimodal function $|F(x)|$. The idea of this transformation is shown in Fig. 2.6.

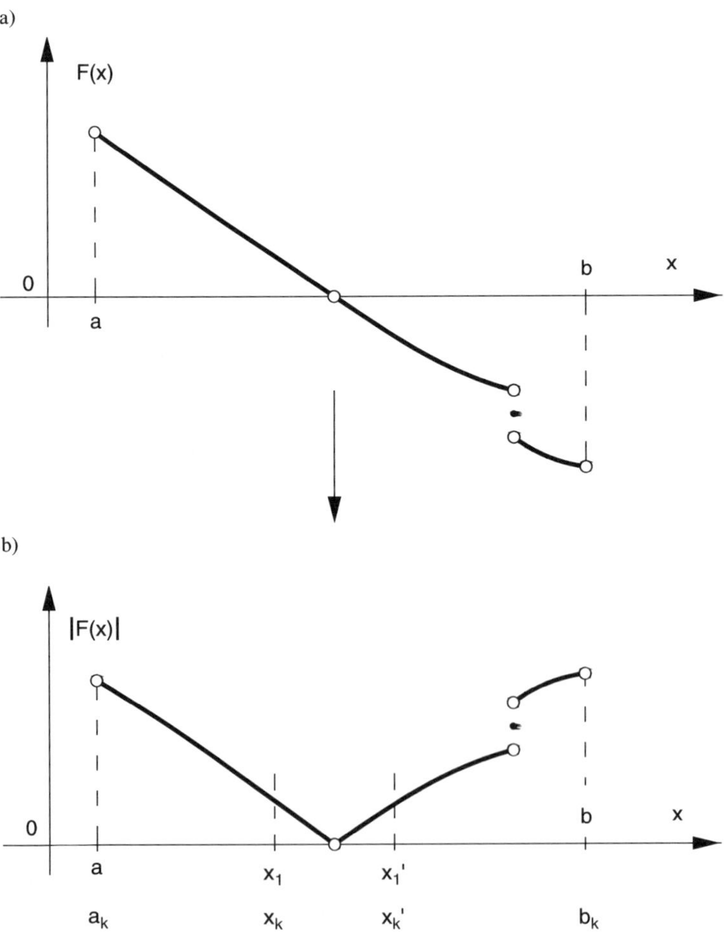

Fig. 2.6

The process of evaluation of the global minimum of the function $|F(x)|$ can be effectively performed by means of the golden or Fibonacci cut methods [6]. As an example, let us consider the algorithm of the golden cut method. In the first iteration, for a given interval $[a, b]$, we determine the interior points (coordinates) x_1 and x_1', see Fig. 2.6(b), from the following formulas:

$$x_1 = a + (b - a)/s^2$$
$$x_1' = a + (b - a)/s \tag{2.28}$$

where

$$s = (1 + \sqrt{5})/2 = 1.618\,033\,989$$

Next, we calculate the values $|F(x_1)|$ and $\left|F(x_1')\right|$. When $|F(x_1)| < \left|F(x_1')\right|$, we take the new search interval as $[a, \; x_1']$. In the opposite case, i.e., when $|F(x_1)| > \left|F(x_1')\right|$, the searching interval is reduced by removing the section $[a, \; x_1]$. When the values $|F(x_1)|$ and $\left|F(x_1')\right|$ of the function are equal, the new interval is chosen as an arbitrary one among the subintervals determined above, that is $[a, \; x_1']$ or $[x_1, \; b]$. In a similar way, we perform each successive iteration. Next, the above process is repeated several times in order to make the search interval narrower. After performing n iterations, the length of this interval is reduced to:

$$|b_n - a_n| = \frac{|b - a|}{s^{n-1}} \tag{2.29}$$

As the criterion for the end of calculation, the following condition can be used:

$$|b_k - a_k| < \varepsilon \tag{2.30}$$

where ε is an arbitrarily small positive number. It is worth emphasizing the fact that in each kth iteration, where $k = 2, 3, \ldots, n$, only one of the coordinates is determined; x_k or x_k' and the corresponding value of the function, i.e., $|F(x_k)|$ or $\left|F(x_k')\right|$. Due to the adopted procedure of determination of the coordinates, see Eq. (2.28), it is possible that one of the coordinates, x_k or x_k', is the same as one of the coordinates determined in the previous iteration. This precious property leads to the reduction of the calculations almost by two. The parameter s, appearing in the relation (2.28), is the inverse of the parameter $t = (-1 + \sqrt{5})/2$ describing the golden cut of a line segment; hence justification of the name of the method. The Fibonacci cut method differs from the one described earlier chiefly due to the manner in which the coordinates x_k and x_k' are evaluated.

Numerous comparative calculations made by means of both methods prove that the golden cut method is only a little less effective. Nevertheless, this method is more frequently used, because the Fibonacci cut method requires previous determination of the number of iterations, which is not always easy to determine.

2.4 Examples of Applications

Example 2.2 As a first example of applications of numerical methods presented above, we consider the problem of designing the lossless slab line. The cross-section of this TEM transmission line is shown in Fig. 2.7.

The distributions of the electrical **E** and magnetic **H** fields are shown in Fig. 2.8.

The line under consideration belongs to the class of the dispersionless, two-conductor waveguides, in which the TEM electromagnetic waves are propagated (Transverse Electro Magnetic Mode). The external conductor of the line is formed by two, parallel equipotential conductive planes. As in the case of other configurations of the TEM transmission lines, circuit parameters, such as the complex amplitudes (phasors) of the voltage **U** and of the current **I**, as well as characteristic impedance Z_0 can be used for a description of them. In order to explain the meaning of these parameters, we assume that a variable voltage (difference of potentials) exists between two arbitrary points A and B lying on the surfaces of the internal and external conductors.

$$u(t) = U_0 \cos(\omega t + \varphi_u) \tag{2.31}$$

At any fixed moment of time t this voltage is equal to:

$$u(t) = \int_A^B \mathbf{E}(t)\mathbf{dl} \tag{2.32}$$

and the integration is performed along an arbitrary line joining the points A and B. The quantity

$$\mathbf{U} = U_0 \exp(j\varphi_u) \tag{2.33}$$

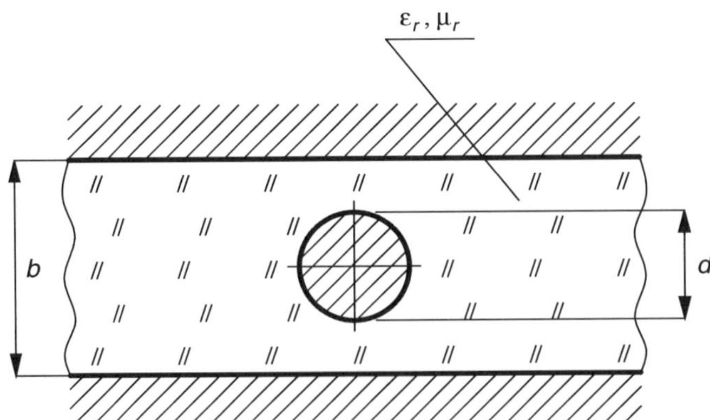

Fig. 2.7

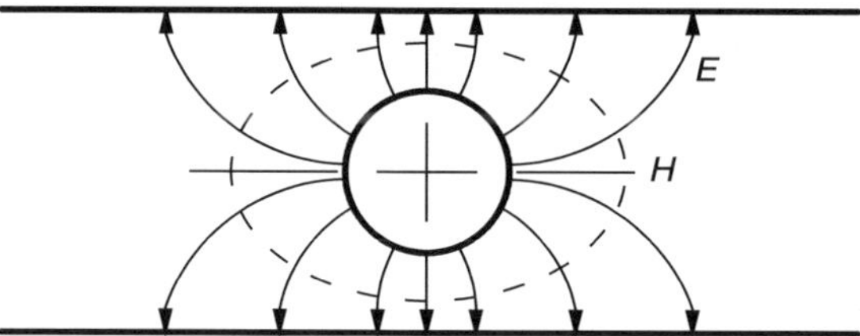

Fig. 2.8

representing the formulas (2.31) and (2.32) is called the complex amplitude (phasor) of the voltage. The variable electric field in the line $\mathbf{E}(t)$ is always accompanied by a variable magnetic field $\mathbf{H}(t)$, having the field lines surrounding the internal conductor of the line. According to the Ampere law, the total current flowing in the internal conductor of the line is:

$$i(t) = I_0 \cos(\omega t + \varphi_i) = \oint_C \mathbf{H}(t)\mathbf{dl} \tag{2.34}$$

where C is an arbitrary closed line surrounding the conductor. The quantity

$$\mathbf{I} = I_0 \exp(j\varphi_i) \tag{2.35}$$

representing the formula (2.34) is called complex amplitude of the current. The ratio of the complex amplitude of the voltage to the complex amplitude of the current is called characteristic impedance of the line:

$$Z_0 = \frac{\mathbf{U}}{\mathbf{I}} \tag{2.36}$$

The characteristic impedance Z_0 should not be confused with the wave impedance defined as:

$$\xi = \frac{\mathbf{E}}{\mathbf{H}} \tag{2.37}$$

The characteristic impedance Z_0 of the slab line depends on its geometrical dimensions b, d, see Fig. 2.7, and the electrical parameters σ, ε_r, μ_r of the dielectric. In general case, the impedance Z_0 can be evaluated by solving the appropriate Laplace's boundary value problem described in detail in Chap. 8 of the present book. Such an approach, however, is rather complicated and strenous. Therefore, for this reason many approximating closed-form design formulas have been elaborated. It is concluded in [9, 10] that the Wheeler's analysis formula is the most accurate and a convenient one for using in an engineering practice. The advantage of this formula lies also in its simple mathematical form, namely:

$$Z_0\left(\frac{d}{b}\right) = 59.952\sqrt{\frac{\mu_r}{\varepsilon_r}}\left(\ln\frac{\sqrt{X}+\sqrt{Y}}{\sqrt{X}-\sqrt{Y}} - \frac{R^4}{30} + 0.014R^8\right), \quad \Omega \qquad (2.38)$$

where

$$R = \frac{\pi}{4}\cdot\frac{d}{b}, \quad X = 1 + 2\,\text{sh}^2(R), \quad Y = 1 - 2\sin^2(R) \quad \pi = 3.141\,592\,653\ldots$$

ε_r – the relative permittivity of the dielectric substrate

μ_r – the relative permeability of the dielectric substrate.

According to the results published in [9, 11], the above formulas make it possible in determining the impedance Z_0 with high accuracy ($\Delta Z_0/Z_0 < 0.001$) for $0.05 \leq d/b < 1$. The design of the slab line on the basis of (2.38) consists in evaluating such a ratio d/b for which the following equation is satisfied:

$$V\left(\frac{d}{b}\right) = Z_0\left(\frac{d}{b}\right) - Z_0 = 0 \qquad (2.39)$$

where Z_0 is the given value of the characteristic impedance. The function $V(d/b)$ assumes its minimum value (zero) at point d/b being sought. This point solution can be effectively evaluated by means of the golden cut method. Some calculation results obtained in this way are presented in Table 2.1.

Example 2.3 Figure 2.9 presents cross-section of the eccentric coaxial line, the inner conductor of which is laterally displaced from its normal position to the axis location.

The characteristic impedance of this TEM transmission line can be evaluated analytically by a field analysis and the resulting expression is:

$$Z_0(x) = 59.952\sqrt{\frac{\mu_r}{\varepsilon_r}}\ln\left(x + \sqrt{x^2 - 1}\right), \quad [\Omega] \qquad (2.40)$$

where
 ε_r – the relative permittivity of the dielectric substrate;
 μ_r – the relative permeability of the dielectric substrate;

$$x = x(b) = \frac{b + \left(a^2 - 4c^2\right)/b}{2a}$$

Table 2.1

Z_0, Ω	ε_r	μ_r	d/b
30	3.78	1	0.4790
50	1	1	0.5486
75	1	1	0.3639

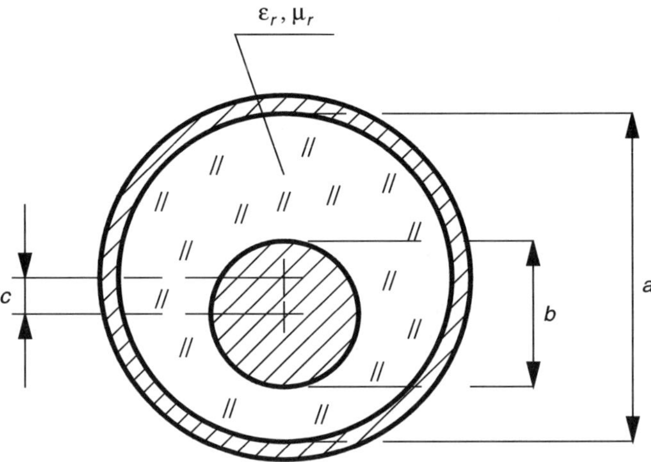

Fig. 2.9

We can easily see that when $c = 0$, the line under discussion becomes the normal coaxial line for which:

$$Z_0\left(\frac{a}{b}\right) = 59.952\sqrt{\frac{\mu_r}{\varepsilon_r}}\ln\left(\frac{a}{b}\right),\ \Omega \tag{2.41}$$

For given values of $a = 7 \times 10^{-3}$ m, $c = 10^{-4}$ m and $Z_0 = 50$, Ω, designing the eccentric coaxial line on the basis of formula (2.40) consists in evaluating such diameter b, see Fig 2.9, for which the following equation is satisfied:

$$|Z_0\,[x(b)] - Z_0| = 0 \tag{2.42}$$

Solving the Eq. (2.42) using the golden cut method we obtain $b = 3.038 \times 10^{-3}$ m. Performing similar calculations for the eccentricity $c = 0$ (the eccentric line becomes the normal coaxial line) we obtain $b = 3.040 \times 10^{-3}$ m. Identical result can be obtained analytically by using the formula:

$$b = a \cdot \exp\left[\frac{-Z_0}{59.952\sqrt{\mu_r/\varepsilon_r}}\right] \tag{2.43}$$

which is the inverted form of (2.41).

References

1. Shoup T.E., Applied numerical methods for the microcomputer. Prentice-Hall, Inc., Englewood Cliffs, NJ, 1984

2. Bender C.M. and S.A. Orszag, Advanced mathematical methods for scientists and engineers. McGraw-Hill, New York, 1978
3. Young D.M. and R.T. Gregory, A survey of numerical mathematics. Addison-Wesley Comp., London, 1973
4. Berezin L.S. and N.P. Zhidkov, Computing methods. Pergamon Press, Elmsford, NY, 1965
5. Bender C.M. and S.A. Orszag, Advanced mathematical methods for scientists and engineers. McGraw-Hill, New York, 1978
6. Bazaraa M.S., Sherali H.D. and C.M. Shetty, Nonlinear programming. Theory and applications. John Wiley and Sons, New York, 1993
7. Mathews J.H., Numerical methods for mathematics, science and engineering. Prentice-Hall Intern. Inc., Englewood Cliffs, NJ, 1992
8. Akai T.J., Applied numerical methods for engineers. John Wiley and Sons, New York, 1994
9. Gunston M.A.R., Microwave transmission line impedance data. Van Nostrandt Reinhold Comp., New York, 1972
10. Wheeler H.A. "The transmission-line properties of a round wire between parallel planes". IRE Trans., AP-3, Antennas and Propagation, October 1955, pp. 203–207
11. Rosłoniec S., Algorithms for computer-aided design of linear microwave circuits. Artech House Inc., Boston, MA, 1990

Chapter 3
Methods for Numerical Solution
of Nonlinear Equations

In this chapter, we consider the systems of n arbitrary equations:

$$
\begin{aligned}
F_1(x_1, x_2, \ldots, x_n) &\equiv F_1(\mathbf{x}) = 0 \\
F_2(x_1, x_2, \ldots, x_n) &\equiv F_2(\mathbf{x}) = 0 \\
&\vdots \\
F_n(x_1, x_2, \ldots, x_n) &\equiv F_n(\mathbf{x}) = 0
\end{aligned}
\tag{3.1}
$$

with n unknowns, creating the vector $\mathbf{x} = [x_1, x_2, \ldots, x_n]$. In the case when at least one of the functions $F_i(\mathbf{x})$ of the system, where $i = 1, 2, 3, \ldots, n$, is nonlinear with respect to at least one unknown (variable) x_j, where $j = 1, 2, 3, \ldots, n$, the system is nonlinear. The nonlinearity of the function $F_i(\mathbf{x})$ with respect to the variable $x_j \subset \mathbf{x}$ should be understood in the following way. Value changes of these functions and related changes of the corresponding variables are not related by means of constant coefficients, independently of the value of variables. Contrary to the linear case, the systems of nonlinear equations cannot be solved by means of direct (simple) methods, because such methods are not elaborated up to now. In consequence, in case of the systems of nonlinear equations, the iterative numerical methods are chiefly used and the most popular among them are:

- Method of direct iterations;
- Iterative parameter perturbation procedure;
- Newton iterative method and
- Equivalent minimization strategies.

3.1 The Method of Direct Iterations

The algorithm of the method of direct iterations is very similar to that of the Gauss–Seidel method used for solving systems of linear equations, see Sect. 1.2.2. During the first stage of the procedure, the system of equations (3.1) is transformed to the following equivalent form:

S. Rosłoniec, *Fundamental Numerical Methods for Electrical Engineering,*
© Springer-Verlag Berlin Heidelberg 2008

$$\begin{aligned}
x_1 &= f_1(x_2, x_3, \ldots, x_n) \\
x_2 &= f_2(x_1, x_3, \ldots, x_n) \\
&\;\;\vdots \\
x_n &= f_n(x_1, x_2, \ldots, x_{n-1})
\end{aligned} \tag{3.2}$$

We begin our calculation by taking the approximate initial solution, i.e.:

$$x_1 = a_1, \quad x_2 = a_2, \quad x_3 = a_3, \ldots, \quad x_n = a_n$$

Then the expressions which permit to find successive, more accurate, approximations of the solution can be written in the following form:

$$\begin{aligned}
x_1 &= f_1(a_2, a_3, \ldots, a_{n-1}, a_n) \\
x_2 &= f_2(x_1, a_3, \ldots, a_{n-1}, a_n) \\
&\;\;\vdots \\
x_i &= f_i(x_1, x_2, \ldots, \; x_{i-1}, a_{i+1}, \ldots, a_n) \\
&\;\;\vdots \\
x_n &= f_n(x_1, x_2, \ldots, x_{n-2}, x_{n-1})
\end{aligned} \tag{3.3}$$

The calculating process, performed according to the formula (3.3) has iterative form, and it means that the approximate solution obtained in the current iteration constitutes the initial approximation (starting point) for the next iteration. These calculations are continued, until the difference

$$R = \sum_{i=1}^{n} |x_i - a_i| \tag{3.4}$$

obtained from the two consecutive solutions (found in the previous and the current iteration) would become sufficiently small. In the limit case (for an infinite number of iterations) the difference R should attain the value equal to zero. The applicability condition for this method is simply identical to the convergence of the vector $\mathbf{x}$, see (3.3), towards a certain limit solution $[\mathbf{x}^*]$. In order to satisfy the above condition, the initial approximation by which the iteration process begins, should be chosen in a possibly close neighborhood of the desired solution. The initial approximation satisfying the above condition is frequently determined by means of the optimization methods. This problem will be explained later in the present chapter. Another indirect way to solve this problem is the application of the iterative parameter perturbation procedure presented in Sect. 3.2.

3.2 The Iterative Parameter Perturbation Procedure

In order to clarify the essence of this procedure, let us reconsider the task of finding the solution of the equation system (3.1) written in the form:

$$F_i(\mathbf{x}) = 0 \tag{3.5}$$

where $i = 1, 2, 3, \ldots, n$. Evaluation of this solution using the procedure of direct iterations is not possible when the initial approximation guaranteeing the convergence of the calculation process is not known. In such cases, it is possible to introduce the second auxiliary equation system of n equations (linear or nonlinear) with n unknowns

$$G_i(\mathbf{x}) \equiv G_i^{(0)}(\mathbf{x}) = 0 \tag{3.6}$$

for which the solution is already known. The systems of equations (3.5) and (3.6) constitute the base of the generalized equation system, defined as follows:

$$G_i^{(k+1)}(\mathbf{x}) = G_i^{(k)}(\mathbf{x}) + [F_i(\mathbf{x}) - G_i^{(k)}(\mathbf{x})]\frac{k}{N} \tag{3.7}$$

Parameter k of this system is an integer taking the values from the interval 0 to N, where N is a fixed integer (e.g., ≥ 10) determining the digitization step. It can be easily shown that for $k = 0$, the system (3.7) is identical to the system described by auxiliary equations (3.6), whose solution is assumed to be known. For the second limiting value of the parameter k, that is for $k = N$, the system (3.7) transforms to the system (3.5). When the value of N is sufficiently great, changing the value of the parameter k with constant step equal to 1 leads to the "almost smooth" transformation of the equation system (3.6) into the system (3.5). For each fixed value of the parameter $k = k'$, beginning from $k = 1$ and ending when $k = N$, the system (3.7) is being solved by the simple iterations method described in the previous section. The solution obtained for this value of the parameter $k = k'$ is used as the initial approximation for the solution to be obtained in the next cycle, i.e., for $k = k'+1$. As the system of equations $G_i^{(k+1)}(\mathbf{x})$ differs small from the system $G_i^{(k)}(\mathbf{x})$ solved in the previous cycle, the convergence probability of the calculations performed by means of the direct iterations is very high. This probability can be increased by diminishing the iteration step; that is by increasing the value of N. Unfortunately, it causes an increase of the number of cycles needed for obtaining the solution of an equation system (3.7) using the method of direct iterations. It does not lead hopefully to the increase of calculation errors, because there is no error accumulation when we use iteration methods. The method we have just explained can be called a "forced method", because the difficult task of finding the solution of the equation system (3.5) has been obtained for the price of big number of auxiliary calculations [1–3].

3.3 The Newton Iterative Method

In this section, we consider the system of n nonlinear equations with n unknowns forming the vector $\mathbf{x} = [x_1, x_2, \ldots, x_n]$:

$$
\begin{aligned}
F_1(x_1, x_2, \ldots, x_n) &\equiv F_1(\mathbf{x}) = 0 \\
F_2(x_1, x_2, \ldots, x_n) &\equiv F_2(\mathbf{x}) = 0 \\
&\vdots \\
F_n(x_1, x_2, \ldots, x_n) &\equiv F_n(\mathbf{x}) = 0
\end{aligned}
\tag{3.8}
$$

Assume that the approximation obtained for the solution of the system (3.8) in the kth iteration is equal to:

$$
x_1 = a_1, \quad x_2 = a_2, \quad x_3 = a_3, \ldots, \quad x_n = a_n
$$

Solution of this system using the Newton method consists in finding such corrections Δx_i, where $i = 1, 2, 3, \ldots, n$, defined for the particular unknowns for which the vector

$$
\mathbf{x} = \left[x_1 = a_1 + \Delta x_1, \; x_2 = a_2 + \Delta x_2, \ldots, \; x_n = a_n + \Delta x_n \right] = \mathbf{a} + \Delta \mathbf{x}
\tag{3.9}
$$

constitutes the solution being sought. Let us develop the functions on the left-side of the system (3.8) into the Taylor series at the known point (vector) $\mathbf{a} = [a_1, a_2, a_3 \ldots, a_n]$

$$
\begin{aligned}
F_1(\mathbf{x}) &\approx F_1(\mathbf{a}) + \frac{\partial F_1}{\partial x_1} \Delta x_1 + \frac{\partial F_1}{\partial x_2} \Delta x_2 + \cdots + \frac{\partial F_1}{\partial x_n} \Delta x_n + \cdots \\
F_2(\mathbf{x}) &\approx F_2(\mathbf{a}) + \frac{\partial F_2}{\partial x_1} \Delta x_1 + \frac{\partial F_2}{\partial x_2} \Delta x_2 + \cdots + \frac{\partial F_2}{\partial x_n} \Delta x_n + \cdots \\
&\vdots \\
F_n(\mathbf{x}) &\approx F_n(\mathbf{a}) + \frac{\partial F_n}{\partial x_1} \Delta x_1 + \frac{\partial F_n}{\partial x_2} \Delta x_2 + \cdots + \frac{\partial F_n}{\partial x_n} \Delta x_n + \cdots
\end{aligned}
\tag{3.10}
$$

According to our previous assumption the vector $\mathbf{x}$, see (3.9), should be the solution of the equation system (3.8). It means that the functions $F_i(\mathbf{x})$, where $i = 1, 2, 3, \ldots, n$, should be equal to zero. After considering this property we obtain:

$$\frac{\partial F_1}{\partial x_1}\Delta x_1 + \frac{\partial F_1}{\partial x_2}\Delta x_2 + \cdots + \frac{\partial F_1}{\partial x_n}\Delta x_n = -F_1(\mathbf{a})$$

$$\frac{\partial F_2}{\partial x_1}\Delta x_1 + \frac{\partial F_2}{\partial x_2}\Delta x_2 + \cdots + \frac{\partial F_2}{\partial x_n}\Delta x_n = -F_2(\mathbf{a}) \qquad (3.11)$$

$$\vdots$$

$$\frac{\partial F_n}{\partial x_1}\Delta x_1 + \frac{\partial F_n}{\partial x_2}\Delta x_2 + \cdots + \frac{\partial F_n}{\partial x_n}\Delta x_n = -F_n(\mathbf{a})$$

In order to assure some clarity to our further considerations, we write the equation system (3.11) in an equivalent matrix form:

$$
\begin{bmatrix}
\dfrac{\partial F_1}{\partial x_1} & \dfrac{\partial F_1}{\partial x_2} & \cdots & \dfrac{\partial F_1}{\partial x_n} \\[2mm]
\dfrac{\partial F_2}{\partial x_1} & \dfrac{\partial F_2}{\partial x_2} & \cdots & \dfrac{\partial F_2}{\partial x_n} \\[2mm]
\vdots & \vdots & \cdots & \vdots \\[2mm]
\dfrac{\partial F_n}{\partial x_1} & \dfrac{\partial F_n}{\partial x_2} & \cdots & \dfrac{\partial F_n}{\partial x_n}
\end{bmatrix}
\cdot
\begin{bmatrix}
\Delta x_1 \\ \Delta x_2 \\ \vdots \\ \Delta x_n
\end{bmatrix}
=
\begin{bmatrix}
-F_1(\mathbf{a}) \\ -F_2(\mathbf{a}) \\ \vdots \\ -F_n(\mathbf{a})
\end{bmatrix}
\qquad (3.12)
$$

In the consecutive developments of the functions $F_i(\mathbf{x})$, where $i = 1, 2, 3, \ldots, n$, at the point $\mathbf{a}$, only the first linear terms of the Taylor series, for all the unknowns x_j, $j = 1, 2, 3, \ldots, n$, have been taken into account. The desired vector $\mathbf{x} = \mathbf{a} + \Delta\mathbf{x}$ represents therefore only a consecutive, better approximation of the desired solution. All partial derivatives of the functions $F_i(\mathbf{x})$ are determined at the point $\mathbf{a}$. It is easy to show that the system of equations (3.12) obtained in this way is linear with respect to the corrections $\Delta\mathbf{x} = [\Delta x_1, \ \Delta x_2, \ \Delta x_3, \ldots, \ \Delta x_n]$ formulated for all the unknown variables.

This system is usually being solved by the Gauss elimination method with the choice of pivoting (principal) element. After determination of the corrections $[\Delta x_1, \Delta x_2, \Delta x_3, \ldots, \Delta x_n]$, new approximation is made by taking $\mathbf{a} = \mathbf{x}$ and the calculations are repeated according to the same algorithm. The iteration process can be stopped only after obtaining the vector $\mathbf{x}$ satisfying equations (3.8) up to the desired accuracy. For each iteration, it is necessary to verify whether the equation system (3.12) is not singular. Thus, it is necessary to verify whether the determinant of the coefficient matrix (Jacobian) is not equal to zero or not infinitely small. In case when the absolute value of this determinant (Jacobian) is too small, the excessive calculation errors may occur. This situation happens most frequently in the final phase of calculations, during which the calculated vector $\mathbf{x}$ becomes very close to the final solution. In such case the derivatives of the functions $F_i(\mathbf{x})$ constituting the coefficient matrix are close to zero. In the engineering practice, we often meet the common problem of solving the system of two nonlinear equations with two unknowns. Therefore, let us consider especially this problem more in detail:

$$F_1(x_1, x_2) = 0$$
$$F_2(x_1, x_2) = 0$$

$$(3.13)$$

Assume first that the nth approximation of the desired solution, that is $x_1^{(n)}$ and $x_2^{(n)}$, is known. According to the theory presented above, the equation system (3.12) formulated for this task has the form:

$$\mathbf{W}(\mathbf{x}^{(n)}) \cdot \Delta \mathbf{x} = -\mathbf{F}(\mathbf{a}^{(n)})$$

$$(3.14)$$

where

$$\mathbf{W}(\mathbf{x}^{(n)}) = \begin{bmatrix} \dfrac{\partial F_1}{\partial x_1} & \dfrac{\partial F_1}{\partial x_2} \\[2mm] \dfrac{\partial F_2}{\partial x_1} & \dfrac{\partial F_2}{\partial x_2} \end{bmatrix}$$

$$(3.15)$$

Solving this equation system with respect to the correction vector $\Delta \mathbf{x}$ we obtain the next $(n+1)$th, better approximation of the desired solution. For that purpose, let us multiply both sides of the equation system (3.14) by the inverse of the coefficient matrix, namely $\mathbf{W}^{-1}(\mathbf{x}^{(n)})$. After this multiplication we obtain:

$$\mathbf{W}^{-1}(\mathbf{x}^{(n)}) \cdot \mathbf{W}(\mathbf{x}^{(n)}) \cdot \Delta \mathbf{x} = -\mathbf{W}^{-1}(\mathbf{x}^{(n)}) \cdot \mathbf{F}(\mathbf{a}^{(n)})$$

$$(3.16)$$

According to the relations (2.24) and (2.25) given in the preceding chapter, Eq. (3.16) can be written as:

$$\Delta \mathbf{x} = -\mathbf{W}^{-1}(\mathbf{x}^{(n)}) \cdot \mathbf{F}(\mathbf{a}^{(n)})$$

$$(3.17)$$

The left-side of the Eq. (3.17) represents the correction vector, which added to the previous approximation forms a new more accurate approximation. The next, $(n+1)$th approximation of the desired solution is:

$$\mathbf{x}^{(n+1)} = \mathbf{x}^{(n)} - \mathbf{W}^{-1}(\mathbf{x}^{(n)}) \cdot \mathbf{F}(\mathbf{a}^{(n)})$$

$$(3.18)$$

The crucial problem we meet when we want to implement the algorithm explained above, in order to determine the next approximation of the needed solution, is finding the inverse of the coefficient matrix described by the relation (3.12). This matrix can be easily found using the algorithm presented in Appendix 2. Finally, we obtain:

$$x_1^{(n+1)} = x_1^{(n)} - \frac{1}{J}\left[F_1 \frac{\partial F_2}{\partial x_2} - F_2 \frac{\partial F_1}{\partial x_2}\right]$$

$$x_2^{(n+1)} = x_2^{(n)} + \frac{1}{J}\left[F_1 \frac{\partial F_2}{\partial x_1} - F_2 \frac{\partial F_1}{\partial x_1}\right]$$

$$(3.19)$$

where

$$J = \begin{bmatrix} \dfrac{\partial F_1}{\partial x_1} & \dfrac{\partial F_1}{\partial x_2} \\[2mm] \dfrac{\partial F_2}{\partial x_1} & \dfrac{\partial F_2}{\partial x_2} \end{bmatrix} = \frac{\partial F_1}{\partial x_1} \cdot \frac{\partial F_2}{\partial x_2} - \frac{\partial F_1}{\partial x_2} \cdot \frac{\partial F_1}{\partial x_2} \neq 0 \tag{3.20}$$

Functions F_1, F_2 appearing in the relations (3.19) and (3.20) and their derivatives are determined at the point $[x_1^{(n)}, x_2^{(n)}]$. During the calculation of the $(n+1)$th approximation of the desired solution it may occur that the Jacobian J turns out to be equal to zero. In such case, the whole calculation process will be interrupted, because the division by zero is prohibited. In order to prevent such events, the corresponding security procedures should be included into the program. We usually adopt the following practical procedure. After finding out that the Jacobian J is equal to zero, a small increment ε (negative or positive) is added to one, freely chosen variable, for example $x_1^{(n)}$. For such incremented value $[x_1^{(n)} + \varepsilon, x_2^{(n)}]$ the functions F_1, F_2, their derivatives and the determinant J is next calculated. When absolute value of the determinant J exceeds the given positive, nonzero value then the new approximation $[x_1^{(n+1)}, x_2^{(n+1)}]$ can be calculated according to formulas (3.19). In the opposite case, we add the increment ε to the different variable and repeat the whole process once more. Sometimes the whole calculation process may be divergent with respect to the limit equal to the solution, because the distance (in the n-dimensional space) of the assumed initial approximation (starting point) from the solution is too large. In such cases, we can transform the problem described by the Eqs. (3.8) into an equivalent optimization problem, according to the method described in Sect. 3.4.

Example 3.1 As an illustration for the algorithm of the Newton method presented above we propose to evaluate the solutions of the following pair of equations:

$$F_1(\mathbf{x}) \equiv x_1^2 + x_2^2 - 5 = 0$$

$$F_2(\mathbf{x}) \equiv x_1^2 - x_2^2 + 3 = 0$$

We assume that the initial approximation of the solution, i.e., the starting point $(x_1^{(0)}, x_2^{(0)}) \equiv (0.5, 1)$ is known. The approximate values of the solution $(1, 2)$, found in the consecutive iterations are given in Table 3.1.

Table 3.1

n	0	1	2	3	4
$x_1^{(n)}$	0.5	1.25	1.025	1.0003	1
$x_2^{(n)}$	1	2.5	2.05	2.0006	2
F_1	−3.75	2.8125	0.2531	3.051×10^{-3}	0
F_2	2.25	−1.6875	−0.1519	-1.830×10^{-3}	0
J	−4	−25	−16.81	−16.0097	−16

The solution (1, 2) evaluated above is only one of the set of four solutions: (1, 2), (−1, 2), (−1, −2) and (1, −2) placed in the particular quadrants of the two-dimensional space (plane) $x_1 x_2$. The consecutive solutions mentioned above can be evaluated in the similar manner by appropriate choice of the initial approximation (starting point). Of course the point (0, 0) is not suitable for this purpose. This restriction is justified by the fact that at this central point the Jacobian (3.20) takes zero value.

Example 3.2 When we use the Bairstow's method to find the roots of a polynomial equation (2.1), the system of equations (2.7) is being solved by means of the Newton iterative method. According to the formulas (3.19) and (3.20), the consecutive, $(n + 1)$th approximation of the coefficients p and q, which we must determine, is calculated by means of the following relations

$$p^{(n+1)} = p^{(n)} - \frac{1}{J} \left(R \cdot \frac{\partial S}{\partial q} - S \cdot \frac{\partial R}{\partial q} \right)$$

$$q^{(n+1)} = q^{(n)} + \frac{1}{J} \left(R \cdot \frac{\partial S}{\partial p} - S \cdot \frac{\partial R}{\partial p} \right)$$

(3.21)

where

$$J = \frac{\partial R}{\partial p} \cdot \frac{\partial S}{\partial q} - \frac{\partial S}{\partial p} \cdot \frac{\partial R}{\partial q} \neq 0$$

During the process of calculation of the partial derivatives appearing in the above formulas it should be remembered that the coefficients R and S are functions of the coefficients $b_1, b_2, b_3, \ldots, b_{n-2}$. These coefficients depend in turn on p and q, see formulas (2.5). Therefore, we have to determine the sequence of partial derivatives of the coefficients $b_1, b_2, b_3, \ldots, b_{n-2}$, R and S with respect to p and q. These derivatives can be found by using relations (2.12–2.15) given in the Sect. 2.1.

3.4 The Equivalent Optimization Strategies

In Sect. 2.3, the problem of finding solution of the one nonlinear equation with one unknown was transformed into an equivalent optimization problem, which was subsequently solved by means of the golden cut method. This approach can be generalized to the form useful in case of the system of n arbitrary equations with n unknowns. For this end, we take the function $F_i(\mathbf{x})$, see the equation system (3.1), and construct the new function which attains the global minimum equal to zero for the vector $\mathbf{x}$, which represents the desired solution. The examples of such functions are:

$$U(\mathbf{x}) = \sum_{i=1}^{n} [F_i(\mathbf{x})]^2 \tag{3.22}$$

$$V(\mathbf{x}) = \sum_{i=1}^{n} |F_i(\mathbf{x})| \tag{3.23}$$

In practice, the function (3.22) is most frequently used, because it is easier to differentiate, than the function (3.23). When applied to the problem considered in Example 3.1, the function $U(\mathbf{x})$ has the form:

$$U(x_1, x_2) = (x_1^2 + x_2^2 - 5)^2 + (x_1^2 - x_2^2 + 3)^2 \tag{3.24}$$

The function (3.24) assumes the minimum value equal to zero at the point [1, 2], which is the desired solution. Outside this point, i.e., for the arbitrary values of the variables x_1, x_2 this function remains positive. This property is shown in Fig. 3.1.

In the close neighborhood of the point (1, 2) the analyzed function describes the surface of the shape similar to the paraboloid of revolution. Therefore, starting at an arbitrary point $P(x_1, x_2)$ lying on that surface and moving in the direction $\mathbf{d}(x_1, x_2)$, pointing to the region in which the function in question decreases, we approach the desired solution. The trajectory from the starting point, up to the point corresponding to the solution may of course consist of several linear sections, positions of which in the two-dimensional space x_1, x_2 (on the plane) define the corresponding directions of improvement, that is $\mathbf{d}_i(x_1, x_2)$. The simplest representative of the functions describing the minimization direction $\mathbf{d}$ is the anti-gradient function defined by:

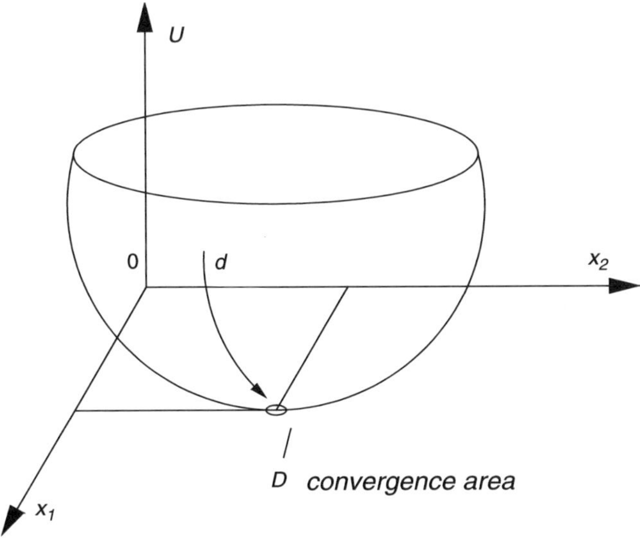

Fig. 3.1

$$-\nabla U(x_1, x_2) = -\mathbf{k}_1 \frac{\partial U}{\partial x_1} - \mathbf{k}_2 \frac{\partial U}{\partial x_2} \tag{3.25}$$

where $\mathbf{k}_1$ and $\mathbf{k}_2$ are the unit vectors (versors) corresponding to the variables x_1 and x_2, respectively. In general case, when we are dealing with a function U depending on n independent variables, the anti-gradient of this function is:

$$-\nabla U(x_1, x_2, x_3, \ldots, x_n) = -\nabla U(\mathbf{x}) = -\sum_{i=1}^{n} \mathbf{k}_i \frac{\partial U(\mathbf{x})}{\partial x_i} \tag{3.26}$$

where $\mathbf{k}_i$ is the versor related to the variable x_i. The direction of minimization, evaluated by (3.26) constitutes the basis for the simplest gradient optimization method, known as the steepest descent method [4, 5]. This method is most effective in the initial phase of searching, when the partial derivatives of the function $U(\mathbf{x})$ are considerably different from zero. As we approach the solution, values of these derivatives decrease, and in consequence the direction $\mathbf{d}$ is determined with decreasing precision. For this reason, the process of searching for the solution in the close neighborhood should be continued by means of more effective methods, as the Fletcher–Reeves and Davidon–Fletcher–Powell method. The algorithms of these methods and corresponding application examples can be found in the literature [4–6].

3.5 Examples of Applications in the Microwave Technique

Example 3.3 As the first example of application of the numerical methods discussed above, let us design the air coupled slab lines for given values of characteristic impedances Z_{0e} and Z_{0o} [7, 8]. The cross-section of these lossless TEM transmission lines is shown in Fig. 3.2.

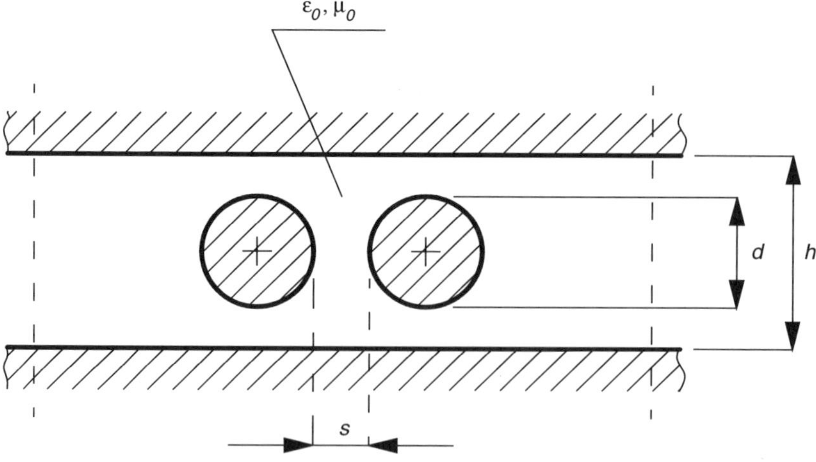

Fig. 3.2

The outer conductor of these lines is formed by two parallel equipotential conducting planes. The lines of this type are used broadly in various microwave devices such as different kinds of filters, directional couplers, etc., chiefly intended to work at medium- and high-power levels. Due to the circular geometry of the inner conductors, they are easy to manufacture and make possible in obtaining the good impedance and constructional matchings with respect to adjacent coaxial lines. Basic electrical parameters of the coupled lines are the two characteristic impedances defined for the symmetrical (in-phase) and antisymmetrical excitations, denoted by Z_{0e} and Z_{0o}, respectively. The coupled lines are excited symmetrically, even-mode $(++)$, if the voltages applied to their inner conductors are:

$$u_1(t) = U_0 \cos(\omega t + \varphi_0)$$
$$u_2(t) = U_0 \cos(\omega t + \varphi_0)$$

(3.27)

The electric field distribution obtained for such symmetrical excitation is shown in Fig. 3.3(a).

The ratio of complex amplitudes of the voltage $\mathbf{U} = U_0 \exp(j\varphi_0)$ and the currents $\mathbf{I}$ flowing in each conductor is called the even-mode characteristic impedance and is denoted as Z_{0e}. In the case when the voltages applied to the inner conductors has the phase lag at an angle $180°$, that is:

$$u_1(t) = U_0 \cos(\omega t + \varphi_0)$$
$$u_2(t) = U_0 \cos(\omega t + \varphi_0 + \pi)$$

(3.28)

(a)

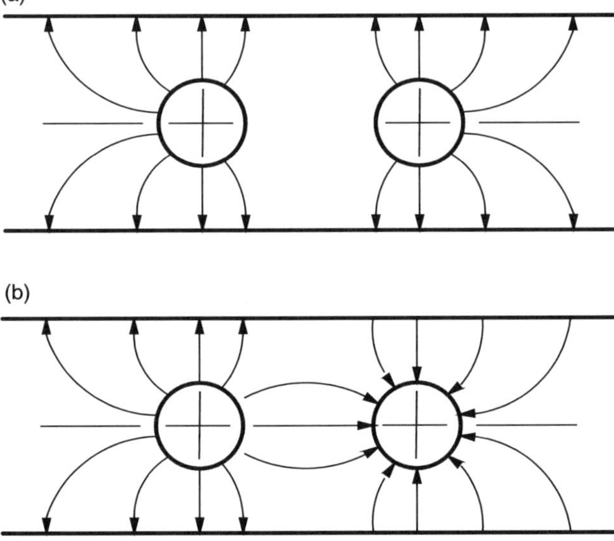

(b)

Fig. 3.3

the excitation is called antisymmetrical, odd-mode $(+-)$. The corresponding electrical field distribution is given in Fig. 3.3(b). In this case there is a strong electrical field between the inner conductors, which may cause a breakdown. The odd-mode characteristic impedance, Z_{0o}, is defined as the ratio of the complex amplitude of the voltage $\mathbf{U} = U_0 \exp(j\varphi_0)$ to the complex amplitude of the currents $\mathbf{I}$ flowing through each inner conductors. For each pair of the design parameters $x = d/h$ and $y = s/h$, see Fig. 3.2, the impedances Z_{0e} and Z_{0o} satisfy the following condition:

$$Z_{0o} \leq Z_0 \leq Z_{0e} \tag{3.29}$$

where Z_0 is the characteristic impedance of the lines without coupling. The impedances Z_{0e} and Z_{0o} are equal, if the distance s between the inner conductors of the line is sufficiently large (theoretically infinite). In the microwave technology, the pair of coupled lines is often characterized by the coupling coefficient defined by:

$$k = \frac{Z_{0e} - Z_{0o}}{Z_{0e} + Z_{0o}} \tag{3.30}$$

It follows from the inequalities (3.29) that $0 \leq k \leq 1$. The characteristic impedances Z_{0e} and Z_{0o} of lines under consideration depend on geometrical dimensions b, h, s, see Fig. 3.2. Similarly for the single slab line, see Example 2.2, the impedances Z_{0e} and Z_{0o} can be evaluated by solving the appropriate Laplace boundary value problem described in detail in Chap. 8. However, such field approach is rather complicated, and therefore inconvenient for engineering practice. Therefore, for this reason many approximating closed-form design formulas have been elaborated. The example of such formulas useful for engineering is presented below [8].

$$Z_{0e}(x, y) = 59.952 \ln\left[\frac{0.523962}{f_1(x)f_2(x, y)f_3(x, y)}\right], \ \Omega$$

$$\tag{3.31}$$

$$Z_{0o}(x, y) = 59.952 \ln\left[\frac{0.523962 f_3(x, y)}{f_1(x)f_4(x, y)}\right], \ \Omega$$

where

$$x = d/h$$

$$y = s/h$$

$$f_1(x) = x \frac{a(x)}{b(x)}$$

$$f_2(x, y) = \begin{cases} c(y) - xd(y) + e(x)g(y) & \text{for } y < 0.9 \\ 1 + 0.004 \exp(0.9 - y) & \text{for } y \geq 0.9 \end{cases}$$

$$f_3(x, y) = \text{th}\left(\pi \frac{x+y}{2}\right)$$

$$f_4(x, y) = \begin{cases} k(y) - xl(y) + m(x)n(y) & \text{for } y < 0.9 \\ 1 & \text{for } y \geq 0.9 \end{cases}$$

$$a(x) = 1 + \exp(16x - 18.272)$$

$$b(x) = \sqrt{5.905 - x^4}$$

$$c(y) = -0.8107y^3 + 1.3401y^2 - 0.6929y + 1.0892 + \frac{0.014002}{y} - \frac{0.000636}{y^2}$$

$$d(y) = 0.11 - 0.83y + 1.64y^2 - y^3$$

$$e(x) = -0.15 \exp(-13x)$$

$$g(y) = 2.23 \exp\left(-7.01y + 10.24y^2 - 27.58y^3\right)$$

$$k(y) = 1 + 0.01\left(-0.0726 - \frac{0.2145}{y} + \frac{0.222573}{y^2} - \frac{0.012823}{y^3}\right)$$

$$l(y) = 0.01\left(-0.26 + \frac{0.6866}{y} + \frac{0.0831}{y^2} - \frac{0.0076}{y^3}\right)$$

$$m(x) = -0.1098 + 1.2138x - 2.2535x^2 + 1.1313x^3$$

$$n(y) = -0.019 - \frac{0.016}{y} + \frac{0.0362}{y^2} - \frac{0.00234}{y^3}$$

The relations given above guarantee the accuracy of the impedances $Z_{0e}(x, y)$ and $Z_{0o}(x, y)$ not worse than 0.7% for $0.1 \leq x \leq 0.8$ and $0.1 \leq y$.

For given values of impedances $Z_{0e} = Z'_{0e}$ and $Z_{0o} = Z'_{0o}$ the design of these lines consists in determining such values of parameters $x = d/h$ and $y = s/h$, see Fig. 3.2, for which the following equations are satisfied:

$$V_1(x, y) = Z_{0e}(x, y) - Z'_{0e} = 0$$

$$V_2(x, y) = Z_{0o}(x, y) - Z'_{0o} = 0 \qquad (3.32)$$

The system of nonlinear equations (3.32) can be effectively solved by using the Newton method described in Sect. 3.3. Thus, the $(n + 1)$th approximation of the desired solution is:

$$x^{(n+1)} = x^{(n)} - \frac{1}{J}\left(V_1 \frac{\partial V_2}{\partial y} - V_2 \frac{\partial V_1}{\partial y}\right)$$

$$y^{(n+1)} = y^{(n)} + \frac{1}{J}\left(V_1 \frac{\partial V_2}{\partial x} - V_2 \frac{\partial V_1}{\partial x}\right) \qquad (3.33)$$

where

$$J = \begin{vmatrix} \dfrac{\partial V_1}{\partial x} & \dfrac{\partial V_1}{\partial y} \\[2ex] \dfrac{\partial V_2}{\partial x} & \dfrac{\partial V_2}{\partial y} \end{vmatrix} \neq 0$$

and $(x^{(n)}, y^{(n)})$ is the nth (previous) approximation. The functions $V_1(x, y)$ and $V_2(x, y)$ appearing in the expressions (3.32) and their derivatives are calculated at point $(x^{(n)}, y^{(n)})$. Before execution of the $(n+1)$th iteration, the value of the Jacobian J should be verified. When its value is different from zero, the calculations can be continued. In the opposite case, we should perform a small shift of the point $(x^{(n)}, y^{(n)})$, by adding a small number ε to one of the variables in order to satisfy the condition $J \neq 0$. After this "small perturbation" the calculations can be continued according to the formulas (3.33). The initial approximation assuring convergence of the calculation process of this algorithm can be found according to the following expressions:

$$x^{(0)} = \frac{4}{\pi} \exp\left(\frac{-Z_0}{59.952\sqrt{0.987 - 0.171k - 1.723k^3}} \right)$$

$$y^{(0)} = \frac{1}{\pi} \ln\left(\frac{r+1}{r-1} \right) - x^{(0)}$$

$$(3.34)$$

where

$$Z_0 = \sqrt{Z_{0e} Z_{0o}}$$

$$k = \frac{Z_{0e} - Z_{0o}}{Z_{0e} + Z_{0o}} \qquad r = \left(\frac{4}{\pi x^{(0)}} \right)^{0.001 + 1.117k - 0.683k^2}$$

As the criterion allowing to stop the calculations, the following condition is most frequently used:

$$V_1^2(x, y) + V_2^2(x, y) \leq Z_{0e} Z_{0o} \times 10^{-6} \qquad (3.35)$$

This inequality guarantees the precision not worse than 0.1%. An example of such calculations is presented in Table 3.2.

Example 3.4 This example deals with a problem concerning design of the non-commensurate four-section impedance transformer with the maximally flat insertion loss function. Calculations will be performed for different values of the ratio $R = Z_{02}/Z_{01}$ of impedances that are to be matched. The electrical scheme of the designed transformer is shown in Fig. 3.4(a).

Table 3.2

$\varepsilon_r = 1, \; \mu_r = 1$			
$Z_{0e}\ \Omega$	$Z_{0o}\ \Omega$	$x = d/h$	$y = s/h$
51.607	48.443	0.5483	0.7882
52.895	47.263	0.5460	0.6119
55.277	45.227	0.5418	0.4439
59.845	41.774	0.5286	0.2802
69.371	36.038	0.4893	0.1460

The insertion loss function of this transformer is:

$$L = \frac{P_{\mathrm{we}}}{P_{\mathrm{wy}}} = 1 + |T_{21}|^2 = 1 + T_{21}^2 \tag{3.36}$$

where

$$T_{21} = C\{D - 2\cos(2a_2\theta) + E\cos(2\theta + 2a_2\theta) - F[2\cos(2\theta) - \cos(2a_2\theta - 2\theta)]\}$$
$$C = (R-1)(R^2+1)(R+1)^2/(16R^2\sqrt{R}), \quad D = 2(R^2+1)/(R+1)^2$$
$$E = (R+1)^2/(R^2+1), \quad F = (R-1)^2/(R^2+1)$$

(a)

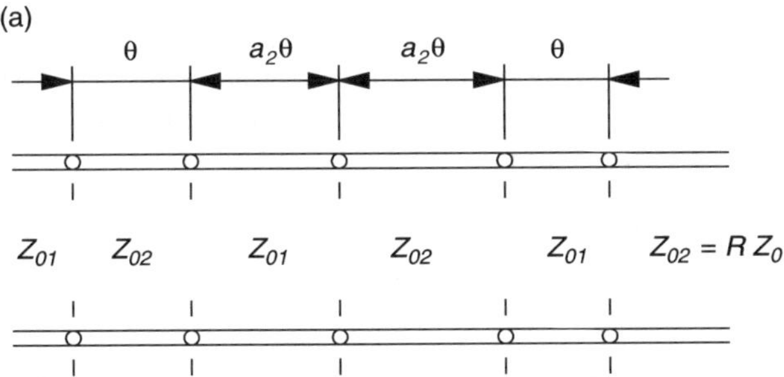

(b)

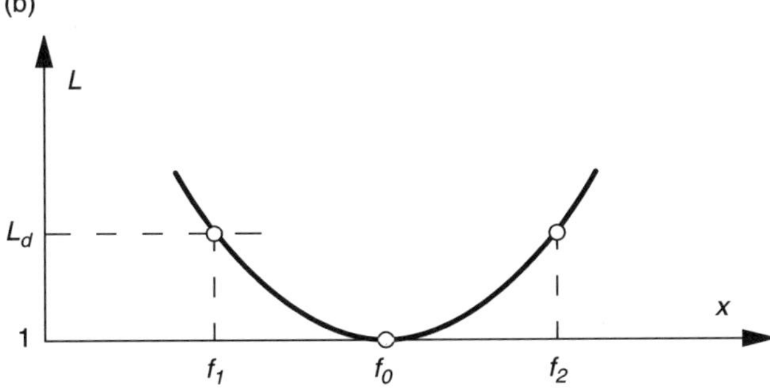

Fig. 3.4

Design of the transformer of this type consists in finding such values of electrical lengths $\theta(f_0)$ and $a_2\theta(f_0)$ of the particular sections, for which the insertion loss function (3.36) will satisfy the following conditions:

$$L[\theta(f_0),\ a_2] = 1$$
$$L^{(k)}[\theta = \theta(f_0),\ a_2] = 0 \tag{3.37}$$

where the index $k = 1, 2, 3$ describes the derivative of the order k with respect to $\theta(f)$ calculated at the point $\theta = \theta(f_0)$. It follows from analysis of the relation (3.36) that conditions (3.37) are satisfied if:

$$\frac{1}{C}T_{21}[\theta(f_0),\ a_2] = 0$$
$$\frac{d}{d\theta}\left\{\frac{1}{C}T_{21}[\theta = \theta(f_0),\ a_2]\right\} = 0 \tag{3.38}$$

The derivative appearing in the system of equations (3.38) can be determined using the following explicit formula:

$$\frac{d}{d\theta}\left[\frac{1}{C}T_{21}(\theta, a_2)\right] = 4a_2\sin(2a_2\theta) - E(2 + 2a_2)\sin(2\theta + 2a_2\theta)$$
$$- F[-4\sin(2\theta) + (2a_2 - 2)\sin(2a_2\theta - 2\theta)] \tag{3.39}$$

where the coefficients E and F are defined by the relation (3.36). The system of equations (3.38) can be effectively solved by means of the Newton method, that is in the similar way as the previous Example 3.2. Convergence of the iteration process is guaranteed, when the following initial approximation is adopted:

$$\theta_0(R) = 0.273\exp(0.188 - 0.131R + 0.004R^2),$$
$$a_2(R) = 2.948 + 0.175R \tag{3.40}$$

Results of some chosen calculations performed by means of the algorithm presented above are given in Table 3.3.

For the three transformers designed in this way (see the columns 1, 4 and 5 of the Table 3.3), the frequency responses of the voltage standing wave ratio (VSWR) have been evaluated. The corresponding plots are presented in Fig. 3.5.

Table 3.3

R	$\theta_0(f_0, R)$, rad	$a_2(R)$	$\theta(f_0, R)$, rad	a_2
1.5	0.2702	3.1395	0.2734	3.1341
2	0.2475	3.2104	0.2592	3.2168
3	0.2304	3.4072	0.2306	3.4104
4	0.2074	3.6093	0.2067	3.6058
5	0.1904	3.7979	0.1876	3.7929

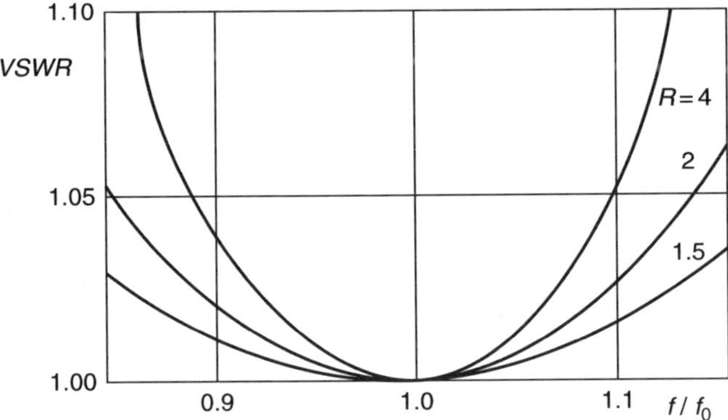

Fig. 3.5

These responses have been evaluated by using the formula:

$$\text{VSWR} = \frac{1 + |\Gamma|}{1 - |\Gamma|} \tag{3.41}$$

where

$$|\Gamma| = \frac{|T_{21}|}{\sqrt{1 + |T_{21}|^2}}$$

The formula (3.41) was determined for the case, when the analyzed transformers are lossless. Such assumption is fully justified in case of the small transformers operating in the UHF band and at lower frequencies of the microwave range.

Example 3.5 Let us consider once again the design problem for the noncommensurate four-section impedance transformer whose electrical scheme is shown in Fig. 3.4(a). The insertion loss function $L(f)$ of this transformer should be similar to that shown in Fig. 3.6. In other words, the insertion loss function $L(f)$ of the designed transformer should be equal to ripples similar as the insertion

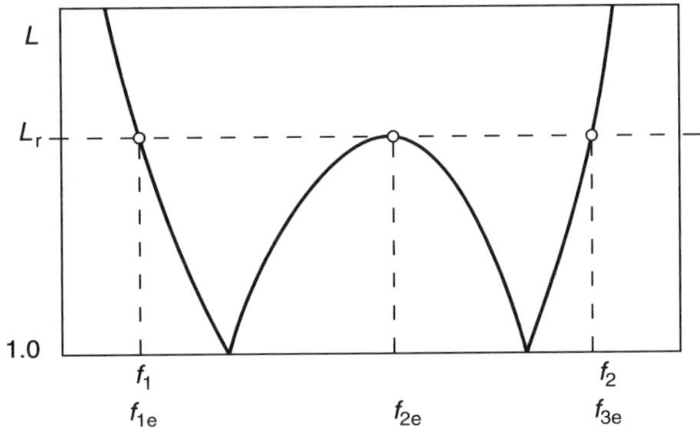

Fig. 3.6

loss function of the corresponding Chebyshev transformer composed of quarter wavelength sections [9].

We assume that the ratio $R = Z_{02}/Z_{01}$ of impedance that are to be matched and the band coverage coefficient $x = f_2/f_1 = \theta(f_2)/\theta(f_1)$ are given. The problem formulated above leads to the conclusion that we should find such electrical length $\theta = \theta(f_1)$ and the coefficient a_2, see Fig. 3.4(a), for which the curve $L(f)$ will be similar to that shown in Fig. 3.6. It is clearly visible that this response should take the identical extreme values at the three points representing different frequencies $f_{1e} = f_1$, f_{2e}, $f_{3e} = f_2$ lying inside the given bandwidth $(f_1 - f_2)$. This requirement can be described by the following nonlinear equations:

$$T_{21}(\theta_a, a_2) - T_{21}(\theta_a \cdot f_{3e}/f_{1e}, a_2) = 0$$
$$T_{21}(\theta_a, a_2) + T_{21}(\theta_a \cdot f_{2e}/f_{1e}, a_2) = 0$$

$$(3.42)$$

where $T_{21}(\theta, a_2)$ is the element of the wave transmission matrix described by the formula (3.36). The system of equations (3.42) can be solved by means of the Newton–Raphson method when the appropriate initial approximation $[\theta^{(0)}(R, x), a_2^{(0)}(R, x)]$ is known. Of course, this initial approximation should ensure convergence of the calculation process. In other words, the initial approximation must lie in the sufficiently close neighborhood of the desired solution. The two-variable functions $\theta^{(0)}(R, x)$ and $\alpha_2^{(0)}(R, x)$ satisfying this requirement can be found in the following way. Assume that the solution of the equation system (3.42) is known at the arbitrary point (R_0, x_0) of our region of interest. This solution should be treated as an initial approximation for the solution of this system of equations for $R = R_0 + \Delta R_1, x = x_0 + \Delta x_1$, where ΔR_1 and Δx_1 represent small increments, for which convergence of the calculation process is assured. The solution obtained in this way serves as initial approximation used subsequently to obtain next solution of the system (3.42) for $R = R_0 + \Delta R_1 + \Delta R_2$ and $x = x_0 + \Delta x_1 + \Delta x_2$. Proceeding in a similar way, we can determine the set of solutions guaranteeing convergence of the calculation process performed by means of the Newton method, in our region of interest $D : [R_{min} \leq R \leq R_{max}, x_{min} \leq x \leq x_{max}]$. The set of discrete solutions evaluated in this manner for $1.3 \leq R \leq 10$ and $1.4 \leq x \leq 2.6$ has been subsequently approximated by using the following uncomplicated continuous functions [9]:

$$\theta^{(0)}(R, x) = \theta(f_1) = \frac{V_4(r, x)}{1 + x}$$

$$(3.43)$$

$$a_2^{(0)}(R, x) = \frac{f_3(r) + f_4(r)(2 - x)}{V_4(r, x)}$$

where

$$r = R - 1.5, \quad V_4(r, x) = f_1(r) + f_2(r)(x - 2)$$

$$f_1(r) = 0.629575 \ \exp(-0.115156r + 0.004939r^2 - 0.000074r^3)$$

$$f_2(r) = 0.105558 \ \exp(-0.046644r - 0.001213r^2 + 0.000267r^3)$$

$$f_3(r) = 1.614779 \ \exp(-0.079409r + 0.003701r^2 - 0.000075r^3)$$

$$f_4(r) = 0.251327 - 0.123151 \ \exp(-0.219819r + 0.016291r^2 - 0.000646r^3)$$

Formulas (3.43) make it possible the effective solution of the system of equations (3.42) for $1.3 \le R \le 10$ and $1.4 \le x \le 2.6$. Some calculation results, obtained by using these approximating formulas are given in Tables 3.4 and 3.5.

The VSWR(f/f_1) responses obtained for two four-section transformers designed for ($R = 2, x = 2$) and ($R = 4, x = 1.9$) are shown in Figs. 3.7 and 3.8, respectively.

Similar to (3.43) closed-form formulas for designing the more broadband nonsynchronous transmission line transformers composed of six and eight noncommensurate sections are described in papers [10, 11]. It deserves noting that the

Table 3.4

$R = Z_{02}/Z_{01}$	$x = f_2/f_1$	$\theta(f_1)$, rad	a_2	VSWR$_{\max}$
4	1.3	0.1845	3.4618	1.0430
4	1.9	0.1642	2.8787	1.2759
4	2.6	0.1500	2.3205	1.6492

Table 3.5

$R = Z_{02}/Z_{01}$	$x = f_2/f_1$	$\theta(f_1)$, rad	a_2	VSWR$_{\max}$
1.5	2	0.2098	2.5659	1.0740
2	2	0.1996	2.6058	1.1340
3	2	0.1790	2.6969	1.2359
4	2	0.1619	2.7849	1.3253
5	2	0.1482	2.8655	1.4074

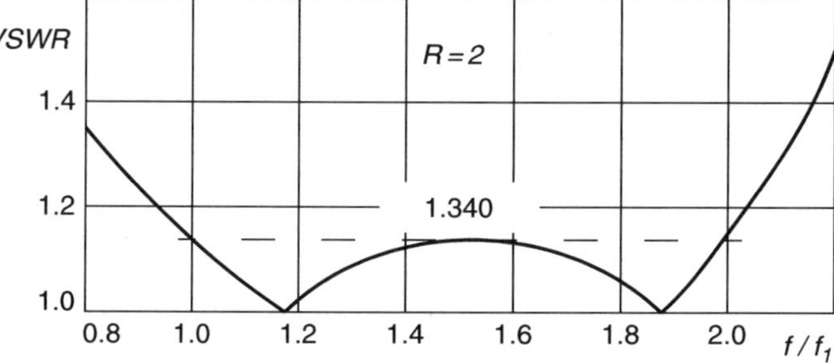

Fig. 3.7

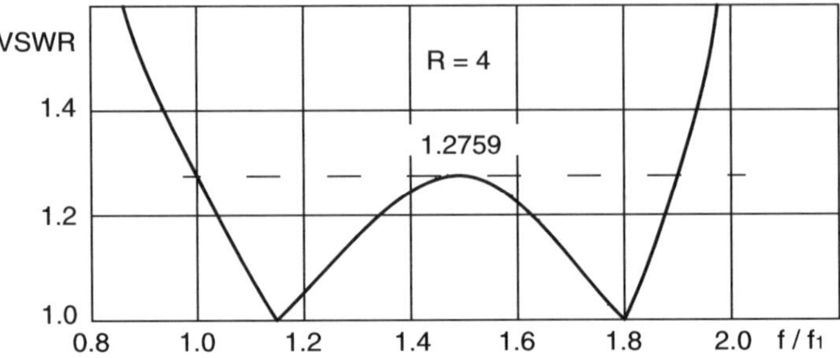

Fig. 3.8

eight-section transformer described in [11] has been used as a prototype circuit for designing the eight-way power divider/combiner implemented in the L-band rotary joint. Moreover, a similar multi-section nonsynchronous noncommensurate transmission line structures can be applied to matching two-frequency dependent complex impedances over the given frequency band. The corresponding design algorithm and examples of applications are described in [12, 13].

References

1. Ortega J.M., and W.C. Rheinboldt, Iterative solution of nonlinear equations in several variables. Academic Press, New York, 1970
2. Young D.M., and R.T. Gregory, A survey of numerical mathematics. Addison-Wesley Comp., London, 1973
3. Pearson C.E., Numerical methods in engineering and science. Van Nostrand Reinhold, New York, 1986
4. Fletcher R., A review of methods for unconstrained optimization. Academic Press, New York, 1969
5. Himmelblau D.M., Applied nonlinear programming. McGraw-Hill, New York, 1972
6. Bazaraa M.S., Sherali H.D. and C.M. Shetty, Nonlinear programming. Theory and applications. John Wiley and Sons, New York, 1993
7. Cristal E.G., "Coupled circular cylindrical rods between parallel ground planes". IEEE Trans., MTT – 12, Microwave Theory and Techniques, July 1964
8. Rosłoniec S., Algorithms for computer-aided design of linear microwave circuits. Artech House Inc., Boston, MA, 1990
9. Matthaei G.L., Young L. and E.M.T. Jones, Microwave filters, impedance matching networks and coupling structures. Artech House Inc., Boston, MA, 1980
10. Rosłoniec S. "Algorithms for the computer-aided design of nonsynchronous, noncommensurate transmission-line impedance transformers". Intern. Jour. of Microwave and Millimeter-Wave Computer-Aided Engineering, vol. 4, July 1994
11. Rosłoniec S. "An optimization algorithm for design of eight-section of nonsynchronous, noncommensurate impedance transformers" Proc. of the Microwave and Optronic Conf. MIOP – 97. Stuttgart/Sindelfingen, April 1997
12. Rosłoniec S., "Design of stepped transmission line matching circuits by optimization methods". IEEE Trans., MTT – 42, Microwave Theory and Techniques, December 1994
13. Rosłoniec S., Linear microwave circuits – analysis and design (in Polish) Published by: Wydawnictwa Komunikacji i Łączności, Warsaw, 1999

Chapter 4
Methods for the Interpolation
and Approximation of One Variable Function

Each professionally active engineer usually has to do with large number of numerical data acquired during the calculation or experimenting process. It is therefore obvious that statistical processing of this data or assignment of the corresponding relations in the form of analytic functions has major importance for practical purposes. The methods of interpolation and approximation discussed in this chapter serve, among other goals, for this purpose. By interpolation we understand the process of assignment, for the given function $y = y(x)$, continuous or discrete, of a continuous function $f(x)$ which for a finite number of values of x_i takes the same values, that is $f(x_i) = y_i = y(x_i)$, where $i = 0, 1, 2, 3, \ldots, n$. The values of x_i and y_i represent the coordinates of the points (see Fig. 4.1), called the interpolation points (nodes). The coordinates x_i are therefore often identified with this names [1–4].

In most practical cases, we are concerned with interpolation of a discrete function $y_i = y(x_i)$ by means of a continuous function $f(x)$. This problem may be solved in many different ways but the methods most frequently used for this purpose are: piecewise linear interpolation, using the Lagrange or Newton–Gregory interpolation polynomial, interpolation by means of cubic spline functions and interpolation using a finite linear combination of Chebyshev polynomials of the first kind. All these methods are discussed in Sect. 4.1. Slightly different, but a more general problem is the approximation of a given, continuous or discrete function $y = y(x)$, by a continuous function $f(x)$. In this case, both functions can take the same values $f(x_i) = y_i = y(x_i)$, for a finite number of points x_i, but it is not a necessary condition. Such a particular case is shown in Fig. 4.2.

In case of approximating the discrete function $y_i = y(x_i)$, where $i = 0, 1, 2, 3, \ldots, n$, by a continuous function $f(x)$, the most frequently used measure of the quality of the approximation (norm) is the following sum:

$$R_1 = \sum_{i=0}^{n} |f(x_i) - y_i| \tag{4.1}$$

which should take the possibly smallest value. When the approximated function $y = y(x)$ satisfies the Dirichlet's conditions in an approximation interval $[a, b]$ (is bounded and has a finite number of discontinuities of the first kind), we can evaluate:

S. Rosłoniec, *Fundamental Numerical Methods for Electrical Engineering,*
© Springer-Verlag Berlin Heidelberg 2008

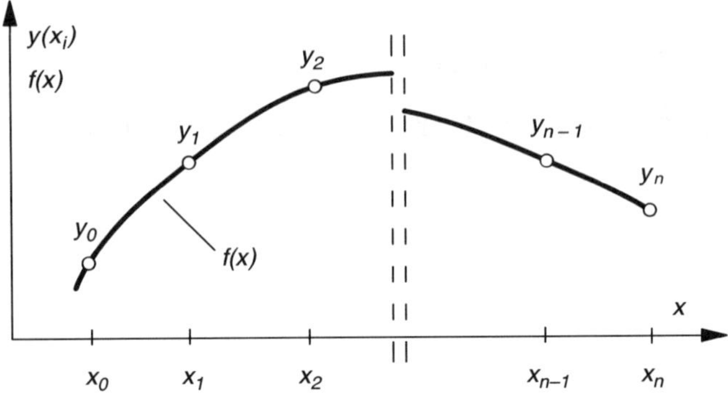

Fig. 4.1

$$R_2 = \frac{1}{|b-a|} \int_a^b \big| f(x) - y(x) \big| \, dx \tag{4.2}$$

The norm given by formula (4.2) has a simple geometrical interpretation. It is equal to the ratio of the hatched area in Fig. 4.3 to the length of the approximation interval.

Another example of the norm used for approximation of a continuous function $y(x)$ by a function $f(x)$ is:

$$R_3 = \frac{1}{|b-a|} \int_a^b \big[f(x) - y(x) \big]^2 \, dx \tag{4.3}$$

The approximation methods used most frequently in the engineering can be discriminated in consideration of the approximating function $f(x)$ or the adopted approximation norm. In general, an approximating function can be a linear combination of linearly independent component functions $\varphi_k(x)$, where $k = 0, 1, 2, 3, \ldots, m$

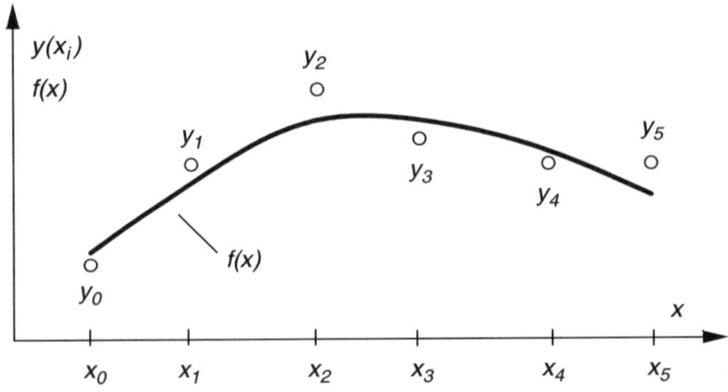

Fig. 4.2

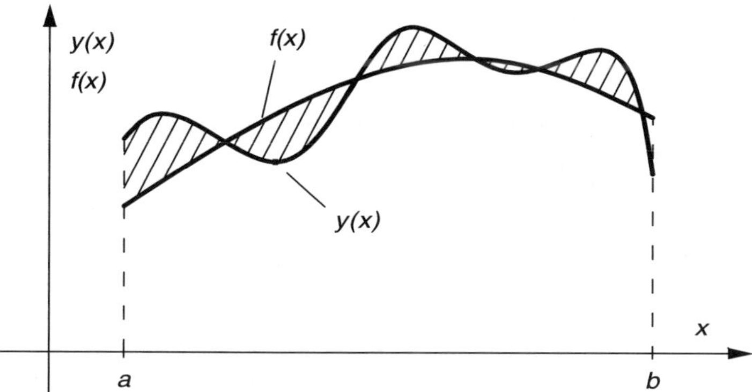

Fig. 4.3

$$f(x) = a_0\varphi_0(x) + a_1\varphi_1(x) + a_2\varphi_2(x) + a_3\varphi_3(x) + \cdots + a_m\varphi_m(x) = \sum_{k=0}^{m} a_k\varphi_k(x)$$

$$(4.4)$$

The set of functions $\varphi_k(x)$ is called linearly independent, when no such function belonging to this set exists that can be expressed as the linear combination of the remaining functions. It means that the identity

$$\sum_{k=0}^{m} \alpha_k\varphi_k(x) = 0 \tag{4.5}$$

cannot be satisfied for an arbitrary set of coefficients $a_0, a_1, a_2, a_3, \ldots, a_m$ different from zero. The simplest example of the linearly independent functions is the set of power functions: $1, x, x^2, x^3, \ldots, x^m$. The problem of approximation by means of the function (4.4) can be solved very easily in case when a given system of linearly independent functions is orthogonal [5, 6]. In order to explain this property, we formulate the relation:

$$\int_a^b \varphi_i(x)\varphi_k(x)\mathrm{d}x \equiv \overline{\varphi_i(x)\varphi_k(x)} = 0 \quad \text{for } i \neq k \tag{4.6}$$

If each arbitrarily chosen pair of functions belonging to the set of linearly independent functions satisfies the condition (4.6), then such a set is orthogonal in an interval $[a, b]$. Moreover, if the following condition is simultaneously satisfied

$$\overline{\varphi_i^2(x)} \equiv N_i = 1 \quad \text{for } i = 0, 1, 2, 3, \ldots, m \tag{4.7}$$

then the set of functions is said to be orthonormal in an interval $[a, b]$.

The approximation methods presented in Sect. 4.2 will be discussed in the following order:

- approximation of a constant function $y(x) = \text{const}$ in the limited interval $[a, b]$ by the Chebyshev polynomial $T_n(x)$;
- approximation of a constant function, $y(x) = \text{const}$ in the interval $[a, b]$ by the Butterworth polynomial $B_n(x)$;
- approximation of a discrete function $y_i = y(x_i)$, where $i = 0, 1, 2, 3, \ldots, n$, by the polynomial

$$\Phi(x) = a_0\varphi_0(x) + a_1\varphi_1(x) + a_2\varphi_2(x) + a_3\varphi_3(x) + \cdots + a_n\varphi_n(x) = \sum_{i=0}^{n} a_i\varphi_i(x)$$

according to the least squares criterion;
- approximation of periodic functions satisfying the Dirichlet's conditions by the trigonometric polynomial

$$\Psi(x) = a_0\psi_0(x) + a_1\psi_1(x) + a_2\psi_2(x) + a_3\psi_3(x) + \cdots + a_n\psi_n(x) = \sum_{i=0}^{n} a_i\psi_i(x)$$

whose basis functions $\psi_0(x), \psi_1(x), \psi_2(x), \psi_3(x), \ldots, \psi_n(x)$, are linearly independent and satisfy the condition (4.6).

4.1 Fundamental Interpolation Methods

4.1.1 The Piecewise Linear Interpolation

The essence of the piecewise linear interpolation of a given discrete function $y_i = y(x_i)$, where $i = 0, 1, 2, 3, \ldots, n$, is illustrated in Fig. 4.4.

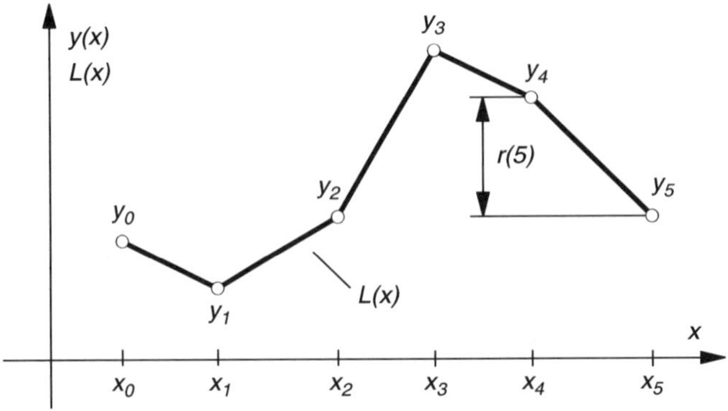

Fig. 4.4

For each value of the variable x lying in the approximation interval, the corresponding value of an interpolating function can be calculated from the formula:

$$L(x) = y_0 + \sum_{i=1}^{k} r(i) + r(k+1) \frac{x - x_k}{x_{k+1} - x_k} \tag{4.8}$$

where $r(i) = y_i - y_{i-1}$ and k is the largest index value for which $x_k < x$.

4.1.2 The Lagrange Interpolating Polynomial

The problem of interpolation by the power polynomial consists in finding a polynomial $P(x)$ of degree not greater than n, which for the given values of x_i, where $i = 0, 1, 2, 3, \ldots, n$, takes the same values as the approximated function $y_i = y(x_i)$. There is a mathematical proof that there exists only one polynomial,

$$P_n(x) = a_0 + a_1 x + a_2 x^2 + a_3 x^3 + \cdots + a_n x^n = \sum_{i=0}^{n} a_i x^i \tag{4.9}$$

having $(n + 1)$ coefficients $a_0, a_1, a_2, a_3, \ldots, a_n$. In order to determine them, we should formulate $(n + 1)$ reasonably stated conditions. According to the idea of the interpolation, see Fig. 4.1, these conditions are:

$$\begin{aligned}
y_0 &= P_n(x_0) = a_0 + a_1 x_0 + a_2 x_0^2 + \cdots + a_n x_0^n \\
y_1 &= P_n(x_1) = a_0 + a_1 x_1 + a_2 x_1^2 + \cdots + a_n x_1^n \\
y_2 &= P_n(x_2) = a_0 + a_1 x_2 + a_2 x_2^2 + \cdots + a_n x_2^n \\
&\vdots \\
y_n &= P_n(x_n) = a_0 + a_1 x_n + a_2 x_n^2 + \cdots + a_n x_n^n
\end{aligned} \tag{4.10}$$

The equation system (4.10) can be written in the following matrix form:

$$\begin{bmatrix} 1 & x_0^1 & x_0^2 & \cdots & x_0^n \\ 1 & x_1^1 & x_1^2 & \cdots & x_1^n \\ 1 & x_2^1 & x_2^2 & \cdots & x_2^n \\ \vdots & \vdots & \vdots & \cdots & \vdots \\ 1 & x_n^1 & x_n^2 & \cdots & x_n^n \end{bmatrix} \cdot \begin{bmatrix} a_0 \\ a_1 \\ a_2 \\ \vdots \\ a_n \end{bmatrix} = \begin{bmatrix} y_0 \\ y_1 \\ y_2 \\ \vdots \\ y_n \end{bmatrix} \tag{4.11}$$

As we see, this equation system is linear with respect to the unknown coefficients $a_0, a_1, a_2, a_3, \ldots, a_n$ and has a unique solution, because the determinant of the matrix of coefficients (called the Vandermonde's determinant) cannot be equal to zero. The system of equations (4.11) can be solved using one of the methods presented in Chap. 1. We must, however, emphasize that in many cases the equation systems formulated in such a way may be ill-conditioned and as a consequence of this fact, considerable computational errors may occur. Therefore, usefulness of a polynomial found in this way may be sometimes doubtful. This effect is particularly dangerous when we are dealing with polynomials of large degree n. It may be considerably eliminated, when we use the algorithm for determination of interpolating polynomial, introduced by Lagrange. The Lagrange interpolation consists in finding such polynomial $P_n(x) \equiv L_n(x)$ of degree not greater than n, which at interpolation points $x_0, x_1, x_2, \ldots, x_n$ takes the same values as the interpolated function $y_i = y(x_i)$, i.e.:

$$L_n(x_i) = y_i = y(x_i) \quad \text{for } i = 0, 1, 2, 3, \ldots, n \tag{4.12}$$

At the beginning of the procedure of evaluating such polynomial, let us consider an auxiliary problem consisting in finding the polynomial $\delta_i(x)$, satisfying the conditions:

$$\begin{aligned} \delta_i(x_j) &= 0 \qquad \text{for } j \neq i \\ \delta_i(x_j) &= 1 \qquad \text{for } j = i \end{aligned} \tag{4.13}$$

where $i = 0, 1, 2, 3, \ldots, n$. An example of such polynomial $\delta_i(x)$ is shown in Fig. 4.5.

Conditions (4.13) can be expressed by means of the Kronecker symbol. The polynomial $\delta_i(x)$ is equal to zero for $(x_0, x_1, x_2, \ldots, x_{i-1}, x_{i+1}, \ldots, x_n)$, and therefore can be written as the following product:

$$\delta_i(x) = c_i \prod_{\substack{j=0 \\ j \neq i}}^{n} (x - x_j) \tag{4.14}$$

where c_i is a certain coefficient. This coefficient may be easily found from the relations (4.14) and (4.13), precisely $\delta_i(x_i) = 1$. After the uncomplicated substitution we obtain:

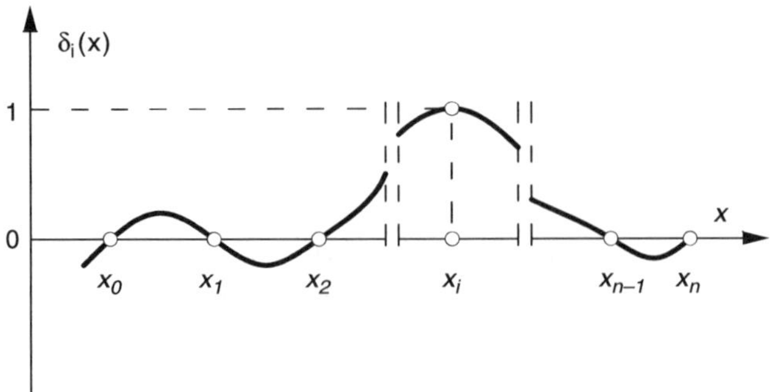

Fig. 4.5

$$c_i = \frac{1}{(x_i - x_0)(x_i - x_1)\ldots(x_i - x_{i-1})(x_i - x_{x+1})\ldots(x_i - x_{n-1})(x_i - x_n)}$$
(4.15)

Introducing formula (4.15) to the relation (4.14) we get:

$$\delta_i(x) = \frac{(x - x_0)(x - x_1)\ldots(x - x_{i-1})(x - x_{i+1})\ldots(x - x_{n-1})(x - x_n)}{(x_i - x_0)(x_i - x_1)\ldots(x_i - x_{i-1})(x_i - x_{i+1})\ldots(x_i - x_{n-1})(x_i - x_n)}$$
(4.16)

The Lagrange interpolation polynomial is defined as the linear combination of the polynomials (4.16), in which all coefficients are equal to the corresponding values of an interpolated function, namely:

$$L_n(x) = \sum_{i=0}^{n} y_i \delta_i(x)$$
(4.17)

We can prove that the polynomial (4.17) represents a unique solution of the problem stated above.

Example 4.1 Let us assume that the discrete function $y_i = y(x_i)$ is defined as: $(x_0 = 1.0, y_0 = 2.0)$, $(x_1 = 3.0, y_1 = 3.5)$, $(x_2 = 5.0, y_2 = 3.7)$ and $(x_3 = 7.0, y_3 = 3.5)$. The Lagrange interpolating polynomial evaluated for this function is:

$$
\begin{aligned}
L_3(x) = 2 \cdot &\frac{(x - 3)(x - 5)(x - 7)}{(1 - 3)(1 - 5)(1 - 7)} && (i = 0) \\
+ 3.5 \cdot &\frac{(x - 1)(x - 5)(x - 7)}{(3 - 1)(3 - 5)(3 - 7)} && (i = 1) \\
+ 3.7 \cdot &\frac{(x - 1)(x - 3)(x - 7)}{(5 - 1)(5 - 3)(5 - 7)} && (i = 2) \\
+ 3.5 \cdot &\frac{(x - 1)(x - 3)(x - 5)}{(7 - 1)(7 - 3)(7 - 5)} && (i = 3)
\end{aligned}
$$

After rearrangement, this polynomial can be written in the traditional power form, namely $L_3(x) = 0.01875x^3 - 0.33125x^2 + 1.83125x + 0.48125$. Identical values of the coefficients $a_0 = 0.48125$, $a_1 = 1.83125$, $a_2 = -0.33125$ and $a_3 = 0.01875$ can be found when we solve the following equation system formulated for this problem:

$$
\begin{bmatrix}
1 & 1^1 & 1^2 & 1^3 \\
1 & 3^1 & 3^2 & 3^3 \\
1 & 5^1 & 5^2 & 5^3 \\
1 & 7^1 & 7^2 & 7^3
\end{bmatrix}
\cdot
\begin{bmatrix}
a_0 \\
a_1 \\
a_2 \\
a_3
\end{bmatrix}
=
\begin{bmatrix}
2.0 \\
3.5 \\
3.7 \\
3.5
\end{bmatrix}
$$

The above equation system corresponds to the system (4.11).

4.1.3 The Aitken Interpolation Method

The Lagrange interpolating polynomial $W_n(x)$ can be evaluated using a series of linear interpolations [7]. Thus, let us assume that the points x_0, x_1, x_2, x_3, $\ldots$, x_n, and values $y_0 = y(x_0)$, $y_1 = y(x_1)$, $y_2 = y(x_2)$, $y_3 = y(x_3)$, $\ldots$, $y_n = y(x_n)$ of the function $y(x)$ being interpolated are given. According to Aitken interpolation method, the interpolating polynomial $W_n(x)$ can be evaluated by using the following *n-iterative* algorithm.

4.1.3.1 First Iteration

At the beginning, n polynomials $W_{0,k}$ of degree 1 are created in the following manner:

$$W_{0,k} = y_0 \frac{x - x_k}{x_0 - x_k} + y_k \frac{x - x_0}{x_k - x_0} = \frac{1}{x_0 - x_k} \begin{vmatrix} y_0 & x - x_0 \\ y_k & x - x_k \end{vmatrix}$$

where $k = 1$, 2, 3, 4, $\ldots$, n. The set of these polynomials is written in third column of the Table 4.1.

4.1.3.2 Second Iteration

Polynomials $W_{0,k}$ evaluated above allow us to create the set of $n - 1$ polynomials of degree 2, i.e.:

$$W_{0,1,l} = W_{0,1} \frac{x - x_l}{x_1 - x_l} + W_{0,l} \frac{x - x_1}{x_l - x_1} = \frac{1}{x_1 - x_l} \begin{vmatrix} W_{0,1} & x - x_1 \\ W_{0,l} & x - x_l \end{vmatrix}$$

where $l = 2, 3, 4, \ldots, n$. These polynomials $W_{0,1,l}$ are written in the fourth column of Table 4.1.

4.1.3.3 Third Iteration

In the same way, the set of $n - 2$ polynomials

$$W_{0,1,2,m} = W_{0,1,2} \frac{x - x_m}{x_2 - x_m} + W_{0,1,m} \frac{x - x_2}{x_m - x_2} = \frac{1}{x_m - x_2} \begin{vmatrix} W_{0,1,2} & x - x_2 \\ W_{0,1,m} & x - x_m \end{vmatrix}$$

of degree 3 is created. In this case, $m = 3, 4, \ldots, n$.

Table 4.1

x_0	y_0						
x_1	y_1	$W_{0,1}$					
x_2	y_2	$W_{0,2}$	$W_{0,1,2}$				
x_3	y_3	$W_{0,3}$	$W_{0,1,3}$	$W_{0,1,,2,3}$			
x_4	y_4	$W_{0,4}$	$W_{0,1,4}$	$W_{0,1,,2,4}$			
$\ldots$	$\ldots$	$\ldots$	$\ldots$	$\ldots$	$\ldots$	$W_{0,1,2,3,\ldots,n-2,n-1}$	
x_n	y_n	$W_{0,n}$	$W_{0,1,n}$	$W_{0,1,,2,n}$	$\ldots$	$W_{0,1,2,3,\ldots,n-2,n}$	$W_{0,1,2,3,\ldots,n-2,n-1,n}$

4.1.3.4 Successive Iterations

In successive iterations the polynomials of higher degrees, see Table 4.1, are evaluated analogously.

4.1.3.5 Last Iteration

Only one polynomial $W_{0,1,2,3,...,n} \equiv W_n(x)$ of degree n is evaluated in the last (n) iteration. This polynomial is identical to the Lagrange polynomial (4.17) being sought.

In order to illustrate the algorithm presented above, let us consider again the interpolation problem described in Example 4.1. The intermediate polynomials evaluated for this interpolation problem are:

$$W_{0,1} = 0.75x + 1.25, \; W_{0,2} = 0.425x + 1.575, \; W_{0,3} = 0.25x + 1.75$$

$$W_{0,1,2} = -0.1625x^2 + 1.4x + 0.7625, \; W_{0,1,3} = -0.125x^2 + 1.25x + 0.875$$

$$W_{0,1,2,3} = 0.01875x^3 - 0.33125x^2 + 1.83125x + 0.48125$$

The polynomial $W_{0,1,2,3}(x)$ is the same as the polynomial $L_3(x)$, which was obtained by Lagrangian interpolation, see Example 4.1. Undoubtedly, this identity confirms the validity of that algorithm.

4.1.4 The Newton–Gregory Interpolating Polynomial

The Newton–Gregory polynomial interpolating the function $y_i = y(x_i)$ at $(n+1)$ points (nodes) has the form:

$$N_n(x) = a_0 + a_1(x - x_0) + a_2(x - x_0)(x - x_1) + a_3(x - x_0)(x - x_1)(x - x_2)$$
$$+ \cdots + a_n(x - x_0)(x - x_1)(x - x_2)\ldots(x - x_{n-1})$$

$$(4.18)$$

The coefficients a_i, for $i = 0, 1, 2, 3, \ldots, n$, of this polynomial can be determined from the system of $(n+1)$ equations written below:

$$
\begin{aligned}
&a_0 = y_0 \\
&a_0 + a_1(x_1 - x_0) = y_1 \\
&a_0 + a_1(x_2 - x_0) + a_2(x_2 - x_0)(x_2 - x_1) = y_2 \\
&\quad \vdots \\
&a_0 + a_1(x_n - x_0) + a_2(x_n - x_0)(x_n - x_1) + \cdots \\
&\qquad + a_n(x_n - x_0)(x_n - x_1)\ldots(x_n - x_{n-1}) = y_n
\end{aligned}
$$

$$(4.19)$$

The matrix $\mathbf{A}$ of coefficients of the equation system (4.19) has the triangular form, which makes of course the solution of this system much easier. We may use here, for example the second stage of Gaussian elimination method. Consequently, we can write:

$$a_0 = y_0$$

$$a_1 = \frac{1}{(x_1 - x_0)}\,[y_1 - a_0]$$

$$a_2 = \frac{1}{(x_2 - x_0)(x_2 - x_1)}\,[y_2 - a_0 - a_1(x_2 - x_0)]$$

$$a_3 = \frac{1}{(x_3 - x_0)(x_3 - x_1)(x_3 - x_2)}\,[y_3 - a_0 - a_1(x_3 - x_0) - a_2(x_3 - x_0)(x_3 - x_1)]$$

$$\vdots$$

$$a_m = \frac{1}{\prod\limits_{i=0}^{m-1}(x_m - x_i)}\left\{ y_m - a_0 - \sum_{k=1}^{m-1}\left[a_k \prod_{j=0}^{k-1}(x_m - x_j)\right]\right\}$$

$$(4.20)$$

The recursive formulas that follows represent the generalized relation (4.20), satisfied for $2 \le m \le n$. The problem of calculating the coefficients a_m can be simplified, if the interpolation points $(x_0,\ x_1,\ x_2, \ldots,\ x_n)$ are equally spaced by the step

$$h = x_{i+1} - x_i \qquad (4.21)$$

Applying the relation (4.21) to the equations (4.19) we get:

$$y_0 = a_0$$
$$y_1 = a_0 + a_1 \cdot h$$
$$y_2 = a_0 + a_1(2h) + a_2 \cdot (2h) \cdot h$$
$$\vdots$$
$$y_n = a_0 + a_1(nh) + a_2(nh)(n-1)h + a_3(nh)(n-1)h(n-2)h + \cdots + a_n(n!)\,h^n$$

$$(4.22)$$

The following coefficients represent the solution of equation system (4.22)

$$a_0 = y_0$$

$$a_1 = \frac{y_1 - y_0}{h} = \frac{\Delta y_0}{h}$$

$$a_2 = \frac{1}{2h^2}\,[(y_2 - y_1) - (y_1 - y_0)] = \frac{\Delta^2 y_0}{2h^2}$$

$$\ldots$$

The coefficients a_m, for $m = 1,\ 2,\ 3, \ldots,\ n$, can be written in the following general form:

$$a_m = \frac{\Delta^m y_0}{(m!)\,h^m} \qquad (4.23)$$

where $\Delta^m y_0$ is the finite difference of order m. Values of the finite differences used for calculation of the coefficients a_m can be evaluated according to the following multi-step algorithm:

- differences of the first order, $m = 1$

$$\Delta y_0 = y_1 - y_0$$
$$\Delta y_1 = y_2 - y_1$$
$$\vdots$$
$$\Delta y_{n-1} = y_n - y_{n-1}$$

- differences of the second order, $m = 2$

$$\Delta^2 y_0 = \Delta y_1 - \Delta y_0$$
$$\Delta^2 y_1 = \Delta y_2 - \Delta y_1$$
$$\vdots$$
$$\Delta^2 y_{n-2} = \Delta y_{n-1} - \Delta y_{n-2}$$

$$(4.24)$$

- differences of the third order, $m = 3$

$$\Delta^3 y_0 = \Delta^2 y_1 - \Delta^2 y_0$$
$$\Delta^3 y_1 = \Delta^2 y_2 - \Delta^2 y_1$$
$$\vdots$$
$$\Delta^3 y_{n-3} = \Delta^2 y_{n-2} - \Delta^2 y_{n-3}$$

- differences of the m – order

$$\Delta^m y_0 = \Delta^{m-1} y_1 - \Delta^{m-1} y_0$$
$$\Delta^m y_1 = \Delta^{m-1} y_2 - \Delta^{m-1} y_1$$
$$\vdots$$
$$\Delta^m y_{n-m} = \Delta^{m-1} y_{n+1-m} - \Delta^{m-1} y_{n-m}$$

Finite differences of an arbitrary order $1 \leq m \leq n$ can be expressed directly by means of the values y_i of the interpolated function.

$$\Delta^m y_0 = y_m - m y_{m-1} + \frac{m(m-1)}{2!} y_{m-2} + \cdots + (-1)^m y_0 \qquad (4.25)$$

The formula (4.25) may be generalized for finite differences calculated at the point x_i

$$\Delta^m y_i = y_{m+i} - m y_{m+i-1} + \frac{m(m-1)}{2!} y_{m+i-2} + \cdots + (-1)^m y_i \qquad (4.26)$$

After introducing relation (4.23) into the polynomial (4.18) we obtain the standard form of Newton–Gregory interpolation polynomial, namely:

$$N(x) = y_0 + \frac{\Delta y_0}{h}(x - x_0) + \frac{\Delta^2 y_0}{(2!)h^2}(x - x_0)(x - x_1) + \cdots$$
$$+ \frac{\Delta^n y_0}{(n!)h^n}(x - x_0)(x - x_1)(x - x_2)\ldots(x - x_{n-1}) \tag{4.27}$$

By introducing a new variable

$$t = \frac{x - x_0}{h} \tag{4.28}$$

for which

$$x = x_0 + th, \quad \frac{x - x_1}{h} = \frac{x - x_0 - h}{h} = t - 1, \quad \frac{x - x_2}{h} = t - 2, \ldots$$
$$\frac{x - x_{n-1}}{h} = t - n + 1$$

the polynomial (4.27) can be written as:

$$N(x_0 + th) = y_0 + t\,\Delta y_0 + \frac{t(t-1)}{2!}\Delta^2 y_0 + \frac{t(t-1)(t-2)}{3!}\Delta^3 y_0 + \cdots$$
$$+ \frac{t(t-1)(t-2)\ldots(t-n+1)}{n!}\Delta^n y_0 \tag{4.29}$$

The polynomial (4.29) can be used for interpolation of a given function $y_i = y(x_i)$ over the whole interval $[x_0, x_n]$. For the sake of computing precision, it is however recommended to reduce the interpolation interval to $[x_0, x_1]$, assuring that $t < 1$. For different values of the variable x, as for example $x_i < x < x_{i+1}$, we should take x_i instead of x_0. In this case, for $i = 1, 2, 3, \ldots, n-1$, this interpolation polynomial has the form:

$$N(x_i + th) = y_i + t\,\Delta y_i + \frac{t(t-1)}{2!}\Delta^2 y_i + \frac{t(t-1)(t-2)}{3!}\Delta^3 y_i + \cdots$$
$$+ \frac{t(t-1)(t-2)\ldots(t-n+1)}{n!}\Delta^n y_i \tag{4.30}$$

which in the literature is known as the first Newton–Gregory polynomial for the forward interpolation. Polynomial (4.30) is used chiefly to determine the values of a given function, lying in the left-half of the interpolation interval $[x_0, x_n]$. Justification of this fact may be explained in the following manner. The finite differences $\Delta^m y_i$ are calculated on the basis of values $y_i, \ y_{i+1}, \ y_{i+2}, \ y_{i+3}, \ldots, \ y_{i+m}$, when $i + m \leq n$. For i close to n, the finite differences of higher orders are not calculated. For example, if $i = n - 3$, the polynomial (4.30) contains only the differences $\Delta y_i, \ \Delta^2 y_i$ and $\Delta^3 y_i$. When the points lying in the right-half of the interpolation interval $[x_0, \ x_n]$ are concerned, it is recommended to use the polynomial

$$N(x_n + th) = y_n + t\Delta y_{n-1} + \frac{t(t+1)}{2!}\Delta^2 y_{n-2} + \frac{t(t+1)(t+2)}{3!}\Delta^3 y_{n-3} + \cdots$$
$$+ \frac{t(t+1)(t+2)\ldots(t+n-1)}{n!}\Delta^n y_0$$

$$(4.31)$$

defined for

$$t = \frac{x - x_n}{h} \le 0 \tag{4.32}$$

This version of polynomial is called the Newton–Gregory polynomial for backward interpolation.

Example 4.2 The function $y_i = y(x_i)$ defined in Example 4.1 interpolates the polynomial (4.18) whose coefficients are: $a_0 = 2,\quad a_1 = 0.75,\quad a_2 = -0.1625$ and $a_3 = 0.01875$. Some values of this polynomial are given in Table 4.2

Example 4.3 As next example, let us calculate the values $N_5(0.1)$ and $N_5(0.9)$ of the first and second Newton–Gregory polynomials interpolating the function $y_i = y(x_i)$ given in the first and second columns of Table 4.3.

In the remaining columns of this table, values of finite differences calculated from the formulas (4.24) are given. For $x = 0.1$ we obtain $t = (x - x_0)/h = (0.1 - 0)/0.2 = 0.5$.

According to the formula (4.30)

$$N_5(0.1) = 1.2715 + 0.5 \times 1.1937 + \frac{0.5(0.5 - 1)}{2}(-0.0146)$$
$$+ \frac{0.5(0.5 - 1)(0.5 - 2)}{6}0.0007 + \frac{0.5(0.5 - 1)(0.5 - 2)(0.5 - 3)}{24}(-0.0001)$$
$$= 1.8702$$

Table 4.2

x	1.5	3.0	4.5	5.5	6	7
$N_3(x)$	2.546094	3.500000	3.722656	3.652344	3.593750	3.500000

Table 4.3

x_i	y_i	Δy_i	$\Delta^2 y_i$	$\Delta^3 y_i$	$\Delta^4 y_i$	$\Delta^5 y_i$
0.0	1.2715					
		1.1937				
0.2	2.4652		−0.0146			
		1.1791		0.0007		
0.4	3.6443		−0.0139		−0.0001	
		1.1652		0.0006		0.0000
0.6	4.8095		−0.0133		−0.0001	
		1.1519		0.0005		
0.8	5.9614		−0.0128			
		1.1391				
1.0	7.1005					

Value of the polynomial $N_5(0.9)$ should be calculated according to the formula (4.31) by introducing $t = (x - x_n)/h = (0.9 - 1)/0.2 = -0.5$. In this case

$$
\begin{aligned}
N_5(0.9) =\, &7.1005 + 0.5 \cdot 1.1391 - \frac{0.5(-0.5 + 1)}{2}(-0.0128) \\
&- \frac{0.5(-0.5 + 1)(-0.5 + 2)}{6}(0.0005) \\
&- \frac{0.5(-0.5 + 1)(-0.5 + 2)(-0.5 + 3)}{24}(-0.0001) = 6.5325
\end{aligned}
$$

4.1.5 Interpolation by Cubic Spline Functions

The spline function used for the interpolation of the function $y_i = y(x_i)$, where $i = 0, 1, 2, 3, \ldots, n$, is defined as a set of n conjugate trinomials. The properties of such trinomials will be discussed below on the example of the cubic spline function, composed of three trinomials, i.e.:

$$
\begin{aligned}
q_1(x) &= k_{10} + k_{11}x + k_{12}x^2 + k_{13}x^3 \\
q_2(x) &= k_{20} + k_{21}x + k_{22}x^2 + k_{23}x^3 \\
q_3(x) &= k_{30} + k_{31}x + k_{32}x^2 + k_{33}x^3
\end{aligned}
\tag{4.33}
$$

Of course, the above approach does not limit the generality of our considerations. The spline function (4.33) interpolates a function $y_i = y(x_i)$ defined at four points, as shown in Fig. 4.6.

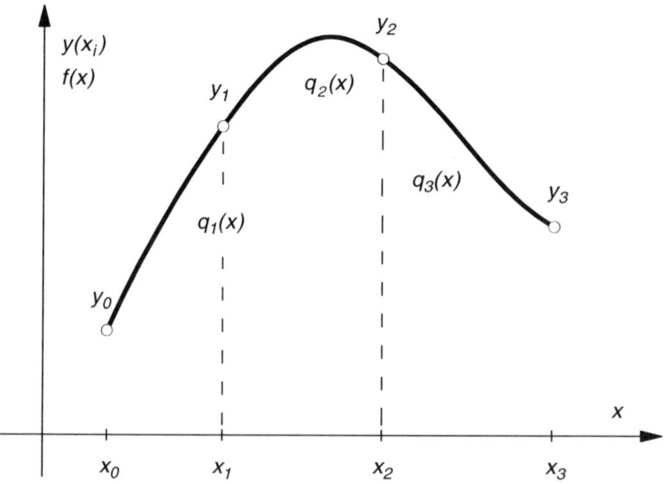

Fig. 4.6

According to the general concept of interpolation, these trinomials should satisfy the following requirements:

$$\begin{aligned}
q_1(x_0) &= y_0 \\
q_1(x_1) &= q_2(x_1) = y_1 \\
q_2(x_2) &= q_3(x_2) = y_2 \\
q_3(x_3) &= y_3
\end{aligned} \tag{4.34}$$

Moreover, it is required that the corresponding first and second derivatives of the trinomials, calculated at internal interpolation points should be equal to each other. This requirement can be expressed as follows:

$$\begin{aligned}
q_1'(x_1) &= q_2'(x_1) \\
q_1''(x_1) &= q_2''(x_1) \\
q_2'(x_2) &= q_3'(x_2) \\
q_2''(x_2) &= q_3''(x_2)
\end{aligned} \tag{4.35}$$

Similar equations can be formulated for the second derivatives of the first and the last (third) trinomials. These derivatives, calculated for the first and last (external) interpolation points, should be equal to zero.

$$q_1''(x_0) = q_3''(x_3) = 0 \tag{4.36}$$

Equations (4.34), (4.35) and (4.36) formulated above, form the system of 12 linear equations with 12 unknown coefficients of the polynomials $q_1(x)$, $q_2(x)$ and $q_3(x)$, see relations (4.33). It can be solved by using one of the methods described in Chap. 1. For example, the Gaussian elimination with the choice of the pivotal element is suitable for this purpose.

Example 4.4 Given the following discrete function $y_i = y(x_i)$ defined at three points: $(x_0 = 1, y_0 = 2)$, $(x_1 = 3, y_1 = 3.5)$ and $(x_2 = 5, y_2 = 3.7)$. The corresponding interpolating spline function composed of two trinomials has the form:

$$q_1(x) = k_{10} + k_{11}x + k_{12}x^2 + k_{13}x^3, \quad q_2(x) = k_{20} + k_{21}x + k_{22}x^2 + k_{23}x^3$$

According to the idea of interpolation we can formulate the set of eight equations for these trinomials, i.e.:

$$\begin{aligned}
q_1(1) &= 2, & q_1'(3) - q_2'(3) &= 0 \\
q_1(3) &= 3.5, & q_1''(3) - q_2''(3) &= 0 \\
q_2(3) &= 3.5, & q_1''(1) &= 0 \\
q_2(5) &= 3.7, & q_2''(5) &= 0
\end{aligned}$$

The equations formulated above can be written in the matrix form with respect to coefficients $k_{10}, k_{11}, k_{12}, k_{13}, k_{20}, k_{21}, k_{22}$ and k_{23} being sought.

$$
\begin{bmatrix}
1 & 1 & 1 & 1 & 0 & 0 & 0 & 0 \\
1 & 3 & 9 & 27 & 0 & 0 & 0 & 0 \\
0 & 0 & 0 & 0 & 1 & 3 & 9 & 27 \\
0 & 0 & 0 & 0 & 1 & 5 & 25 & 125 \\
0 & 1 & 6 & 27 & 0 & -1 & -6 & -27 \\
0 & 0 & 2 & 18 & 0 & 0 & -2 & -18 \\
0 & 0 & 2 & 6 & 0 & 0 & 0 & 0 \\
0 & 0 & 0 & 0 & 0 & 0 & 2 & 30
\end{bmatrix}
\cdot
\begin{bmatrix}
k_{10} \\ k_{11} \\ k_{12} \\ k_{13} \\ k_{20} \\ k_{21} \\ k_{22} \\ k_{23}
\end{bmatrix}
=
\begin{bmatrix}
2 \\ 3.5 \\ 3.5 \\ 3.7 \\ 0 \\ 0 \\ 0 \\ 0
\end{bmatrix}
$$

The solution of this equation system is: $k_{10} = 1.128125$, $k_{11} = 0.790625$, $k_{12} = 0.121875$, $k_{13} = -0.040625$, $k_{20} = -1.065624$, $k_{21} = 2.984375$, $k_{22} = -0.609375$ and $k_{23} = 0.040625$. It follows from our above considerations that, in case $(m + 1)$ interpolation points, the number of interpolating trinomials equals m and the number of the coefficients in all trinomials is $4m$. The total number of coefficients can be considerably reduced by using the set of properly correlated trinomials [2, 8]. Here, we have some examples of such trinomials:

$$q_i\,[t\,(x)] = t \cdot y_i + \bar{t} \cdot y_{i-1} + \Delta x_i \left[(k_{i-1} - d_i) \cdot t \cdot \bar{t}^2 - (k_i - d_i) \cdot t^2 \cdot \bar{t}\right] \quad (4.37)$$

where

$$\Delta x_i = x_i - x_{i-1}, \quad \Delta y_i = y_i - y_{i-1}, \quad d_i = \frac{\Delta y_i}{\Delta x_i}, \quad t = \frac{x - x_{i-1}}{\Delta x_i}, \quad \bar{t} = 1 - t$$

$$i = 1, 2, \ldots, m$$

Each trinomial (4.37) has only two coefficients k_{i-1} and k_i. When the form of the first trinomial $q_1\,[t(x)]$ (having the coefficients k_0 and k_1) is given, then only one new unknown coefficient is added when we pass to the next trinomial. The following conclusions result from the relations (4.37):

1. For $x = x_{i-1}$

$$t = 0, \quad \bar{t} = 1, \quad q_i[t(x)] = y_{i-1} \quad (4.38)$$

2. For $x = x_i$

$$t = 1, \quad \bar{t} = 0, \quad q_i[t(x)] = y_i$$

Moreover, we can find in the literature the proof of the fact that trinomials (4.37) satisfy automatically the requirement of equality of the corresponding first derivatives at the internal interpolation points. Similar requirements formulated for second derivatives have the form:

$$k_{i-1}\Delta x_{i+1} + 2k_i\,(\Delta x_i + \Delta x_{i+1}) + k_{i+1}\Delta x_i = 3\,(d_i\Delta x_{i+1} + d_{i+1}\Delta x_i) \quad (4.39)$$

In case of the external interpolation points the requirement:

$$q_1''[t(x_0)] = 0, \quad q_m''[t(x_m)] = 0$$

is satisfied when

$$
\begin{aligned}
2k_0 + k_1 &= 3d_1 \\
k_{m-1} + 2k_m &= 3d_m
\end{aligned}
\tag{4.40}
$$

Equations (4.38), (4.39) and (4.40) form the following system of m linear equations

$$
\begin{bmatrix}
2 & 1 & 0 & 0 & \dots & 0 & 0 & 0 \\
\Delta x_2 & 2(\Delta x_1 + \Delta x_2) & \Delta x_1 & 0 & \dots & 0 & 0 & 0 \\
0 & \Delta x_3 & 2(\Delta x_2 + \Delta x_3) & \Delta x_2 & \dots & 0 & 0 & 0 \\
\vdots & \vdots & \vdots & \vdots & \vdots & \vdots & \vdots & \vdots \\
0 & 0 & 0 & 0 & \dots & \Delta x_m & 2(\Delta x_{m-1} + \Delta x_m) & \Delta x_{m-1} \\
0 & 0 & 0 & 0 & \dots & 0 & 1 & 2
\end{bmatrix}
\begin{bmatrix}
k_0 \\ k_1 \\ k_2 \\ \vdots \\ k_{m-1} \\ k_m
\end{bmatrix}
$$

$$
= 3
\begin{bmatrix}
d_1 \\
d_1 \Delta x_2 + d_2 \Delta x_1 \\
d_2 \Delta x_3 + d_3 \Delta x_2 \\
\vdots \\
d_{m-1} \Delta x_m + d_m \Delta x_{m-1} \\
d_m
\end{bmatrix}
$$

Solution of this system is identical to the vector $[k_0, k_1, k_2, \dots, k_m]$ of the desired coefficients. The matrix $\mathbf{A}$ of coefficients of the above equation system is the particular case of the sparse square matrix known in the literature under the name of the ribbon or, more specifically, tri-diagonal matrix. When we solve large ($n \geq 10$) systems of linear equations having tri-diagonal matrices of coefficients $\mathbf{A}$, a method of fast elimination proves to be very useful. It may be interpreted as a modification of the Gaussian elimination method described in Chap. 1. In the Russian language literature, this method is known under the name of "progonka". Similarly, as in the case of the Gaussian elimination, we distinguish here two stages, namely the forward and backward movement. Due to the ribbon structure of the coefficient matrix $\mathbf{A}$, computing formulas describing this method are not complicated and can be expressed in the simple recursive form. The algorithm of this special numerically effective method is described in Appendix C.

Example 4.5 Consider an application of the cubic spline function (4.37) to interpolation of the following discrete function: ($x_0 = 1, y_0 = 2$), ($x_1 = 3, y_1 = 3.5$), ($x_2 = 5, y_2 = 3.8$) and ($x_3 = 7, y_3 = 3$). The differences Δx_i, Δy_i and coefficients d_i calculated for this function are:

$$
\begin{aligned}
\Delta x_1 &= 2, & \Delta x_2 &= 2, & \Delta x_3 &= 2 \\
\Delta y_1 &= 1.5, & \Delta y_2 &= 0.3, & \Delta y_3 &= -0.8 \\
d_1 &= 0.75, & d_2 &= 0.15, & d_3 &= -0.4
\end{aligned}
$$

The quantities given by (4.38), (4.39) and (4.40), can be used to formulate the following system of four linear equations:

$$
\begin{bmatrix}
2 & 1 & 0 & 0 \\
2 & 2(2+2) & 2 & 0 \\
0 & 2 & 2(2+2) & 2 \\
0 & 0 & 1 & 2
\end{bmatrix}
\cdot
\begin{bmatrix}
k_0 \\ k_1 \\ k_2 \\ k_3
\end{bmatrix}
=
\begin{bmatrix}
2.25 \\ 5.4 \\ -1.5 \\ -1.2
\end{bmatrix}
$$

with respect to the desired coefficients. Solution of this system is equal to:

$$
k_0 = 0.873333, \quad k_1 = 0.503333, \quad k_2 = -0.186666, \quad k_3 = -0.506666.
$$

4.1.6 Interpolation by a Linear Combination of Chebyshev Polynomials of the First Kind

Let $f(x)$ be a real function of one variable defined at least in the interval $[-1, 1]$. This function can be interpolated over the interval $[-1, 1]$ by means of finite, linear sum of the Chebyshev polynomials $T_j(x)$ of the first kind

$$
P_N(x) = c_0 T_0(x) + c_1 T_1(x) + c_2 T_2(x) + \cdots + c_N T_N(x) = \sum_{j=0}^{N} c_j T_j(x) \quad (4.41)
$$

The coefficients c_j, $j = 0, 1, 2, 3, \ldots, N$, of the interpolating polynomial $P_N(x)$ can be found using the following relations:

$$
\sum_{k=0}^{N} T_i(x_k) T_j(x_k) = 0 \quad \text{for } i \neq j
$$

$$
\sum_{k=0}^{N} T_i(x_k) T_j(x_k) = \frac{N+1}{2} \quad \text{for } i = j \neq 0 \quad (4.42)
$$

$$
\sum_{k=0}^{N} T_0(x_k) T_0(x_k) = N + 1
$$

that are satisfied only when

$$
x_k = \cos\left(\pi \frac{2k+1}{2N+2}\right), \quad k = 0, 1, 2, 3, \ldots, N \quad (4.43)
$$

The property expressed by relations (4.42) is often called conditional orthogonality, because it takes place only for the set of discrete values x_k defined above. For arbitrary values of the variable x the Chebyshev polynomials, see relations (4.52), do not form the set of orthogonal functions with the weights $p(x) = 1$. In order to determine the coefficients c_j, see (4.41), let us consider the products of the discrete values of the interpolating polynomial $P_N(x_k) = f(x_k)$ and the Chebyshev polynomial $T_i(x_k)$, where $k = 0, 1, 2, 3, \ldots, N, 0 \leq i \leq N$.

$$f(x_0)T_i(x_0) = c_0 T_i(x_0)T_0(x_0) + c_1 T_i(x_0)T_1(x_0) + \cdots + c_N T_i(x_0)T_N(x_0)$$
$$f(x_1)T_i(x_1) = c_0 T_i(x_1)T_0(x_1) + c_1 T_i(x_1)T_1(x_1) + \cdots + c_N T_i(x_1)T_N(x_1)$$
$$f(x_2)T_i(x_2) = c_0 T_i(x_2)T_0(x_2) + c_1 T_i(x_2)T_1(x_2) + \cdots + c_N T_i(x_2)T_N(x_2)$$
$$\vdots$$
$$f(x_i)T_i(x_i) = c_0 T_i(x_i)T_0(x_i) + c_1 T_i(x_i)T_1(x_{i1}) + \cdots + c_N T_i(x_i)T_N(x_i)$$
$$\vdots$$
$$f(x_N)T_i(x_N) = c_0 T_i(x_N)T_0(x_N) + c_1 T_i(x_N)T_1(x_N) + \cdots + c_N T_i(x_N)T_N(x_N)$$

$$(4.44)$$

Taking the sums of the terms on the left and right sides of the relations (4.44) we obtain:

$$\sum_{k=0}^{N} f(x_k)T_i(x_k) = c_0 \sum_{k=0}^{N} T_i(x_k)T_0(x_k) + c_1 \sum_{k=0}^{N} T_i(x_k)T_1(x_k) + c_2 \sum_{k=0}^{N} T_i(x_k)T_2(x_k)$$
$$+ \cdots + c_i \sum_{k=0}^{N} T_i(x_k)T_i(x_k) + \cdots + c_N \sum_{k=0}^{N} T_i(x_k)T_N(x_k)$$

$$(4.45)$$

Using the property (4.42), the sums written above simplify to the form:

$$\sum_{k=0}^{N} f(x_k)T_i(x_k) = c_i \sum_{k=0}^{N} T_i(x_k)T_i(x_k) = c_i \frac{N+1}{2}, \quad i \neq 0 \qquad (4.46)$$

According to the relation (4.46), the desired coefficient c_j is:

$$c_i = \frac{2}{N+1} \sum_{k=0}^{N} f(x_k)T_i(x_k) \quad \text{for } 1 \leq i \leq N \qquad (4.47)$$

In case when $i = 0$:

$$c_0 = \frac{1}{N+1} \sum_{k=0}^{N} f(x_k)T_0(x_k) \qquad (4.48)$$

The interpolation method described above can be generalized to the case of an arbitrary real function $f(t)$ defined over an interval $[t_a, t_b]$, where $t_a \neq -1$ and $t_b \neq 1$. To this end, we write the variable x as a function of a new variable t using the transformation:

$$x \equiv x(t) = -1 + 2\frac{t - t_a}{t_b - t_a} \qquad (4.49)$$

The changes of the variable t over the interval $[t_a, t_b]$ correspond here to respective changes of the variable x over the interval $[-1, 1]$.

Example 4.6 As an illustration example, let us evaluate a polynomial of third degree $P_3[x(t)]$ interpolating the function $f(t) = 3\exp(-0.1t)$ over an interval $[0, 10]$. The values of x_k, calculated for $k = 0, 1, 2, 3$, are given in the second column of Table 4.4.

The third column of this table presents values of t_k calculated for $t_a = 0$ and $t_b = 10$ according to the following relation $t \equiv t(x) = 0.5[(1 - x)t_a + (1 + x)t_b]$, which is the inverse of the transformation (4.49). In the last two columns of the table, only the values of $T_2(x_k)$ and $T_3(x_k)$ are given, because $T_0(x_k) = 1$ and $T_1(x_k) = x_k$. The coefficients c_j, where $j = 0, 1, 2, 3$, calculated by means of (4.48) and (4.47), are equal to: $c_0 = 1.935105711$, $c_1 = -0.938524669$, $c_2 = 0.116111001$ and $c_3 = -9.595734359 \times 10^{-3}$. Hence the interpolating polynomial being evaluated has the form:

$$\begin{aligned} P_3[x(t)] = {}& 1.935105711 - 0.938524669x + 0.116111001(2x^2 - 1) \\ & -9.595734359 \times 10^{-3}(4x^3 - 3x) \end{aligned}$$

The maximum absolute deviation $R_{3\,\mathrm{max}} = \max|f(t) - P_3[x(t)]|$, computed for $0 \leq t \leq 10$, does not exceed 6.629×10^{-4}. This deviation can be reduced using the interpolating polynomial of higher degree. As an example, the interpolating polynomial of the fifth degree ($n = 5$) has been found:

$$\begin{aligned} P_5[x(t)] = {}& 1.935105840 - 0.938524911x + 0.116112773(2x^2 - 1) \\ & -9.626569571 \times 10^{-3}(4x^3 - 3x) + 6.001497450 \times 10^{-4}(8x^4 - 8x^2 + 1) \\ & -3.061124322 \times 10^{-5}(16x^5 - 20x^3 + 5x) \end{aligned}$$

In this case, the maximum deviation $R_{5\,\mathrm{max}} = \max|f(t) - P_5[x(t)]|$ is not greater than 2×10^{-6} and about 320 times less than the deviation $R_{3\,\mathrm{max}}$. On the basis of the function $f(t)$ and the corresponding interpolating polynomial $P_N[x(t)]$, it is possible to determine the deviation function $R_N(t) = f(t) - P_N[x(t)]$, which is undoubtedly the quality measure of the performed interpolation. For majority of the real functions $f(t)$, the deviation function $R_N(t)$ has similar shape as the equal

Table 4.4

k	$x_k = x(t_k)$	t_k	$f(t_k)$	$T_2(x_k)$	$T_3(x_k)$
0	0.923879564	9.619397819	1.146452546	0.707106897	0.382683659
1	0.382683456	6.913417279	1.502710580	-0.707106745	-0.923879562
2	-0.382683426	3.086582869	2.203294992	-0.707106790	0.923879525
3	-0.923879504	0.380624789	2.887964725	0.707106676	-0.382683227

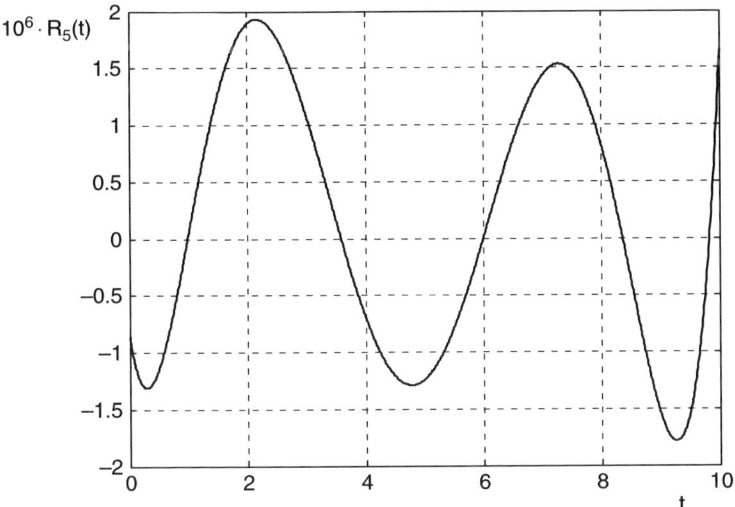

Fig. 4.7

ripple waveform over the interpolation interval $t_a \leq t \leq t_b$, $-1 \leq x(t) \leq 1$, and absolute values of its local extreme values are close to the maximum equal to $R_{N\,max}$. This conclusion is well confirmed by the numerical results obtained for the function $R_5(t) = f(t) - P_5[x(t)]$ shown in Fig. 4.7.

The properties mentioned just above have considerably decided about numerous applications of the interpolation method for various scientific and engineering problems. The simplicity of the algorithm constructed for evaluation of the interpolating polynomial $P_N(x)$ has also contributed to its popularity, see relations (4.41), (4.47) and (4.48).

4.2 Fundamental Approximation Methods for One Variable Functions

4.2.1 The Equal Ripple (Chebyshev) Approximation

Approximation of the constant function over a given limited interval, by a polynomial providing the equal ripple (Chebyshev) deviations, has found many applications for solving a great number of various engineering problems. As an example, let us consider the design problem for the low-pass filter (LPF) with the insertion loss function $L(f)$ [dB] similar to that shown in Fig. 4.8(b).

The insertion loss function (expressed in dB) of an any passive two-port electrical circuit is defined as:

a)

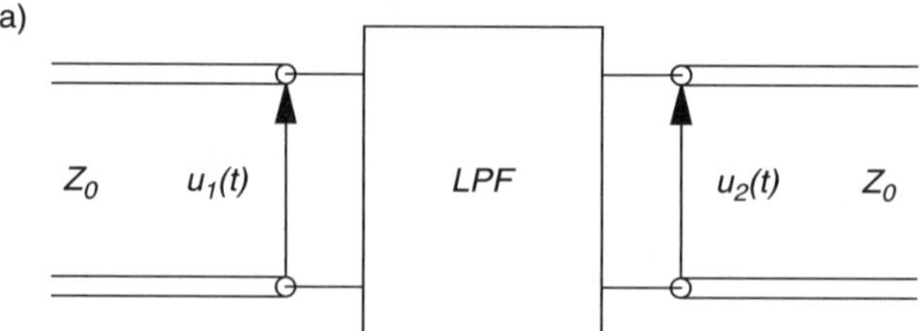

b)

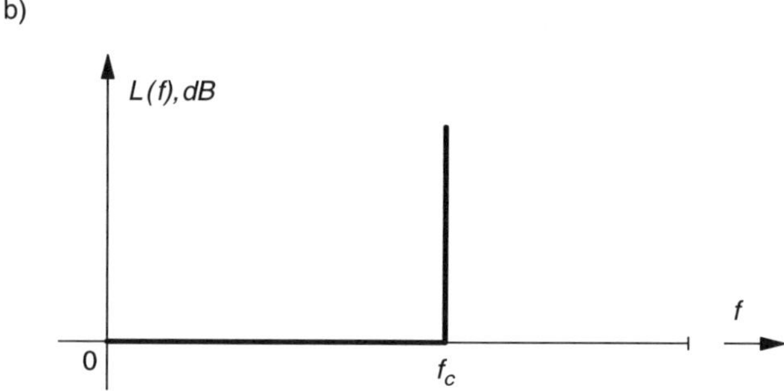

c)

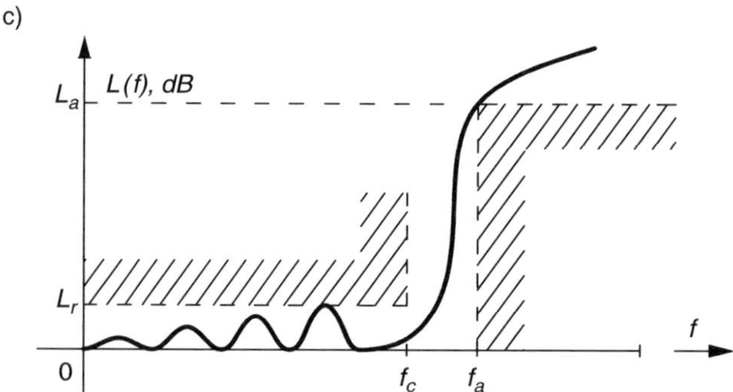

Fig. 4.8

$$L(f)[\text{dB}] = 20 \log \left| \frac{\mathbf{U}_1(f)}{\mathbf{U}_2(f)} \right| \tag{4.50}$$

where $\mathbf{U}_1(f)$ and $\mathbf{U}_2(f)$ denote complex amplitudes of the input and output voltages, respectively. The design of a LPF with the "ideal" insertion loss function $L(f)$ [dB] similar to that shown in Fig. 4.8(b) is not possible, because such filter would be composed of an infinite number of reactive elements, i.e., capacitors and inductors. In other words, this function would not satisfy the condition of physical realizability. In this situation, it is necessary to replace (interpolate or approximate) this function by another one, satisfying simultaneously the problem requirements and conditions of a physical realizability. Limitations imposed on the permissible (acceptable) function $L(f)$ [dB] are illustrated in Fig. 4.8(c). This curve representing the assumed function $L(f)$ [dB] should pass through the nondashed area

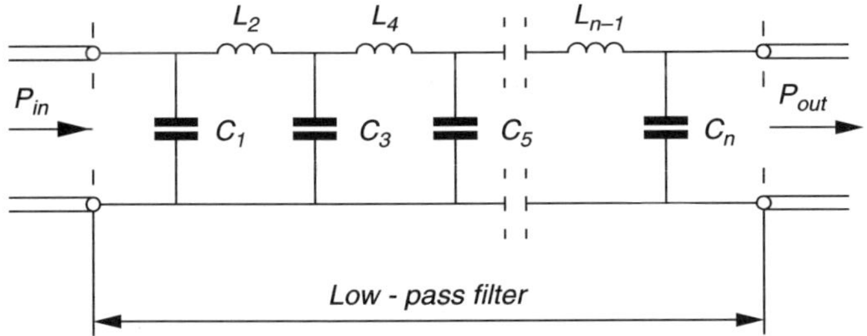

Fig. 4.9

defined by the cut-off frequency f_c, permissible attenuation L_r in the pass-band, lower frequency f_a of the stop-band and minimum value L_a of attenuation in the stop-band. The requirements formulated above can be satisfied by many different functions, satisfying simultaneously the condition of physical realizability. In electrical engineering, the functions $L(f)$ [dB] having the form of polynomials of a possible lowest degree are used most frequently. The LPF realized on a basis the polynomial function can take the ladder structure similar to that shown in Fig. 4.9, [9, 10].

One example of such classical function is:

$$L(f)[\text{dB}] = 10 \, \log \left[1 + \varepsilon T_n^2(x) \right] \tag{4.51}$$

where $x = f/f_c$ is a normalized frequency, ε is the parameter determining permissible losses in the pass-band and $T_n(x)$ is the Chebyshev polynomial of the first kind of degree n. The Chebyshev polynomials $T_n(x)$ can be easily computed from the following recursive formulas:

$$
\begin{aligned}
T_0(x) &= 1 \\
T_1(x) &= x \\
T_2(x) &= 2x^2 - 1 \\
&\;\;\vdots \\
T_n(x) &= 2x T_{n-1}(x) - T_{n-2}(x), \quad n = 2, 3, 4, \ldots
\end{aligned}
\tag{4.52}
$$

Fig. 4.10 presents the Chebyshev polynomials $T_n(x)$ of degrees 0, 1, 2 and 3.

It follows from Fig. 4.10 that over the interval $[-1, 1]$, polynomials $T_n(x)$ take the extreme values equal to 1 or -1. Moreover, polynomials $T_n(x)$ are even functions for even degrees n and respectively odd functions for odd degrees n. This property is expressed by the formula:

$$T_n(-x) = (-1)^n T_n(x) \tag{4.53}$$

The Chebyshev polynomials $T_n(x)$ are often written in their trigonometric form presented below:

$$T_n(x) = \cos[n \arccos(x)] \quad \text{when } |x| < 1 \text{ and } n = 0,\ 1,\ 2,\ 3,\ldots \qquad (4.54)$$

$$T_n(x) = \text{ch}[n\,\text{arch}(x)] \quad \text{when } |x| > 1 \text{ and } n = 0,\ 1,\ 2,\ 3,\ldots \qquad (4.55)$$

In an interval $[-1, 1]$ the polynomials $T_n(x)$ take the zero value for

$$x_k = \cos\left[\frac{(2k+1)\pi}{2n}\right] \qquad (4.56)$$

where $k = 0, 1, 2, 3, \ldots, n - 1$ and $n \geq 1$. The values (4.56) are sometimes called the Chebyshev interpolating points, because they represent the coordinates of intersection points of the polynomial $T_n(x)$ and the approximated function $y(x) = 0$.

If we want to determine the parameters ε and n of the function (4.51), we must know the values of f_c, f_a, L_r and L_a, see Fig. 4.8(c). Thus, let us assume now that $f_c = 10\,\text{MHz}$, $f_a = 13\,\text{MHz}$, $L_r = 0.3\,\text{dB}$ and $L_a = 15\,\text{dB}$. From the formula (4.51) we obtain:

$$L(f = f_c) = 10\,\log[1 + \varepsilon T_n^2(1)] = L_r,\ \ \text{dB}$$

$$L(f = f_a) = 10\,\log[1 + \varepsilon T_n^2(1.3)] \geq L_a,\ \ \text{dB}$$

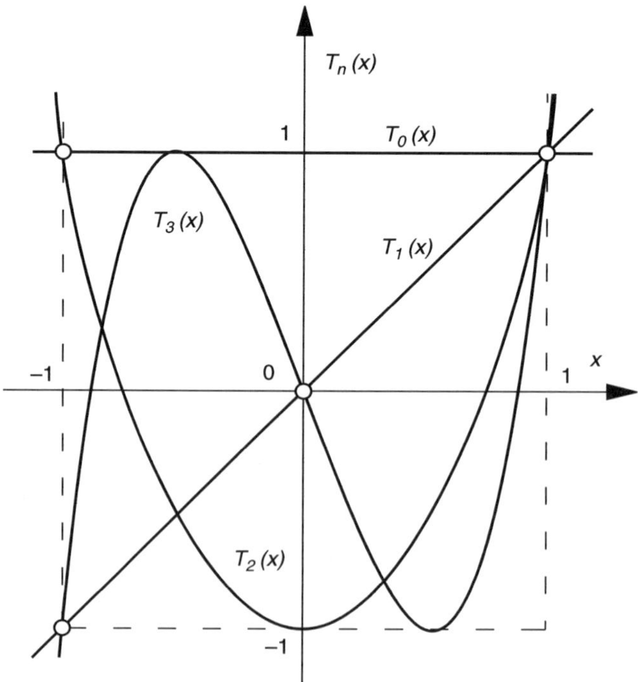

Fig. 4.10

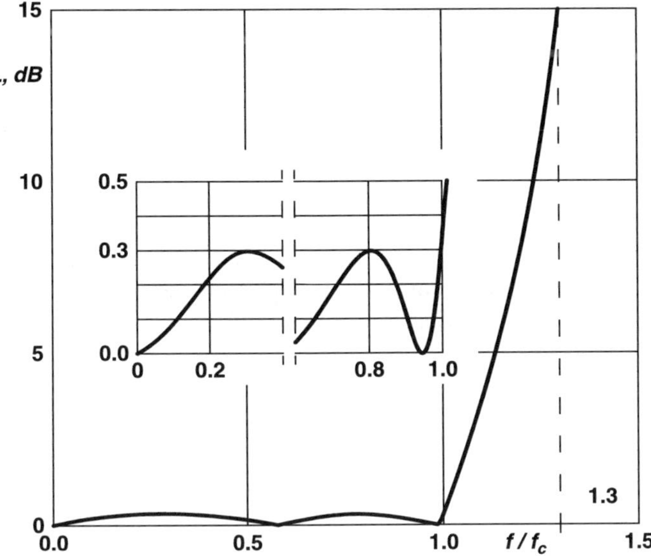

Fig. 4.11

These expressions can be transformed to the form:

$$10^{L_r/10} - 1 = \varepsilon T_n^2(1) = \varepsilon = 0.071519305$$

$$10^{L_a/10} - 1 = 30.622776 \leq \varepsilon T_n^2(1.3)$$

From the above relations follows the inequality $20.692389 \leq T_n(1.3)$. This inequality is satisfied by the polynomial at least of fifth degree ($n = 5$), i.e., taking the value $T_5(1.3) = 21.96688$. The insertion loss function $L(f)[\text{dB}] = 10\log\left[1 + \varepsilon T_n^2(x)\right]$ evaluated in this way ($n = 5$, $\varepsilon = 0.071519305$, $L_r = 10\log(1 + \varepsilon) = 0.3$ dB) is shown in Fig. 4.11. It deserves noting that maximum values of all ripples of the evaluated function $L(f)[\text{dB}]$ are equal to each other over the frequency range $1 \leq (f/f_c) \leq 1$. Equality of amplitudes of these deviations justifies the title of present section.

The LPF implementing this insertion loss function contains five reactive elements creating the LC ladder structure shown in Fig. 4.12.

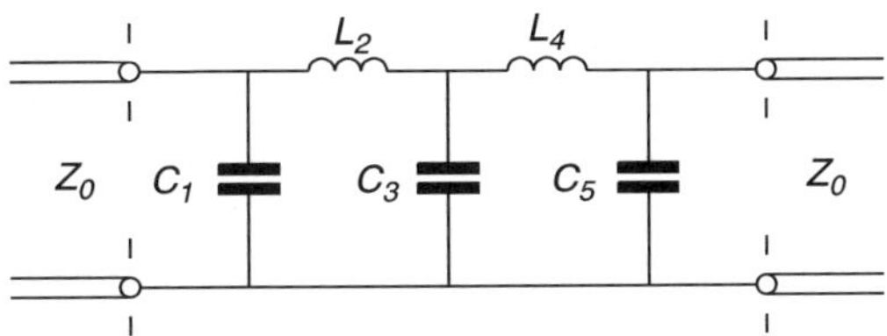

Fig. 4.12

The values of particular capacitances and inductances can be computed from the formulas:

$$C_i = \frac{1}{Z_0 2\pi f_c} g_i \quad \text{for } i = 1,\ 3,\ 5 \tag{4.57}$$

$$L_j = \frac{Z_0}{2\pi f_c} g_j \quad \text{for } j = 2, 4$$

where $g_k, k = 1, 2, \ldots, 5$ are parameters of the prototype LPF, determined on the basis of the evaluated insertion loss function. According to [9]

$$g_1 = \frac{2a_1}{\gamma}$$

$$g_k = \frac{4a_{k-1}a_k}{b_{k-1}g_{k-1}} \quad \text{for } k = 2,\ 3,\ 4 \text{ and } 5 \tag{4.58}$$

where

$$x = \ln\left[\operatorname{cth}\left(\frac{L_r}{17.37}\right)\right]$$

$$\gamma = \operatorname{sh}\left(\frac{x}{2n}\right) \quad n = 5$$

$$a_k = \sin\left[\frac{(2k-1)\pi}{2n}\right] \quad \text{for } k = 1,\ 2,\ 3,\ 4,\ 5,\quad n = 5$$

$$b_k = \gamma^2 + \sin^2\left(\frac{k\pi}{n}\right) \quad \text{for } k = 1,\ 2,\ 3,\ 4,\ 5,\quad n = 5$$

For $n = 5$ and $L_r = 0.3$ dB we obtain: $g_1 = g_5 = 1.4817$, $g_2 = g_4 = 1.2992$ and $g_3 = 2.3095$. Let us assume that the designed filter is loaded on both sides by the real impedances $Z_0 = 50\ \Omega$. Thus, the filter capacitances and inductances calculated from formulas (4.57) are equal to: $C_1 = C_5 = 471.633$ pF, $C_3 = 735.140$ pF, $L_2 = L_4 = 1.034\ \mu$H.

4.2.2 The Maximally Flat (Butterworth) Approximation

In the literature on the linear electrical circuits, including different kinds of filters and impedance matching circuits, the kind of approximation mentioned in the title of this section is called maximally flat or Butterworth approximation [9]. Approximation of the constant function $y = y(x) = \text{const}$ over a limited interval, according to the criterion of maximum flatness, consists in determination of the polynomial $B_n(x)$ of degree n, satisfying the following conditions:

$$B_n(x_0) = y(x_0)$$
$$B_n^{(k)}(x_0) = 0 \quad \text{for } k = 1, 2, 3, \ldots, n - 1 \tag{4.59}$$

where x_0 is an arbitrary point of the approximation interval and the index (k) denotes the derivative of degree k with respect to x calculated at x_0. Moreover, the approximating polynomial $B_n(x)$ should satisfy the condition of the physical realizability defined for the circuit being designed. In order to concretize our further considerations, let us consider once again the problem of designing the LPF with the insertion loss function similar to that shown in Fig. 4.8(b). As in the previous case, see Sect. 4.2.1, this function should be approximated by another one satisfying the requirements specified by f_c, f_a, L_r and L_a in Fig. 4.13.

The insertion loss function satisfying the above requirements, criterion (4.59) and the condition of physical realizability determined for a ladder LPF, see Fig. 4.9, should have the form:

$$B_n(f) = \frac{P_{we}}{P_{wy}} = 1 + \varepsilon \left(\frac{f}{f_c}\right)^{2n} \tag{4.60}$$

The function (4.60) is known in the literature as the Butterworth polynomial of degree n. It is most frequently presented in the logarithmic form:

$$B_n(\omega)[\text{dB}] = 10 \log \left[1 + \varepsilon \left(\frac{\omega}{\omega_c}\right)^{2n} \right] \tag{4.61}$$

where $\omega = 2\pi f$ is an angular frequency. A number of curves, calculated from formula (4.61) for $\varepsilon = 1$ and some values of n, are shown in Fig. 4.14.

The function $B_n(\omega)[\text{dB}]$ similar to that shown in Fig. 4.8(b) (marked in Fig. 4.14 by a broken line) corresponds to the limit case $n \to \infty$. Let us assume that function (4.61) satisfies the requirements specified by $f_c = 10\,\text{MHz}$, $f_a = 13\,\text{MHz}$, $L_r = 3.1\,\text{dB}$ and $L_a = 15\,\text{dB}$ in Fig. 4.13. These requirements are satisfied by the polynomial $B_n(\omega)[\text{dB}]$ of the seventh degree $(n = 7)$ taking for $f_a/f_c = 1.3$ value $B_7(f_a)[\text{dB}] = 10 \log[1 + (1.3)^{14}] = 16.06$ which exceeds the value of $L_a = 15\,\text{dB}$, Fig. 4.15.

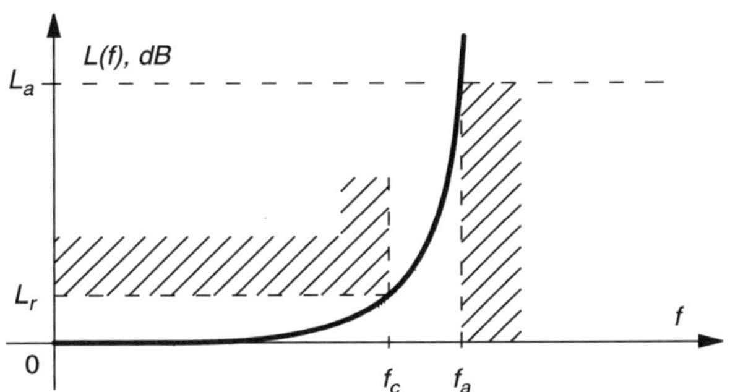

Fig. 4.13

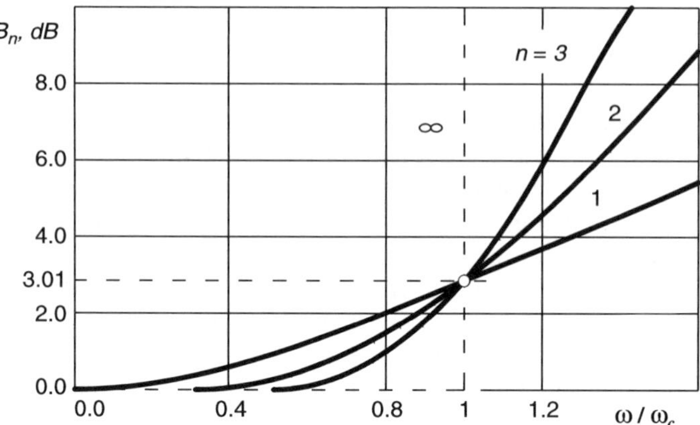

Fig. 4.14

The LPF, see Fig. 4.9, designed on the basis of the insertion loss function evaluated above ($n = 7$, $\varepsilon = 1$) contains seven reactive elements, namely four capacitors and three inductors. Values of these elements can be calculated from

$$C_i = \frac{1}{Z_0 2\pi f_c} g_i \quad \text{for } i = 1, 3, 5, 7 \tag{4.62}$$

$$L_j = \frac{Z_0}{2\pi f_c} g_j \quad \text{for } j = 2, 4, 6$$

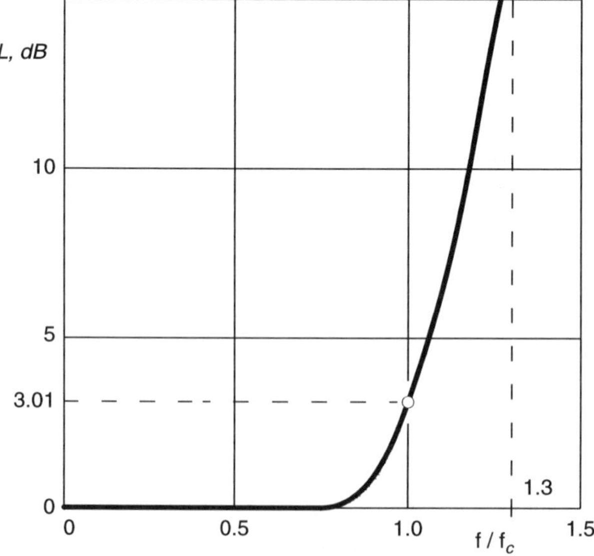

Fig. 4.15

where g_k, $k = 1, 2, 3, \ldots, 7$ are parameters of the corresponding prototype LPF, [9]. These parameters for $\varepsilon = 1$ and $L_r = 3.01$ dB can be calculated by using the following simple formula:

$$g_k = 2 \sin \left[\frac{(2k - 1)\pi}{2n} \right] \tag{4.63}$$

where $k = 1, 2, 3, \ldots, 7$ and $n = 7$. Let us assume that the designed filter is loaded on both sides by real impedances $Z_0 = 75\ \Omega$. With this assumption the filter capacitances and inductances are: $C_1 = C_7 = 94.431\ \text{pF}$, $C_3 = C_5 = 382.396\ \text{pF}$, $L_2 = L_6 = 1.488\ \mu\text{H}$ and $L_4 = 2.387\ \mu\text{H}$. As it was mentioned at the beginning of this section, the Butterworth approximation is used also for designing the impedance matching circuits. This fact is well confirmed by the Example 3.4, which presents the design algorithm for the broadband nonsynchronous noncommensurate impedance transformer composed of four noncommensurate TEM transmission line segments.

4.2.3 Approximation (Curve Fitting) by the Method of Least Squares

In order to explain an idea of the least squares method, let us assume that the approximated function $y_i = y(x_i)$ is defined for $(n + 1)$ points x_i, where $i = 0, 1, 2, 3, \ldots, n$. Let us assume also that the approximating polynomial

$$f(x) = a_0 + a_1 x + a_2 x^2 + \cdots + a_m x^m \tag{4.64}$$

is of degree m. The approximation quality measure for the function $y_i = y(x_i)$ approximated by the function $f(x)$ according to the least squares criterion is the norm:

$$R_S(a_0, a_1, a_2, \ldots, a_m) = \sum_{i=0}^{n} [f(x_i) - y_i]^2 \tag{4.65}$$

The essence of this approximation form, called frequently the least squares approximation, is the evaluation of such values of coefficients $a_0, a_1, a_2, a_3, \ldots, a_m$ for which the norm (4.65) achieves its minimum. According to the corresponding theorems of the differential calculus, concerning the functions of many variables, the norm (4.65) achieves the minimum when

$$\frac{\partial R_S}{\partial a_0} = \frac{\partial R_S}{\partial a_1} = \frac{\partial R_S}{\partial a_2} = \cdots = \frac{\partial R_S}{\partial a_m} = 0 \tag{4.66}$$

After introducing the function $y_i = y(x_i)$ and polynomial (4.64) into conditions (4.66) we obtain the following equations:

$$\frac{\partial R_S}{\partial a_0} = 2 \sum_{i=0}^{n} (a_0 + a_1 x_i + a_2 x_i^2 + \cdots + a_m x_i^m - y_i) \cdot 1 = 0$$

$$\frac{\partial R_S}{\partial a_1} = 2 \sum_{i=0}^{n} (a_0 + a_1 x_i + a_2 x_i^2 + \cdots + a_m x_i^m - y_i) \cdot x_i = 0$$

$$\frac{\partial R_S}{\partial a_2} = 2 \sum_{i=0}^{n} (a_0 + a_1 x_i + a_2 x_i^2 + \cdots + a_m x_i^m - y_i) \cdot x_i^2 = 0 \qquad (4.67)$$

$$\vdots$$

$$\frac{\partial R_S}{\partial a_m} = 2 \sum_{i=0}^{n} (a_0 + a_1 x_i + a_2 x_i^2 + \cdots + a_m x_i^m - y_i) \cdot x_i^m = 0$$

Rearrangement of the equation system (4.67) yields:

$$
\begin{bmatrix}
(n+1) & \sum_{i=0}^{n} x_i & \sum_{i=0}^{n} x_i^2 & \cdots & \sum_{i=0}^{n} x_i^m \\
\sum_{i=0}^{n} x_i & \sum_{i=0}^{n} x_i^2 & \sum_{i=0}^{n} x_i^3 & \cdots & \sum_{i=0}^{n} x_i^{m+1} \\
\sum_{i=0}^{n} x_i^2 & \sum_{i=0}^{n} x_i^3 & \sum_{i=0}^{n} x_i^4 & \cdots & \sum_{i=0}^{n} x_i^{m+2} \\
\vdots & \vdots & \vdots & \cdots & \vdots \\
\sum_{i=0}^{n} x_i^m & \sum_{i=0}^{n} x_i^{m+1} & \sum_{i=0}^{n} x_i^{m+2} & \cdots & \sum_{i=0}^{n} x_i^{m+m}
\end{bmatrix}
\cdot
\begin{bmatrix}
a_0 \\ a_1 \\ a_2 \\ \vdots \\ a_m
\end{bmatrix}
=
\begin{bmatrix}
\sum_{i=0}^{n} y_i \\
\sum_{i=0}^{n} x_i y_i \\
\sum_{i=0}^{n} x_i^2 y_i \\
\vdots \\
\sum_{i=0}^{n} x_i^m y_i
\end{bmatrix}
\qquad (4.68)
$$

Solution of the system of $(m + 1)$ linear equations obtained above can be performed by means of one of the direct methods described in Chap. 1, as for example the Gauss elimination method with the choice of the pivotal element.

The polynomial (4.64) is only a particular version of the generalized one, namely:

$$g(x) = a_1 q_1(x) + a_2 q_2(x) + \cdots + a_m q_m(x) \qquad (4.69)$$

composed of linearly independent basis functions $q_1(x), q_2(x), q_3(x), \ldots, q_m(x)$. Coefficients $a_1, a_2, a_3, \ldots, a_m$ of this generalized polynomial can also be obtained from relations (4.66), expressed as:

$$
\begin{bmatrix}
\sum\limits_{i=0}^{n} q_1^2(x_i) & \sum\limits_{i=0}^{n} q_1(x_i)q_2(x_i) & \dots & \sum\limits_{i=0}^{n} q_1(x_i)q_m(x_i) \\[2ex]
\sum\limits_{i=0}^{n} q_2(x_i)q_1(x_i) & \sum\limits_{i=0}^{n} q_2^2(x_i) & \dots & \sum\limits_{i=0}^{n} q_2(x_i)q_m(x_i) \\[2ex]
\sum\limits_{i=0}^{n} q_3(x_i)q_1(x_i) & \sum\limits_{i=0}^{n} q_3(x_i)q_2(x_i) & \dots & \sum\limits_{i=0}^{n} q_3(x_i)q_m(x_i) \\[2ex]
\vdots & \vdots & \dots & \vdots \\[2ex]
\sum\limits_{i=0}^{n} q_m(x_i)q_1(x_i) & \sum\limits_{i=0}^{n} q_m(x_i)q_2(x_i) & \dots & \sum\limits_{i=0}^{n} q_m^2(x_i)
\end{bmatrix}
\begin{bmatrix} a_1 \\ a_2 \\ a_3 \\ \vdots \\ a_m \end{bmatrix}
=
\begin{bmatrix}
\sum\limits_{i=0}^{n} q_1(x_i)y_i \\[2ex]
\sum\limits_{i=0}^{n} q_2(x_i)y_i \\[2ex]
\sum\limits_{i=0}^{n} q_3(x_i)y_i \\[2ex]
\vdots \\[2ex]
\sum\limits_{i=0}^{n} q_m(x_i)y_i
\end{bmatrix}
$$

Also in this more general case the obtained equation system is linear with respect to desired coefficients $a_1, a_2, a_3, \dots, a_m$. Due to this linearity, this system can be solved by using one of the direct methods described in Chap. 1. When the basis functions of the polynomial (4.69) constitute an orthogonal set of functions, for which

$$
\sum_{i=0}^{n} q_k(x_i)q_j(x_i) = 0 \quad \text{for } j \neq k \tag{4.70}
$$

the matrix of coefficients of the equation system formulated above is the diagonal matrix. In this special case

$$
a_j = \frac{\sum\limits_{i=0}^{n} q_j(x_i)y_i}{\sum\limits_{i=0}^{n} q_j^2(x_i)} \tag{4.71}
$$

where $j = 1, 2, 3, \dots, m$. Of course, in this special case the computation process becomes much simpler.

Historically, the first and the most useful sets of functions, orthogonal over the interval $0 \leq x \leq \pi$, are the following sets of trigonometric functions:

$$
1, \cos(x), \cos(2x), \cos(3x), \dots, \cos(nx), \dots \tag{4.72}
$$

$$
\sin(x), \sin(2x), \sin(3x), \dots, \sin(nx), \dots \tag{4.73}
$$

Combination of functions (4.72) and (4.73), i.e.:

$$
1, \cos(x), \sin(x), \cos(2x), \sin(2x), \cos(3x), \sin(3x), \dots, \cos(nx), \sin(nx), \dots \tag{4.74}
$$

gives a new set of orthogonal functions over the interval $-\pi \leq x \leq \pi$. Variable x appearing in these functions can be treated as linear function of a new variable t, for example:

$$x = \frac{\pi}{t_m} t \tag{4.75}$$

where t_m represents a fixed maximum value of the variable t. Thus, the sets of functions (4.72) and (4.73) can be written as:

$$1, \cos\left(\frac{\pi}{t_m} t\right), \cos\left(2\frac{\pi}{t_m} t\right), \cos\left(3\frac{\pi}{t_m} t\right), \ldots, \cos\left(n\frac{\pi}{t_m} t\right), \ldots \tag{4.76}$$

$$\sin\left(\frac{\pi}{t_m} t\right), \sin\left(2\frac{\pi}{t_m} t\right), \sin\left(3\frac{\pi}{t_m} t\right), \ldots, \sin\left(n\frac{\pi}{t_m} t\right), \ldots \tag{4.77}$$

It is easy to prove that functions (4.76) and (4.77) are orthogonal over the interval $[0, t_m]$. The set of functions (4.74) transformed in the same way, i.e.:

$$1, \cos\left(\frac{\pi}{t_m} t\right), \sin\left(\frac{\pi}{t_m} t\right), \cos\left(2\frac{\pi}{t_m} t\right), \sin\left(2\frac{\pi}{t_m} t\right), \ldots, \cos\left(n\frac{\pi}{t_m} t\right),$$
$$\sin\left(n\frac{\pi}{t_m} t\right), \ldots \tag{4.78}$$

is orthogonal for $-t_m \leq t \leq t_m$. There, it should be pointed out that not only the trigonometric functions have the property of orthogonality. Some polynomials can also be orthogonal over the interval $[-1, 1]$. In order to determine one such set of polynomials, let us consider the functions

$$1, x, x^2, x^3, \ldots, x^n \tag{4.79}$$

First two functions of this set are orthogonal because

$$\int_{-1}^{1} 1 \cdot x \, dx = \frac{1}{2}[1^2 - (-1)^2] = 0 \tag{4.80}$$

Consequently, we can assume that $P_0(x) \equiv 1$ and $P_1(x) \equiv x$. The function x^2 of the set (4.79) is not orthogonal with respect to $P_0(x) \equiv 1$. Therefore, we assume the polynomial $P_2(x)$ as linear combination of first three functions of the set (4.79), namely $P_2(x) = ax^2 + bx + c$. The coefficients a, b and c of this polynomial should ensure its orthogonality with respect to $P_0(x) \equiv 1$ and $P_1(x) \equiv x$. This requirement is expressed by the following equations:

$$\int_{-1}^{1} (ax^2 + bx + c) \cdot 1 \, dx = \frac{2}{3}a + 2c = 0$$

$$\int_{-1}^{1} (ax^2 + bx + c) \cdot x \, dx = \frac{2}{3}b = 0 \tag{4.81}$$

The system of equations (4.81) is satisfied for $b = 0$ and $a = -3c$. Therefore, the desired polynomial is $P_2(x) = c(-3x^2 + 1)$, where c is an arbitrary constant. Value of this constant can be evaluated from an additional condition which is usually taken in the form $P_2(1) = 1$. The normalized polynomial $P_2(x)$ becomes:

$$P_2(x) = \frac{1}{2}(3x^2 - 1) \tag{4.82}$$

Similarly, we evaluate other polynomials, orthogonal over the interval $[-1, 1]$. Some of them are given below:

$$P_3(x) = \frac{1}{2}(5x^3 - 3x)$$

$$P_4(x) = \frac{1}{8}(35x^4 - 30x^2 + 3) \tag{4.83}$$

$$P_5(x) = \frac{1}{8}(63x^5 - 70x^3 + 15x)$$

$$\vdots$$

Polynomials of higher degrees, for $n \geq 3$, can be evaluated from the following recursive formula:

$$P_{n+1}(x) = \frac{2n + 1}{n + 1} x P_n(x) - \frac{n}{n + 1} P_{n-1}(x)$$

In the literature polynomials $P_n(x)$, where $n = 0, 1, 2, 3, \ldots$ are called spherical functions or the Legendre polynomials. Also in this case, variable x can be treated as a linear function of a new variable t, for example:

$$x = \frac{1}{t_m} t \tag{4.84}$$

where t_m is a fixed maximum value of the variable t. The Legendre polynomials expressed in terms of the variable t, i.e.:

$$P_0[x(t)] = 1$$

$$P_1[x(t)] = \frac{1}{t_m}t$$

$$P_2[x(t)] = \frac{1}{2}\left[3\left(\frac{1}{t_m}t\right)^2 - 1\right] = \frac{1}{2}\left[\frac{3}{t_m^2}t^2 - 1\right] \qquad (4.85)$$

$$P_3[x(t)] = \frac{1}{2}\left[5\left(\frac{1}{t_m}t\right)^3 - 3\left(\frac{1}{t_m}t\right)\right] = \frac{1}{2}\left[\frac{5}{t_m^3}t^3 - \frac{3}{t_m}t\right]$$

$$\vdots$$

are orthogonal over the interval $[-t_m, t_m]$.

Example 4.7 As an illustration of the least squares method we evaluate the polynomial of the second degree ($m = 2$) approximating the function $y_i = y(x_i)$ given in the second and third columns of Table 4.5.

According to the algorithm described above we obtain the following equation system:

$$\begin{bmatrix} 5 & 11.2500 & 30.9375 \\ 11.2500 & 30.9375 & 94.9218 \\ 30.9375 & 94.9218 & 309.7617 \end{bmatrix} \cdot \begin{bmatrix} a_0 \\ a_1 \\ a_2 \end{bmatrix} = \begin{bmatrix} 11.4800 \\ 29.2875 \\ 90.8719 \end{bmatrix}$$

whose solution is: $a_0 = 4.717444$, $a_1 = -3.733289$ and $a_2 = 0.966215$. Hence the desired approximating polynomial is: $f(x) = 4.717444 - 3.733289x + 0.966215x^2$.

Values of this polynomial $f(x_i)$ and the corresponding differences (deviations) $f(x_i) - y_i$ are given in the fourth and fifth columns of Table 4.5. The more precise approximating polynomial of the third degree ($m = 3$) evaluated in the similar manner has a form: $f(x) = 5.165683 - 4.572444x + 1.392736x^2 - 0.063169x^3$. In this case max $\{|f(x_i) - y_i|\} \leq 0.041125$, where $i = 0, 1, 2, 3$ and 4.

4.2.4 Approximation of Periodical Functions by Fourier Series

The orthogonal series in one variable is defined as the following linear sum:

Table 4.5

i	x_i	$y_i = y(x_i)$	$f(x_i)$	$f(x_i) - y_i$
0	0.75	2.50	2.460973	−0.039027
1	1.50	1.20	1.291494	0.091494
2	2.25	1.25	1.209008	−0.040992
3	3.00	2.25	2.213513	−0.036486
4	3.75	4.28	4.305012	0.025012

$$\Psi(x) = a_0\psi_0(x) + a_1\psi_1(x) + a_2\psi_2(x) + a_3\psi_3(x) + \cdots + a_n\psi_n(x) + \cdots$$

$$= \sum_{i=0}^{\infty} a_i\psi_i(x) \tag{4.86}$$

whose basis functions $\psi_0(x)$, $\psi_1(x)$, $\psi_2(x)$, $\psi_3(x)$, ..., $\psi_n(x)$, ... form the orthogonal set for $a \leq x \leq b$. When we consider the approximation of an arbitrary, bounded function $f(x)$ by the series (4.86) over a limited interval $[a, b]$, it is first of all necessary to answer the following questions:

- whether an arbitrary limited function $f(x)$ can be approximated with sufficient accuracy by the series (4.86) over the interval $[a, b]$, and
- how to determine the coefficients $a_0, a_1, a_2, a_3, \ldots$ of this series.

The answer for the first above question is positive, because the basis functions of the approximating series form the complete orthogonal set. Some examples of such sets are given by (4.72), (4.73), (4.74), (4.76), (4.77) and (4.78), each within a corresponding properly chosen orthogonality interval. Assume that the series (4.86) approximate a function $f(x)$ defined over the interval $[a, b]$. According to this assumption

$$\int_a^b f(x)\psi_i(x)\mathrm{d}x = \int_a^b \Psi(x)\psi_i(x)\mathrm{d}x$$

$$= a_0 \int_a^b \psi_0(x)\psi_i(x)\mathrm{d}x + a_1 \int_a^b \psi_1(x)\psi_i(x)\mathrm{d}x$$

$$+ a_2 \int_a^b \psi_2(x)\psi_i(x)\mathrm{d}x \tag{4.87}$$

$$+ a_3 \int_a^b \psi_3(x)\psi_i(x)\mathrm{d}x + \cdots + a_i \int_a^b \psi_i^2(x)\mathrm{d}x + \cdots$$

Owing to the orthogonality of basic functions of the series (4.86), the integral (4.87) reduces to the following form:

$$\int_a^b f(x)\psi_i(x)\mathrm{d}x = \int_a^b \Psi(x)\psi_i(x)\mathrm{d}x = a_i \int_a^b \psi_i^2(x)\mathrm{d}x \tag{4.88}$$

Consequently, we can write:

$$a_i = \frac{\int\limits_a^b f(x)\psi_i(x)dx}{\int\limits_a^b \psi_i^2(x)dx} \tag{4.89}$$

where $i = 0, 1, 2, 3, \ldots$ Relation (4.89) can be used to determine coefficients $a_0, a_1, a_2, a_3, \ldots$ of the series (4.86) whose basis functions create the orthogonal and complete set. As an illustration example, we choose the series:

$$\Psi(x) = a_0 + \sum_{i=1}^{\infty} a_i \cos(ix) \tag{4.90}$$

of the functions (4.72), approximating the function $f(x)$ over the interval $[0, \pi]$. According to the relation (4.89) coefficients of this series are equal to:

$$a_0 = \frac{\int\limits_0^\pi f(x) \cdot 1 \, dx}{\int\limits_0^\pi 1 \cdot 1 \, dx} = \frac{1}{\pi} \int\limits_0^\pi f(x)dx$$

$$a_i = \frac{\int\limits_0^\pi f(x)\cos(ix)dx}{\int\limits_0^\pi \cos^2(ix)dx} = \frac{2}{\pi} \int\limits_0^\pi f(x)\cos(ix)dx \quad \text{for } i \geq 1.$$

Another approximation series composed of basis functions (4.73) has the form:

$$\Psi(x) = \sum_{i=1}^{\infty} b_i \sin(ix) \tag{4.91}$$

The coefficients of this series are calculated from the formula:

$$b_i = \frac{\displaystyle\int_0^{\pi} f(x)\sin(ix)\,dx}{\displaystyle\int_0^{\pi} \sin^2(ix)\,dx} = \frac{1}{\pi}\int_0^{\pi} f(x)\sin(ix)\,dx \quad \text{for } i \geq 1$$

In the same manner, we determine the coefficients of the series:

$$f(x) \equiv \Psi(x) = a_0 + \sum_{i=1}^{\infty}\left[a_i\cos(ix) + b_i\sin(ix)\right] \tag{4.92}$$

involving the functions (4.74). According to the relation (4.89) we obtain:

$$a_0 = \frac{\displaystyle\int_{-\pi}^{\pi} f(x)\cdot 1\,dx}{\displaystyle\int_{-\pi}^{\pi} 1\cdot 1\,dx} = \frac{1}{2\pi}\int_{-\pi}^{\pi} f(x)\,dx$$

$$a_i = \frac{\displaystyle\int_{-\pi}^{\pi} f(x)\cos(ix)\,dx}{\displaystyle\int_{-\pi}^{\pi} \cos^2(ix)\,dx} = \frac{1}{\pi}\int_{-\pi}^{\pi} f(x)\cos(ix)\,dx$$

$$b_i = \frac{\displaystyle\int_{-\pi}^{\pi} f(x)\sin(ix)\,dx}{\displaystyle\int_{-\pi}^{\pi} \sin^2(ix)\,dx} = \frac{1}{\pi}\int_{-\pi}^{\pi} f(x)\sin(ix)\,dx$$

The approximating series given by relations (4.90), (4.91) and (4.92) are commonly known as the Fourier series and serve as theoretical basis for the frequency (spectral) analysis of signals [11, 12]. Another Fourier series often used in the spectral analysis is:

$$f(t) \equiv \Psi(t) = a_0 + \sum_{i=1}^{\infty}\left[a_i\cos\left(i\frac{\pi}{t_m}t\right) + b_i\sin\left(i\frac{\pi}{t_m}t\right)\right] \tag{4.93}$$

The basis functions of this series are orthogonal over the interval $[-t_m, t_m]$. Coefficients $a_0, a_1, a_2, a_3, \ldots$ of this "stretched" series can be calculated according to the formulas:

$$a_0 = \frac{\displaystyle\int_{-t_m}^{t_m} f(t) \cdot 1 \, dt}{\displaystyle\int_{-t_m}^{t_m} 1 \cdot 1 \, dt} = \frac{1}{2t_m} \int_{-t_m}^{t_m} f(t) \, dt$$

$$a_i = \frac{\displaystyle\int_{-t_m}^{t_m} f(t) \cdot \cos\left(i\frac{\pi}{t_m}t\right) dt}{\displaystyle\int_{-t_m}^{t_m} \cos^2\left(i\frac{\pi}{t_m}t\right) dt} = \frac{1}{t_m} \int_{-t_m}^{t_m} f(t) \cdot \cos\left(i\frac{\pi}{t_m}t\right) dt$$

$$b_i = \frac{\displaystyle\int_{-t_m}^{t_m} f(t) \cdot \sin\left(i\frac{\pi}{t_m}t\right) dt}{\displaystyle\int_{-t_m}^{t_m} \sin^2\left(i\frac{\pi}{t_m}t\right) dt} = \frac{1}{t_m} \int_{-t_m}^{t_m} f(t) \cdot \sin\left(i\frac{\pi}{t_m}t\right) dt$$

where $i \geq 1$. The series (4.93) can also be presented in the following form having a univocal physical interpretation

$$\Psi(t) = c_0 + \sum_{i=1}^{\infty} c_i \cos\left[\left(i\frac{\pi}{t_m}t - \varphi_i\right)\right] \tag{4.94}$$

where

$$c_0 = a_0, \quad c_i = \sqrt{a_i^2 + b_i^2}, \quad tg(\varphi_i) = \frac{b_i}{a_i}, \quad i \geq 1$$

The coefficient c_0 denotes the constant component and c_i, φ_i are the amplitude and phase angle of the ith harmonic component, respectively.

Example 4.8 Let us assume that the output current of the half-wave rectifier is described by the function:

$$
i(t) \equiv f(t) = \begin{cases} 0 & \text{for} & -T/2 \le t \le 0 \\[2mm] I_m \sin\left(\dfrac{2\pi}{T}t\right) & \text{for} & 0 \le t \le T/2 \end{cases}
$$

where T is the period, as in Fig. 4.16.

The current function $i(t)$ can be approximated (replaced) by a sum of the constant component and an infinite number of harmonic components. In order to determine values of the constant component and amplitudes of the particular harmonic components, we should approximate the current function $i(t)$ by the series (4.93), assuming that $t_m = T/2$. Using relations given above we obtain:

$$
a_0 = \frac{1}{2T/2} \int_{-T/2}^{T/2} i(t)\mathrm{d}t = 0 + \frac{I_m}{T} \int_{-T/2}^{T/2} \sin\left(\frac{2\pi}{T}t\right)\mathrm{d}t
$$

$$
= \frac{I_m}{2\pi}[-\cos(\pi) + \cos(0)] = \frac{I_m}{\pi}
$$

$$
a_i = \frac{1}{T/2} \int_{-T/2}^{T/2} i(t)\cos\left(i\frac{\pi}{T/2}t\right)\mathrm{d}t = 0 + \frac{2I_m}{T} \int_{0}^{T/2} \sin\left(\frac{2\pi}{T}t\right)\cos\left(i\frac{2\pi}{T}t\right)\mathrm{d}t
$$

$$
= \frac{2I_m}{T} \int_{0}^{T/2} \frac{1}{2} \sin\left[\frac{2\pi}{T}(1+i)t\right]\mathrm{d}t + \frac{2I_m}{T} \int_{0}^{T/2} \frac{1}{2} \sin\left[\frac{2\pi}{T}(1-i)t\right]\mathrm{d}t
$$

$$
= \frac{I_m}{2\pi(1+i)}\{-\cos[\pi(1+i)] + 1\} + \frac{I_m}{2\pi(1-i)}\{-\cos[\pi(1-i)] + 1\}
$$

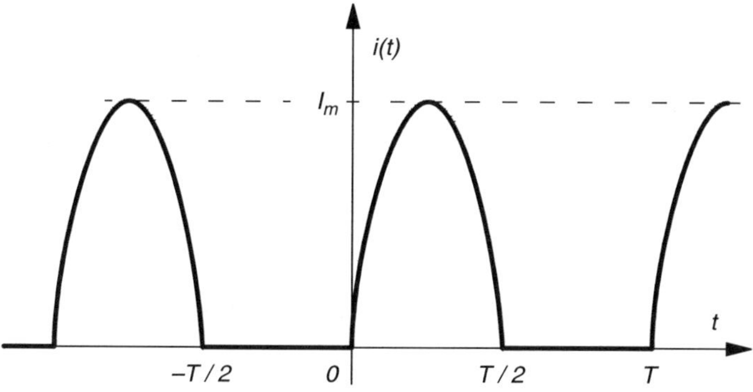

Fig. 4.16

Finally, we have

$$a_i = 0 \quad \text{for } i = 1, 3, 5, \ldots$$

$$a_i = \frac{-2I_m}{\pi} \cdot \frac{1}{i^2 - 1} \quad \text{for } i = 2, 4, 6, \ldots$$

$$b_i = \frac{1}{T/2} \int_{-T/2}^{T/2} i(t) \sin\left(i\frac{\pi}{T/2}t\right) dt = 0 + \frac{2I_m}{T} \int_0^{T/2} \sin\left(\frac{2\pi}{T}t\right) \sin\left(i\frac{2\pi}{T}t\right) dt$$

$$= \frac{2I_m}{T} \int_0^{T/2} \frac{1}{2} \cos\left[\frac{2\pi}{T}(1-i)t\right] dt - \frac{2I_m}{T} \int_0^{T/2} \frac{1}{2} \cos\left[\frac{2\pi}{T}(1+i)t\right] dt$$

$$= \frac{I_m}{2\pi(1-i)}\{\sin[\pi(1-i)] - 0\} - \frac{I_m}{2\pi(1+i)}\{\sin[\pi(1+i)] - 0\}$$

For $i = 1$

$$b_1 = \frac{I_m}{2} \lim_{i \to 1} \frac{\sin[\pi(1-i)]}{\pi(1-i)} = \frac{I_m}{2}$$

For $i > 1$, $b_i = 0$.

The relations determined above make possible the approximation of the current function $i(t)$ by the following infinite series:

$$i(t) \equiv \Psi(t) = \frac{I_m}{\pi} + \frac{I_m}{2} \sin(\omega_0 t) - \frac{2I_m}{\pi} \sum_{k=1}^{\infty} \frac{\cos(2k\omega_0 t)}{4k^2 - 1}$$

where

$$\omega_0 = 2\pi/T \text{ and } k = 1, 2, 3, \ldots$$

Example 4.9 The purpose of this example is to find the Fourier series approximating a periodic function $f(t)$ presented in Fig. 4.17.

The function $f(t)$ is characterized by the coefficients α, β, γ and τ, which must satisfy the following conditions:

$$\alpha > 0, \quad \beta \geq 0, \quad \gamma > 0, \quad \alpha + \beta + \gamma + \tau \leq 1$$

The function under consideration can be approximated by series (4.93), i.e.:

$$\Psi(t) = a_0 + \sum_{i=1}^{\infty} \left[a_i \cos\left(i\frac{2\pi}{T}t\right) + b_i \sin\left(i\frac{2\pi}{T}t\right)\right]$$

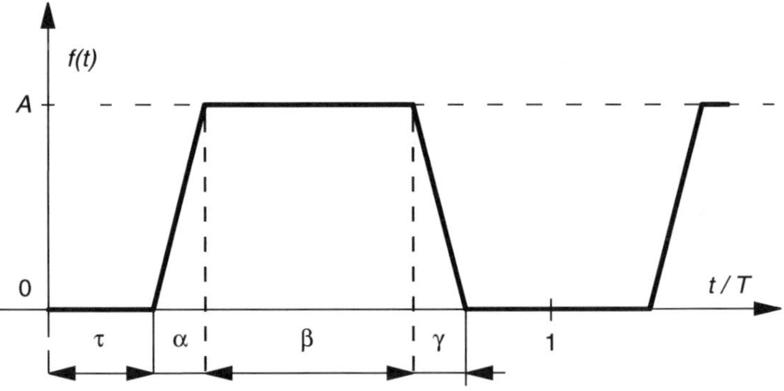

Fig. 4.17

in which $t_m = T/2$. Coefficients of this series can be calculated from the following formulas:

$$a_0 = \frac{1}{2} A \left(\alpha + 2\beta + \gamma \right)$$

$$a_i = \frac{A}{2i^2\pi^2} \left[\frac{1}{\alpha} \left[\cos\, i\omega_0 \left(\tau + \alpha \right) T - \cos\, i\omega_0 \tau T \right] \right.$$
$$\left. - \frac{1}{\gamma} \left[\cos\, i\omega_0 \left(\tau + \alpha + \beta + \gamma \right) T - \cos\, i\omega_0 \left(\tau + \alpha + \beta \right) T \right] \right]$$

$$b_i = \frac{A}{2i^2\pi^2} \left[\frac{1}{\alpha} \left[\sin\, i\omega_0 \left(\tau + \alpha \right) T - \sin\, i\omega_0 \tau T \right] \right.$$
$$\left. - \frac{1}{\gamma} \left[\sin\, i\omega_0 \left(\tau + \alpha + \beta + \gamma \right) T - \sin\, i\omega_0 \left(\tau + \alpha + \beta \right) T \right] \right]$$

where $\omega_0 = 2\pi/T$, $i = 1, 2, 3, \ldots$. By an appropriate choice of the coefficients α, β, γ, τ and adding the constant component we may shape the function $f(t)$. Examples of periodical functions formed in this way are shown in Fig. 4.18.

These functions can be approximated by using the relations given above. For example, in the case when $A = 1$, $\tau = 0.1$, $\beta = 0.2$, $\alpha = 0.0001$ and $\gamma = 0.0002$ we obtain:

$$a_0 = 0.20015, \quad a_1 = 0.115428, \quad b_1 = 0.356203, \ldots, \quad a_3 = -0.162921,$$
$$b_3 = -0.118956, \ldots, \quad a_5 = -0.000299, \quad b_5 = -0.000012, \ldots.$$

The Fourier series $\Psi(t)$ discussed in the Examples 4.8 and 4.9 approximate the given functions $i(t)$ and $f(t)$, see Figs. 4.16 and 4.17, within their appropriate intervals equal to one period T. Validity of this approximation for $-\infty < t < \infty$ can be justified by notifying that basis functions of these series $\Psi(t)$ are themselves periodic functions, whose periods are also equal to the periods of the functions approximated by them. Exact evaluation of the series $\Psi(t)$ is not possible because they

a)

b)

c)

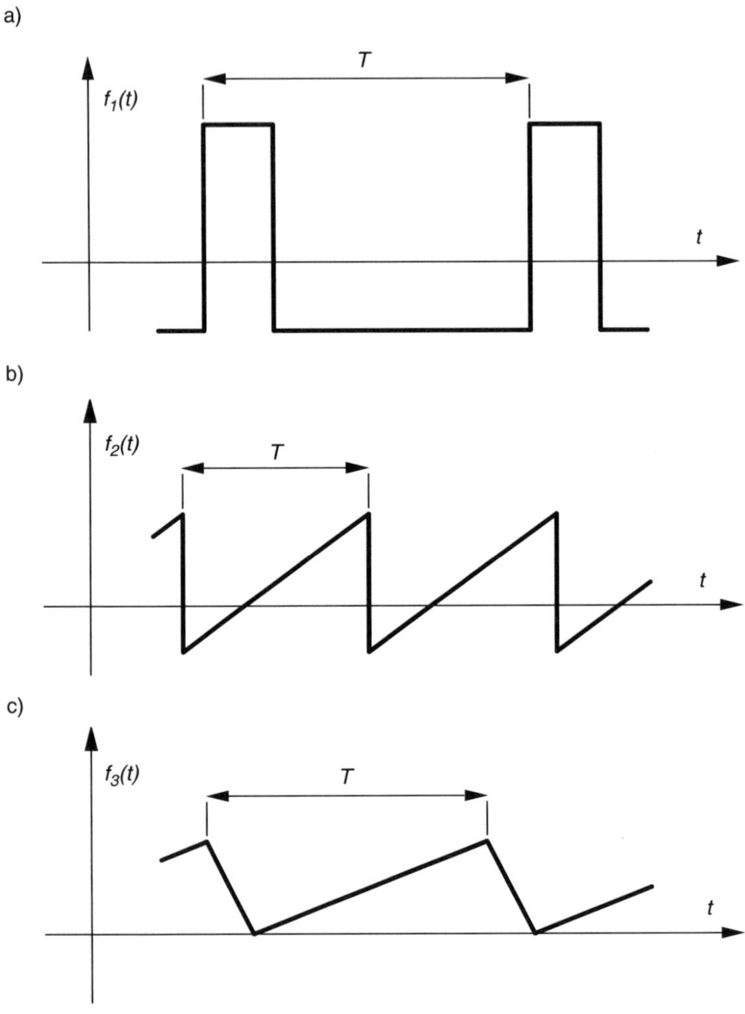

Fig. 4.18

contain an infinite number of terms. Due to that reason, a given function can be approximated only by a series having finite number of terms. Consequently, it yields to a certain approximation error. This error is most clearly visible in the neighborhood of the points of discontinuity of the first kind, in which the approximated series has the waveform similar to that shown in Fig. 4.19(b).

If the number of terms increases, the "oscillations" of the approximating series tend to concentrate in the smaller and smaller neighborhood of the discontinuity point t_0. The decay of this oscillation process is possible only after taking into account the infinite number of terms. In electronic engineering, this effect of the sharp oscillations at the "sharp" edges of the pulses is known under the name of the

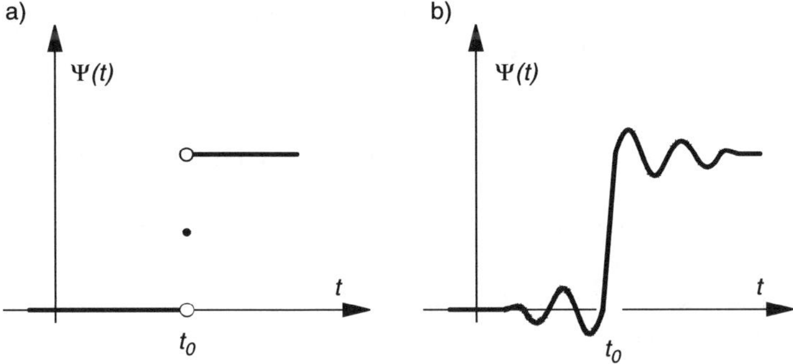

Fig. 4.19

Gibbs effect. It is the direct consequence of the finite frequency bandwidth of the
electronic systems processing these pulse signals.

4.3 Examples of the Application of Chebyshev Polynomials in Synthesis of Radiation Patterns of the In-Phase Linear Array Antenna

By the antenna array we understand a system composed of many identical radiating elements (simple antennas) equally spaced along the straight line, as shown in
Fig. 4.20.

In the case when all radiating elements of the antenna array are supplied by currents of equal amplitudes and equal phases (uniform and in-phase excitation) the
level of the side lobes is equal to $-13.2\,\mathrm{dB}$. Such relatively high-side lobes lead
in case of a transmitting antenna to scattering a considerable part of the electromagnetic energy in the undesired directions and in consequence to reduction of the
antenna directivity. In case of a receiving antenna, the high-side lobes reduce its
immunity against different electromagnetic disturbances. In both cases it is recommended to reduce the side lobes to the possibly minimum value. For this purpose, it

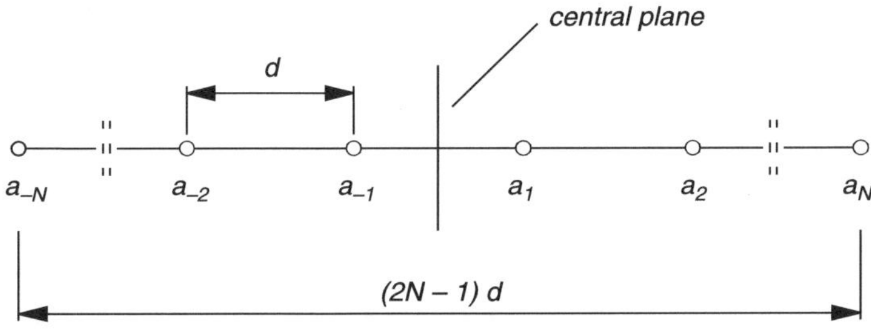

Fig. 4.20

is necessary to change the amplitudes of currents supplying the particular radiating elements. As a rule, these amplitudes decrease with growing distance between the central plane of the antenna and radiating element, see Fig. 4.20. However, this leads in turn to broadening the main lobe of the radiation pattern. It is therefore possible to find such current amplitude distribution for which an optimum will be achieved according to one of the criterions given below.

Criterion 1 The optimum amplitude distribution of the currents supplying the in-phase linear array, having a fixed L/λ ratio, where L is the total length of the antenna, is understood as the distribution for which the side lobe level attains the minimum for a given 3 dB width of the main lobe and λ is the length of the radiated/received wave.

Criterion 2 By the optimal amplitude distribution of the currents supplying the in-phase linear array antenna, having a fixed L/λ ratio and a given side lobe level, we understand such distribution, for which the 3 dB width of the main lobe attains the minimum.

The essence of the design algorithm for a linear array antenna, optimum according to criterion 2, is presented below as an example of the linear antenna including even number $(2N)$ of equally spaced radiating elements, Fig. 4.21.

Let us assume that all $2N$ radiating elements of this antenna array are isotropic. Moreover, let the complex amplitude of the electrical field initiated at fixed point P of the distant zone, by the first radiating element, is:

$$\mathbf{E}_1 = a_1 e^{-j\beta \cdot r_1} = a_1 e^{-j\beta \cdot r_0} e^{j\beta \cdot \frac{d}{2} \cdot \cos(\theta)} \tag{4.95}$$

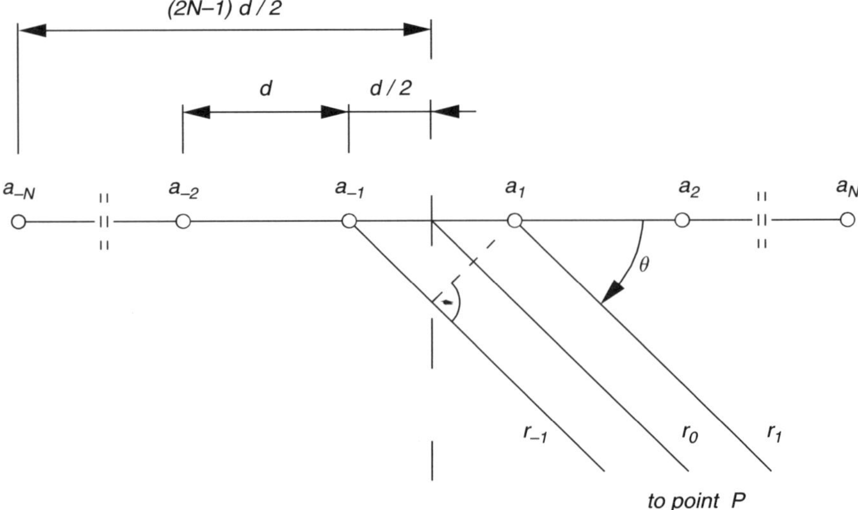

Fig. 4.21

where a_1 is the coefficient proportional to the absolute value of complex amplitude of the current $\mathbf{I}_1$. Similarly, we can determine the complex amplitude of electrical field initiated by a radiating element situated in symmetric position denoted in Fig. 4.21 by the index -1. In this case

$$\mathbf{E}_{-1} = a_{-1}e^{-j\beta \cdot r_{-1}} = a_{-1}e^{-j\beta \cdot r_0}e^{-j\beta \cdot \frac{d}{2} \cdot \cos(\theta)} \tag{4.96}$$

Assuming further that the current amplitude distribution is symmetric with respect to the central plane of the antenna. Thus, for $a_1 = a_{-1}$ we can write the expression:

$$\mathbf{E}_1 + \mathbf{E}_{-1} = a_1 e^{-j\beta \cdot r_0} \cdot \left[e^{-j\beta \cdot \frac{d}{2} \cdot \cos(\theta)} + e^{j\beta \cdot \frac{d}{2} \cdot \cos(\theta)} \right]$$
$$= a_1 e^{-j\beta \cdot r_0} \cdot 2\cos\left[\beta \cdot \frac{d}{2} \cdot \cos(\theta) \right] \tag{4.97}$$

Using the same approach to the remaining pairs of radiating elements, complex amplitude of the resulting electrical field, initiated at point P of the distant zone, can be found as:

$$\mathbf{E} = e^{-j\cdot\beta\cdot r_0} \sum_{k=1}^{N} 2a_k \cos\left[\frac{(2k-1)}{2}\beta \cdot d \cdot \cos(\theta) \right] \tag{4.98}$$

It follows from expression (4.98) that the not normalized radiation pattern of the antenna array under discussion is:

$$f(\theta) = 2\sum_{k=1}^{N} a_k \cos\left[(2k-1) \cdot u\right] \tag{4.99}$$

where $u = \beta \cdot (d/2) \cdot \cos(\theta) = (\pi \cdot d/\lambda)\cos(\theta)$. Consequently, the radiation pattern (4.99) can be presented in the form of a polynomial of degree $(2N-1)$ with respect to variable $x = \cos(u)$, i.e.:

$$f(\theta) = 2\sum_{k=1}^{N} a_k \cos\left[(2k-1) \cdot u\right] = \sum_{i=1}^{2N-1} B_i x^i \tag{4.100}$$

Justification
Each term of the sum (4.100) can be treated as a polynomial of the variable $x = \cos(u)$. This conclusion results from the trigonometric identities given below [13, 14].

$$\cos(2\alpha) = 2\cos^2(\alpha) - 1$$

$$\cos(3\alpha) = 4\cos^3(\alpha) - 3\cos(\alpha)$$

$$\cos(n \cdot \alpha) = 2^{n-1}\cos^n(\alpha) - \frac{n}{1!}2^{n-3}\cos^{n-2}(\alpha) + \frac{n(n-3)}{2!}2^{n-5}\cos^{n-4}(\alpha)$$
$$- \frac{n(n-4)(n-5)}{3!}2^{n-7}\cos^{n-6}(\alpha)$$
$$+ \frac{n(n-5)(n-6)(n-7)}{4!}2^{n-9}\cos^{n-8}(\alpha) - \cdots$$

Using some simple arithmetic operations one can easily show that the sum of polynomials in one variable is itself the polynomial in this variable, what naturally proves validity of the expression (4.100). Coefficients B_i, where $i = 1, 2, 3, \ldots, 2N - 1$, are simply sums of the current coefficients a_i of the terms containing variable $x = \cos(u)$ in the same power i. The essence of the synthesis method, known in the literature as the Dolph–Chebyshev method, consists in approximation of the function (4.100) by the Chebyshev polynomial of the first kind and of degree $(2N - 1)$ [15, 16]. We can write therefore

$$f(\theta) = \sum_{i=1}^{2N-1} B_i x^i = T_{2N-1}(\alpha \cdot x) \tag{4.101}$$

Basic properties of Chebyshev's polynomials $T_n(x)$ of the first kind are described in Sect. 4.2.1. As an illustration, the curve $|T_9(x)|$ is represented in Fig. 4.22.

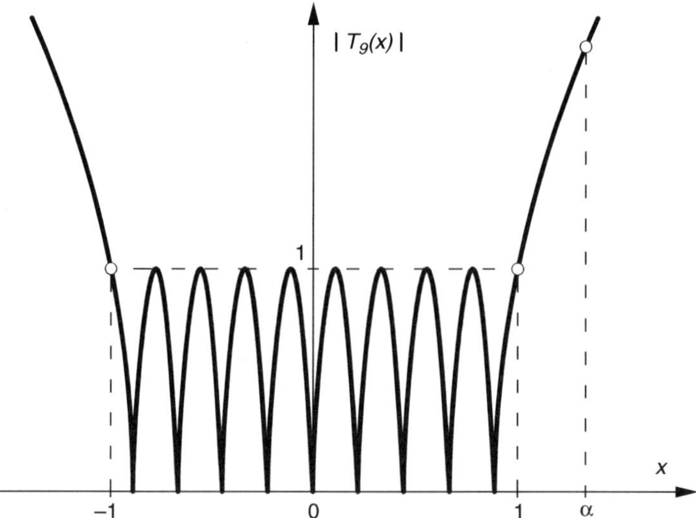

Fig. 4.22

For an arbitrary n the polynomial $T_n(x)$ takes values from the interval $[1, -1]$ if $-1 \leq x \leq 1$. According to (4.101), the function $|f(\theta)|$ takes its maximum value $|T_{2N-1}(\alpha)|$, because the variable $x = \cos(u)$ takes the maximum value equal to 1 for

$$u = \frac{\pi \cdot d}{\lambda} \cos(\theta) = 0 \qquad (4.102)$$

It means that value $x = 1$ corresponds to the direction of the maximum radiation $\theta = 90°$, see Fig. 4.21, because for this value of θ the Eq. (4.102) is satisfied. It is not difficult to prove that for all angles of θ for which $-1 \leq x \leq 1$, the amplitudes of side lobes are equal to 1. According to the above conclusion, we can write the ratio of $\mathbf{E}_{b\,\mathrm{max}}$ to the amplitude of the main lobe $\mathbf{E}_{\mathrm{max}}$, i.e.:

$$\left| \frac{\mathbf{E}_{b\,\mathrm{max}}}{\mathbf{E}_{\mathrm{max}}} \right| = \left| \frac{1}{T_{2N-1}}(\alpha) \right| \qquad (4.103)$$

where $\mathbf{E}_{b\cdot\mathrm{max}}$ and $\mathbf{E}_{\mathrm{max}}$ denote the maximum values of the main and side lobes, respectively. For a given ratio $L = |\mathbf{E}_{b\,\mathrm{max}}/\mathbf{E}_{\mathrm{max}}|$, from Eq. (4.103) we find the value of the parameter α. Next, using the developed trigonometric form of the polynomial $T_{2N-1}(\alpha \cdot x)$, we determine the coefficients B_i, where $i = 1, 2, 3, \ldots, 2N-1$. These coefficients B_i are related to the desired current coefficients $a_1, a_2, a_3, \ldots, a_N$ by equation system which can be written in the following general form:

$$B_1 = f_1(a_1, a_2, a_3, \ldots, a_N)$$
$$B_2 = f_2(a_1, a_2, a_3, \ldots, a_N)$$
$$B_3 = f_3(a_1, a_2, a_3, \ldots, a_N) \qquad (4.104)$$
$$\vdots$$
$$B_N = f_N(a_1, a_2, a_3, \ldots, a_N)$$

Solving this equation system we obtain the desired coefficients $a_1, a_2, a_3, \ldots, a_N$, those according to our assumption are proportional to amplitudes of currents driving particular radiating elements.

Example 4.10 As an illustration of the Dolph–Chebyshev algorithm, let us design the linear in-phase antenna array for the following data: $2N = 8$, $|\mathbf{E}_{\mathrm{max}}/\mathbf{E}_{b\,\mathrm{max}}| = 100$ and $d = 0.7\lambda$. Solving the equation: $1/T_7(\alpha) = 0.01$ we obtain the parameter $\alpha = 1.30038731$. Next, according to the relation (4.100), we determine the system of equations relating the current coefficients $a_1, a_2, a_3, \ldots, a_N$ with coefficients B_1, B_3, B_5 and B_7. In this case

$$f(\theta) = 2\sum_{n=1}^{4} a_n \cos\left[(2n-1)u\right] = 2a_1 \cos(u) + 2a_2\left[4\cos^3(u) - 3\cos(u)\right]$$

$$+ 2a_3\left[16\cos^5(u) - 20\cos^3(u) + 5\cos(u)\right]$$
$$+ 2a_4\left[64\cos^7(u) - 112\cos^5(u) + 56\cos^3(u) - 7\cos(u)\right]$$
$$= B_1\cos(u) + B_3\cos^3(u) + B_5\cos^5(u) + B_7\cos^7(u)$$

where

$$B_1 = 2a_1 - 6a_2 + 10a_3 - 14a_4,$$
$$B_3 = 8a_2 - 40a_3 + 112a_4,$$
$$B_5 = 32a_3 - 224a_4,$$
$$B_7 = 128a_4$$

From the above linear equations it results that

$$a_4 = B_7/128$$
$$a_3 = (B_5 + 224a_4)/32$$
$$a_2 = (B_3 - 112a_4 + 40a_3)/8$$
$$a_1 = (B_1 + 6a_2 - 10a_3 + 14a_4)/2$$

Developing the polynomial $T_7(\alpha \cdot x) = T_7[\alpha \cdot \cos(u)]$ with respect to $x = \cos(u)$ we obtain:

$$B_7 = 64(\alpha)^7, \quad B_5 = -112(\alpha)^5, \quad B_3 = 56(\alpha)^3, \quad B_1 = -7\alpha$$

After introducing the above coefficients into the equation system formulated above we obtain the desired solution: $a_4 = 3.143974851$, $a_3 = 8.993198987$, $a_2 = 16.34309675$, $a_1 = 21.51976369$. Hence the current amplitudes normalized with respect to the maximum component a_1 are equal to:

$$\mathbf{I}_1 = \mathbf{I}_{-1} = 1, \quad \mathbf{I}_2 = \mathbf{I}_{-2} = 0.759445,$$
$$\mathbf{I}_3 = \mathbf{I}_{-3} = 0.417904, \quad \text{and} \quad \mathbf{I}_4 = \mathbf{I}_{-4} = 0.146097.$$

The normalized radiation pattern $F(\theta) = f(\theta)/f_{\max}(\theta)$ of the designed antenna is shown in Fig. 4.23.

According to our assumption $|\mathbf{E}_{b\,\max}/\mathbf{E}_{\max}| = 0.01$, the level of side lobes is equal to $20\log(0.01) = -40\,\mathrm{dB}$ what well confirms the validity of the performed design process.

Example 4.11 This example presents the results of the synthesis performed for a linear in-phase antenna array characterized by the following data: $2N = 16$, $|\mathbf{E}_{b\,\max}/\mathbf{E}_{\max}| = 0.01$ and $d = 0.70\lambda$. Similar as in the previous example, we shall calculate the parameter α by solving the following equation:

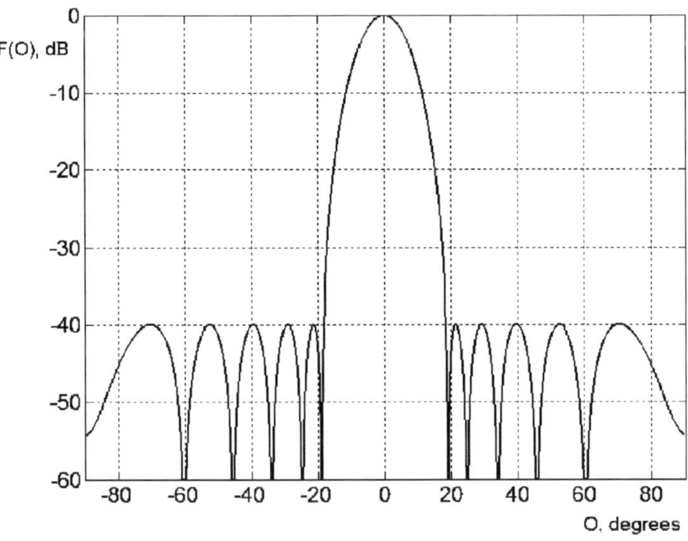

Fig. 4.23

$$T_{2n-1}(\alpha) = \left| \frac{\mathbf{E}_{\max}}{\mathbf{E}_{b\,\max}} \right| = L = 100 \tag{4.105}$$

For large degrees $(2N - 1)$ of the Chebyshev polynomial, Eq. (4.105) can be solved more easily, if we use the following mathematical identity valid for $|x| > 1$

$$T_N(x) = \mathrm{ch}\,[N\,\mathrm{arch}(x)] \tag{4.106}$$

The identity (4.106) makes possible to write Eq. (4.105) in the form:

$$\mathrm{ch}\,[(2N - 1)\,\mathrm{arch}(\alpha)] - L = 0 \tag{4.107}$$

The solution of equation (4.107) may be simplified when we introduce two auxiliary variables:

$$\begin{aligned} W &= (2N - 1)\mathrm{arch}(\alpha) \\ P &= \exp(W) \end{aligned} \tag{4.108}$$

Equation (4.107) expressed in terms of the auxiliary variable P has the simple form of a quadratic equation $P^2 - 2LP + 1 = 0$ whose first root (greater than 1) is equal to:

$$P_1 = L + \sqrt{L^2 - 1} = 100 + \sqrt{100^2 - 1}$$

Having the root P_1 we can find the parameter α from the formula:

$$\alpha = \frac{1}{2}\left(s + \frac{1}{s}\right) \tag{4.109}$$

where

$$s = \exp\left[\frac{1}{2N-1}\ln(P_1)\right]$$

The formula (4.109) represents the inverse relations with respect to (4.108). Thus, for the data assumed in this design example we obtain $P_1 = 199.995002747$, $s = 1.423643589$ and $\alpha = 1.063033294$. For the second stage of the synthesis procedure, we shall formulate the system of N linear equations with N unknown variables a_k, where $k = 1, 2, 3, \ldots, N$, representing amplitudes of the currents driving the particular radiating elements of the designed antenna array. When the number of radiating elements is very large ($2N \geq 10$), this task becomes very difficult and cumbersome. In such case, it is recommended to use the specially developed iterative computational formulas. One of them, namely:

$$a_k = \sum_{q=k}^{N}(-1)^{N-q}\alpha^{2q-1}\frac{(2N-1)(q+N-2)!}{(q-k)!(q+k-1)!(N-q)!} \tag{4.110}$$

proved itself to be very useful for this purpose [16, 17]. The current amplitudes $a_k \equiv \mathbf{I}_k$ calculated according to (4.110) and normalized with respect to the maximum value $a_1 = 21.98914$ are equal to:

$$\mathbf{I}_1 = \mathbf{I}_{-1} = 1.000000, \quad \mathbf{I}_2 = \mathbf{I}_{-2} = 0.935381, \quad \mathbf{I}_3 = \mathbf{I}_{-3} = 0.816304,$$
$$\mathbf{I}_4 = \mathbf{I}_{-4} = 0.661368, \quad \mathbf{I}_5 = \mathbf{I}_{-5} = 0.492615, \quad \mathbf{I}_6 = \mathbf{I}_{-6} = 0.331950,$$
$$\mathbf{I}_7 = \mathbf{I}_{-7} = 0.196367, \quad \mathbf{I}_8 = \mathbf{I}_{-8} = 0.113761$$

The group radiation pattern $F(\theta)$ corresponding to the current distribution given above is shown in Fig. 4.24.

Examples presented in this section illustrate the design methodology for the linear, regular and in-phase antenna arrays with even ($2N$) number of radiating elements. Design of a linear, regular and in-phase array composed of an odd ($2N - 1$) number of radiating elements can be performed in the similar manner [16].

Many linear arrays spaced parallely on the common plane create a planar array antenna. An example of application of such planar array antenna in a mobile radar equipment is shown in Fig. 4.25.

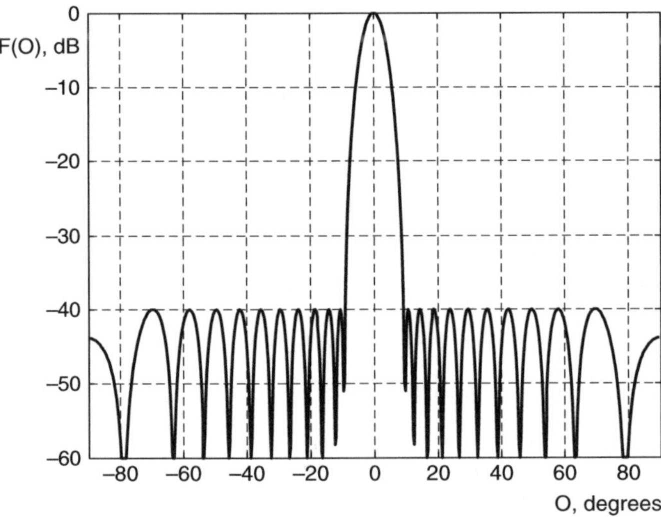

Fig. 4.24

Fig. 4.25

References

1. Achieser N.I., Theory of approximation. Frederick Ungar Pub. Co., New York, 1956
2. Forsythe G.E., Malcolm M.A. and C.B. Moler, Computer methods for mathematical computations, Prentice-Hall, Englewood Cliffs, NJ, 1977
3. Mathews J.H., Numerical methods for mathematics, science and engineering. Prentice-Hall, Inc., Englewood Cliffs, NJ, 1987
4. Mathews J.H., Numerical methods for mathematics, science and engineering. Prentice-Hall Int. Inc., Englewood Cliffs, NJ, 1992
5. Bjorek A.G. Dahlquist G. and A. Bjorck, Numerical methods. Prentice-Hall, New York, 1974
6. Bender C.M. and S.A. Orszag, Advanced mathematical methods for scientists and engineers. McGraw-Hill, New York, 1978
7. Young D.M. and R.T. Gregory, A survey of numerical mathematics. Addison-Wesley Comp., London, 1973
8. Shoup T.E., A practical guide to computer methods for engineers. Prentice-Hall, Englewood Cliffs, NJ, 1979
9. Matthaei G.L., Young L. and E.M.T. Jones, Microwave filters, impedance matching networks and coupling structures. Artech House Inc., Boston, MA, 1980
10. Rosoniec S., Algorithms for computer-aided design of linear microwave circuits. Artech House Inc., Boston, MA, 1990
11. Bracewell R.M., The Fourier integral and its applications. McGraw-Hill Comp., New York, 1965
12. Otnes R.K., and L. Enochson, Digital time series analysis. John Wiley and Sons, New York, 1972
13. Abramowitz M. and I.A. Stegun, Handbook of mathematical functions. Dover, New York, 1954
14. Dwight H.B., Tables of integrals and other mathematical data. The Macmillan Comp., New York, 1961
15. Dolph L.C., "A current distribution for broadside arrays which optimizes the relationship between beam width and side-lobe level". Proc. IRE., vol. 34, June 1946, pp. 335–348
16. Rosłoniec S., Fundamentals of the antenna technique (in Polish). Publishing House of the Warsaw University of Technology, Warsaw, 2006
17. Kuhn R., Mikrowellenantennen. Veb Verlag Technik, Berlin, 1964

Chapter 5
Methods for Numerical Integration of One and Two Variable Functions

The subject of considerations in the first part of this chapter is the definite integral of one variable given in the closed interval $[a, b]$, namely:

$$\int_a^b f(x)dx \tag{5.1}$$

This integral can be easily calculated when an integrable function $f(x)$ is bounded and continuous over this interval and when the primitive function $F(x)$, such that $f(x) \equiv F'(x)$ is known. In this fortunate case, the fundamental Newton formula can be used, namely:

$$\int_a^b f(x)dx \equiv F(b) - F(a) \tag{5.2}$$

In other cases, however, determination of the primitive function $F(x)$ may be very difficult or even impossible. Such situation may occur, for example, if only discrete values $y_i = f(x_i)$, for $i = 0, 1, 2, 3, \ldots, n$, of the integrand are known. In this case, we cannot speak about the primitive function. In other words, relation (5.2) is useless in such cases. Hence there is a necessity for calculating approximate values of definite integrals by means of appropriate numerical methods. In this chapter, these methods have been divided into three groups, see Fig. 5.1, only for didactic reasons.

First group (I) includes methods in which the integrand $f(x)$ is replaced (interpolated or approximated) by a series of elementary functions which are easy to integrate by means of analytical methods. After performing integration of individual terms of the series, we obtain a new series composed of finite or infinite number of terms. This series makes possible calculation of an integral with an arbitrary prescribed accuracy. The integration understood as defined above will be discussed in Sect. 5.1. Most numerous is the second group (II), see Fig. 5.2, which includes the methods commonly known and is most frequently applied in the electrical engineering.

S. Rosłoniec, *Fundamental Numerical Methods for Electrical Engineering*,
© Springer-Verlag Berlin Heidelberg 2008

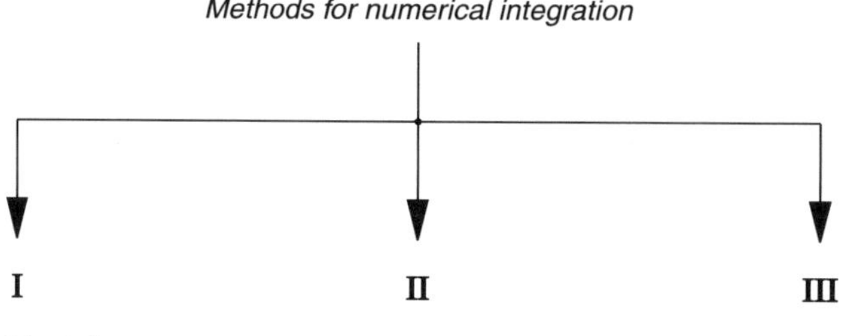

Fig. 5.1

Algorithms of these methods are described below in the same order as in Fig. 5.2, in the Sects. 5.2.1–5.2.5, respectively. The third group (III) contains the methods known in general as the Gaussian methods. Basic feature of these methods is the interpolation (approximation) of integrand $f(x)$ by the orthogonal Legendre, Jacobi or Chebyshev polynomials [1–3]. The essential problem consists here in determining the interpolation points (nodes) x_i and coefficients A_i of the following expression:

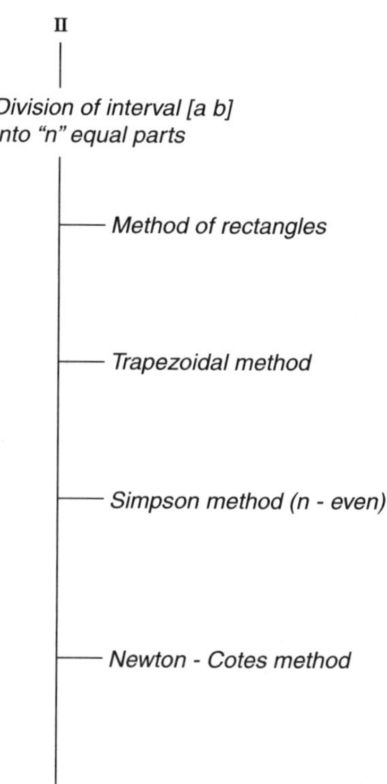

Fig. 5.2

$$\int\limits_a^b f(x)dx \approx \sum_{i=0}^n A_i f(x_i), \tag{5.3}$$

called the mechanical Gaussian quadrature or briefly the quadrature. The algorithms used for evaluation of interpolation points x_i and coefficients A_i, according to the Gauss and Chebyshev quadratures, are described in Sect. 5.2.6. The problem of numerical calculation of double integral, i.e., an integral of a function of two variables $f(x, y)$, defined on the two-dimensional area, is considered in Sect. 5.3. The simplest practical numerical algorithms, such as the algorithm of small (elementary) cells and the Simpson algorithm are described here. In the final section of this chapter, Sect. 5.4, we can find an example of numerical integration applied to determine the counted position of a moving object.

5.1 Integration of Definite Integrals by Expanding the Integrand Function in Finite Series of Analytically Integrable Functions

In case when the integrand $f(x)$ defined over an interval $[a, b]$ can be represented by a series of elementary functions, easily integrable by means of analytical methods, the definite integral of this function can also be represented by the similar series. This property is illustrated in the following examples.

Example 5.1 In the spectral analysis of signals, there is a constant need of calculating the integral sine function defined as:

$$Si(x) = \int\limits_0^x \frac{\sin(t)}{t} dt$$

Corresponding integrand can be represented by the following infinite series:

$$\frac{\sin(t)}{t} = \frac{1}{t}\left[t - \frac{t^3}{3!} + \frac{t^5}{5!} - \frac{t^7}{7!} + \ldots + (-1)^n \frac{t^{2n+1}}{(2n+1)!} + \ldots \right]$$

which is convergent for $-\infty \leq t \leq \infty$. After introducing this series into the integral and performing integration of its individual terms we obtain:

$$Si(x) = x - \frac{x^3}{3!3} + \frac{x^5}{5!5} - \frac{x^7}{7!7} + \ldots + (-1)^n \frac{x^{2n+1}}{(2n+1)!(2n+1)} + \ldots$$

$$= \sum_{n=0}^\infty (-1)^n \frac{x^{2n+1}}{(2n+1)!(2n+1)}$$

This series of functions makes possible calculation of the function $Si(x)$ with assumed accuracy. For example, for $x = 2$ and $n = 5$, we obtain:

$$Si(2) = 2 - \frac{8}{18} + \frac{32}{600} - \frac{128}{35280} + \frac{512}{3265920} - \frac{2048}{439084800}$$

$$+ \sum_{n=6}^{\infty} (-1)^n \frac{2^{2n+1}}{(2n+1)!(2n+1)} \approx 1.605412$$

Consecutive seventh term (not taken into account in the above calculation) takes the value $\sim 8.192/(8.095 \times 10^7)$ which is much smaller than 10^{-6}.

Example 5.2 The following definite integral is very useful for the probabilistic calculations:

$$\frac{1}{\sqrt{2\pi}} \int_0^x e^{-t^2/2} dt$$

It can be computed using the approach similar to that used in Example 5.1. For this end, we represent the integrand by a series similar to the series representing the exponential function, i.e.:

$$e^{-x} = 1 - \frac{x}{1!} + \frac{x^2}{2!} - \frac{x^3}{3!} + \ldots + (-1)^n \frac{x^n}{n!} + \ldots$$

that is valid for $-\infty \leq x \leq \infty$. After introducing the variable $x = t^2/2$ into the above series we obtain:

$$e^{-t^2/2} = 1 - \frac{t^2}{2} + \left(\frac{t^2}{2}\right)^2 \frac{1}{2!} - \left(\frac{t^2}{2}\right)^3 \frac{1}{3!} + \ldots + (-1)^n \left(\frac{t^2}{2}\right)^n \frac{1}{n!} + \ldots$$

Thus

$$\frac{1}{\sqrt{2\pi}} \int_0^x e^{-t^2/2} dt = \frac{1}{\sqrt{2\pi}} \int_0^x 1 \cdot dt - \frac{1}{\sqrt{2\pi}} \int_0^x \frac{t^2}{2} dt + \frac{1}{\sqrt{2\pi}} \int_0^x \left(\frac{t^2}{2}\right)^2 \frac{1}{2!} dt$$

$$- \frac{1}{\sqrt{2\pi}} \int_0^x \left(\frac{t^2}{2}\right)^3 \frac{1}{3!} dt + \ldots$$

$$= \frac{1}{\sqrt{2\pi}} \left[x - \frac{x^3}{2 \cdot 3} + \frac{x^5}{2! 2^2 5} - \frac{x^7}{3! 2^3 7} + \ldots \right.$$

$$\left. + (-1)^n \frac{x^{2n+1}}{n! 2^n (2n+1)} + \ldots \right]$$

For $x = 2$ and $n = 19$, the calculated probability integral is represented by the sum

$$\frac{1}{\sqrt{2\pi}} \int_0^2 e^{-t^2/2} dt = \frac{1}{\sqrt{2\pi}} \left[2 - \frac{8}{6} + \frac{32}{40} - \frac{128}{336} + \frac{512}{3456} \right.$$
$$\left. - \frac{2084}{42240} + .. + (-1)^n \frac{2^{2n+1}}{n!2^n(2n+1)} + \cdots \right]$$

equal to 0.477249868. The next term of the above series is less than 10^{-14}.

Although the approach presented above seems to be very attractive, it should be applied very carefully. In other words, the series representing the computed integral have to converge to the limit equal to exact value of this integral. This obligatory condition may be verified by using the suitable mathematical criterions or numerical simulations described in the more advanced literature.

5.2 Fundamental Methods for Numerical Integration of One Variable Functions

5.2.1 Rectangular and Trapezoidal Methods of Integration

Given an integrand $y = f(x)$ defined over a closed interval $[a, b] \equiv [x_0, x_n]$. In order to calculate the definite integral of this function, we divide the interval $[a, b]$ into n different segments (subintervals) $\Delta x_i = x_i - x_{i-1}$, for $i = 1, 2, 3, \ldots, n$. Theoretical basis for the methods of numerical calculation of definite integrals presented in this section is founded by the theorem about the limit of integral sum defined by the formula:

$$S = S_1 + S_2 + S_3 + \ldots + S_n = \sum_{i=1}^{n} S_i = \sum_{i=1}^{n} f(\xi_i)\Delta x_i \qquad (5.4)$$

where ξ_i is an arbitrarily chosen value of the variable x, taken from the subinterval i, i.e., $x_{i-1} \leq \xi_i \leq x_i$.

5.2.1.1 Theorem About the Limit of an Integral Sum

If an integrand $f(x)$ is bounded and continuous over the closed interval $[a, b]$, then there exists a limit of the integral sum (5.4)

$$\lim_{\Delta x_{i(max)} \to 0} \sum_{i=1}^{n} f(\xi_i)\Delta x_i = \int_a^b f(x)dx \qquad (5.5)$$

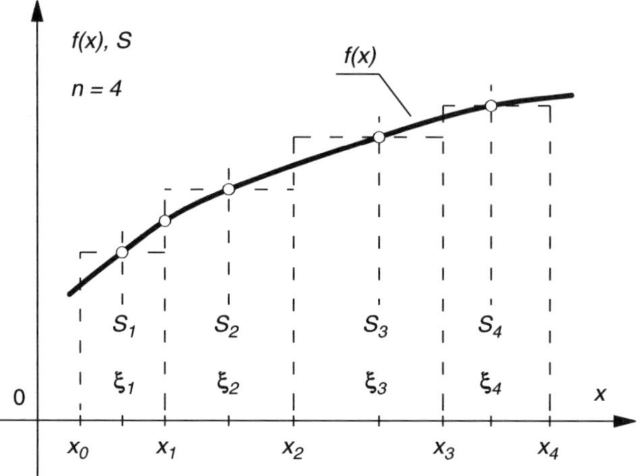

Fig. 5.3

The value of this limit is independent from the partition of the interval $[a, b]$ into elementary subintervals $\Delta x_i = x_i - x_{i-1}$ and from the choice of ξ_i. Geometrical interpretation of the integral sum (5.4), for the case $f(x) > 0$ with $x_0 \leq x \leq x_n$, is shown in Fig. 5.3.

5.2.1.2 The Rectangular Method of Integration

The simplest computation method used for definite integrals is the method of rectangles. Different manners of calculating elementary terms (areas) S_i, $i = 1, 2, 3, \ldots, n$, used in this method are illustrated in Fig. 5.4.

According to the notation used in Fig. 5.4(a, b, c), the approximate values of the integral $\int_a^b f(x)dx$ are, respectively:

$$I_1 = \sum_{i=1}^{n} f(x_{i-1})(x_i - x_{i-1}),$$

$$I_2 = \sum_{i=1}^{n} f(x_i)(x_i - x_{i-1}), \tag{5.6}$$

$$I_3 = \sum_{i=1}^{n} f\left(\frac{x_{i-1} + x_i}{2}\right)(x_i - x_{i-1})$$

The common limit of these approximations when $\max\{x_i - x_{i-1}\} \to 0$ is equal to the accurate value of the integral under computation.

a)

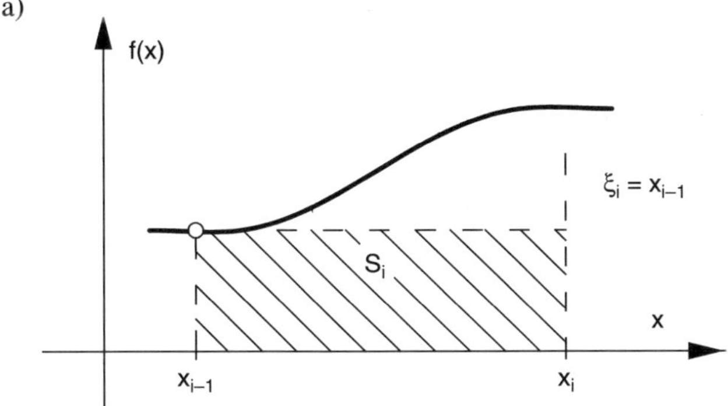

b)

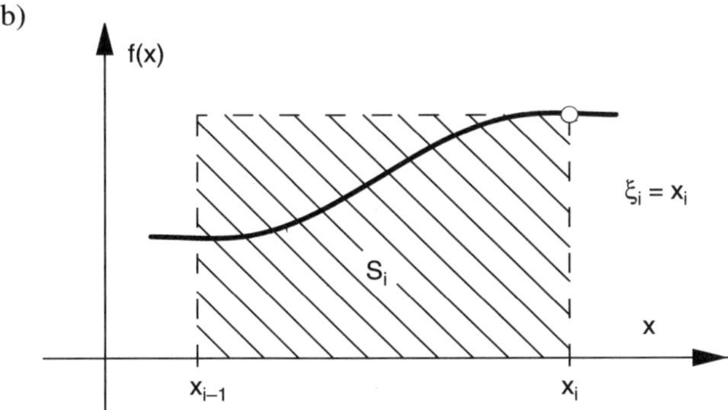

c)

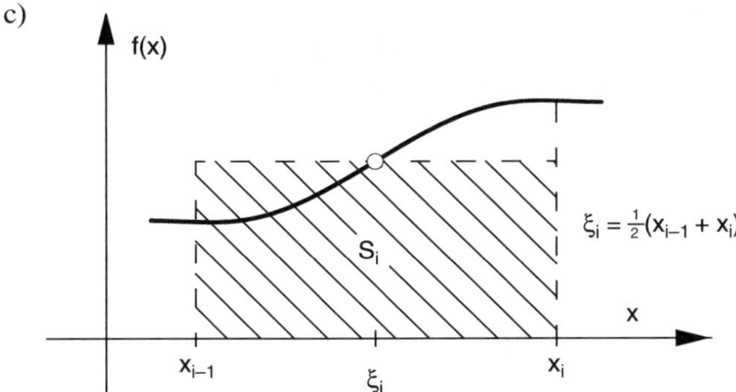

Fig. 5.4

5.2.1.3 The Trapezoid Method of Integration

The essence of the trapezoid method of integration is explained in Fig. 5.5.

In this case, elementary terms are calculated according to the formula defining the surface of a trapezoid, namely:

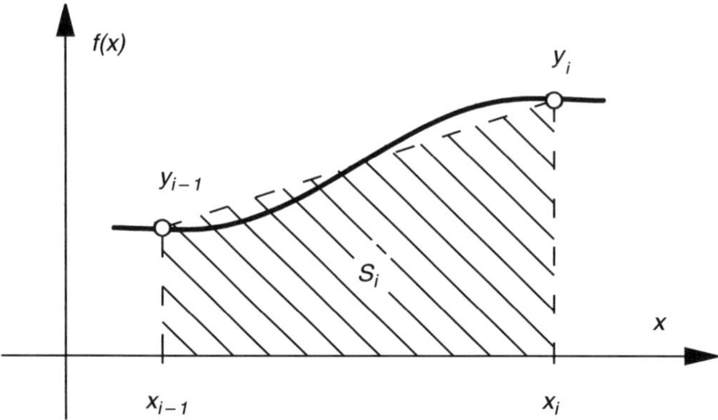

Fig. 5.5

$$S_i = \frac{y_{i-1} + y_i}{2}(x_i - x_{i-1}) \quad i = 1, 2, 3, \ldots, n \tag{5.7}$$

Of course, this fact justifies the name of that method. When all subintervals $\Delta x_i = x_i - x_{i-1}$ have the same lengths equal to $h = (b - a)/n$, then

$$\int_a^b f(x)dx \approx \frac{1}{2}\sum_{i=1}^n h \cdot (y_{i-1} + y_i) = I \tag{5.8}$$

Formula (5.8), called the complete formula of the trapezoidal method, serves for calculation of an approximate value of definite integral over the given interval $[a, b]$. Naturally, the difference between the exact and approximate values of the integral under consideration depends on the integrand $f(x)$ and on the number n of subintervals n and their lengths $|\Delta x_i = x_i - x_{i-1}|$. This difference is a measure of a quality of integration and for this reason it is often called the integration error [4]. This error is a subject of considerations presented below. Let ΔI_i denotes the following partial integral:

$$\Delta I_i = \int_{x_i}^{x_i+h} f(x)dx = F(x_i + h) - F(x_i) \tag{5.9}$$

The approximate value of integral (5.9) calculated by means of the trapezoidal method is

$$\Delta I_i^{(t)} = \frac{h}{2}[f(x_i + h) + f(x_i)] \tag{5.10}$$

Taking the difference of relations (5.9) and (5.10) we obtain a partial error:

$$(Er)_i = F(x_i + h) - F(x_i) - \frac{h}{2}[f(x_i + h) + f(x_i)] \tag{5.11}$$

The primitive function $F(x_i + h)$ included in relation (5.11) can be replaced by its Taylor series written as:

$$F(x_i + h) = F(x_i) + hF'(x_i) + \frac{h^2}{2}F''(x_i) + \frac{h^3}{6}F'''(x_i) + O(h^4) \qquad (5.12)$$

Also the integrand $f(x_i + h)$ can be written in an analogous form, i.e.:

$$f(x_i + h) = f(x_i) + hf'(x_i) + \frac{h^2}{2}f''(x_i) + \frac{h^3}{6}f'''(x_i) + O(h^4) \qquad (5.13)$$

As it has been assumed at the beginning that the function $F(x)$ is a primitive function with respect to integrand $f(x)$ and due to this fact the following relations are valid:

$$f(x_i) = F'(x_i) \equiv \frac{dF(x_i)}{dx},$$
$$f'(x_i) = F''(x_i) \equiv \frac{d^2 F(x_i)}{dx^2}, \qquad (5.14)$$
$$f''(x_i) = F'''(x_i) \equiv \frac{d^3 F(x_i)}{dx^3}$$

After introducing the relations (5.12), (5.13) and (5.14) into formula (5.11) we obtain:

$$(Er)_i = -\frac{h^3}{12}f''(x_i) + O(h^4)$$

Finally, the total error of integration determined over the whole interval $[a, b]$ is:

$$Er = \sum_{i=0}^{n-1}\left[-\frac{h^3}{12}f''(x_i) + O(h^4)\right] = -\frac{h^3}{12}\sum_{i=0}^{n-1}f''(x_i) + O(h^4) \qquad (5.15)$$

It follows from relation (5.15) that the integral sum approximates the definite integral more accurately, if the largest of elementary subintervals $\Delta x_i = x_i - x_{i-1}$ is smaller. It is easy to explain, because in this case the integrand is interpolated more accurately by the step function in case of the rectangle method, or by the broken line in the trapezoid method. In other words, reducing the length of subintervals reduces interpolation error. We must, however, remember that the partition of integration interval into smaller and smaller elementary subintervals Δx_i increases the number of subintervals and the amount of necessary computations. Another negative consequence of such strategy is therefore an increasing of computation error (computer processing error). It follows from the relations explained above that for each particular case there exists such optimal partition number n, for which the sum of interpolation and computation errors would attain the minimum.

5.2.2 The Romberg Integration Rule

The Romberg integration rule known also as Romberg method can be treated as a numerically effective combination of the trapezoid method and the Richardson's extrapolation procedure [4]. In order to explain the corresponding algorithm, let us consider once more the formula (5.8) written in a slightly different form for the function $f(x)$, integrable over an interval $[a, b]$

$$I = \frac{h}{2} \left[f(a) + f(b) + 2 \sum_{j=1}^{n-1} f(x_j) \right] \tag{5.16}$$

where $h = (a - b)/n$, $x_j = a + j \cdot h$. It is not difficult to explain the fact that when the number n is increased, the approximation error becomes smaller.

$$\text{error} \equiv \left| \int_a^b f(x)dx - I \right| \tag{5.17}$$

In case of the Romberg method, this error increases two times in each subsequent iteration, according to the formula:

$$n = 2^{k-1} \tag{5.18}$$

where $k = 1, 2, 3, \ldots, m$ and m is a positive integer. It means that in each consecutive iteration, the integration step is reduced twice, as shown in Fig. 5.6.

$$h_k = \frac{b - a}{2^{k-1}} = \frac{1}{2} h_{k-1} \tag{5.19}$$

Numerical value of the integral calculated according to (5.9) for a fixed (given) parameter k, is denoted in the English language literature usually by $R_{k,1}$ [5, 6]. According to this notation

$$R_{1,1} = \frac{h_1}{2} [f(a) + f(b)]$$

$$R_{2,1} = \frac{h_2}{2} [f(a) + f(b) + 2f(a + h_2)] \tag{5.20}$$

$$= \frac{1}{2} \frac{h_1}{2} [f(a) + f(b)] + \frac{h_1}{2} f\left(a + \frac{h_1}{2}\right) = \frac{1}{2} R_{1,1} + \frac{1}{2} h_1 f\left(a + \frac{h_1}{2}\right)$$

Fig. 5.6

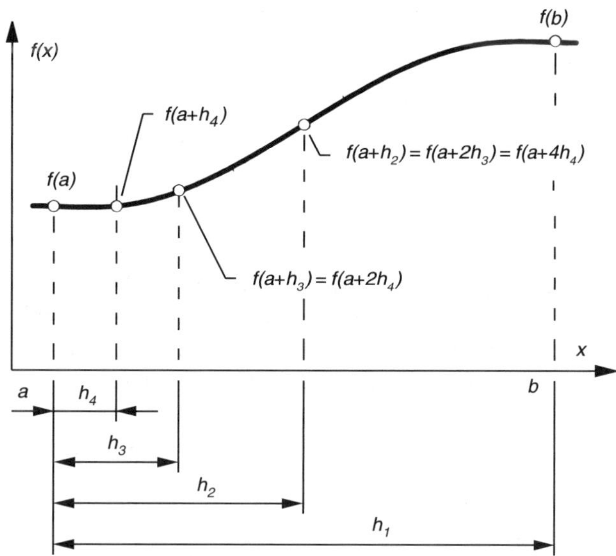

$$R_{3,1} = \frac{h_3}{2}\left[f(a) + f(b) + 2f(a + h_3) + 2f(a + 2h_3) + 2f(a + 3h_3)\right]$$

$$= \frac{1}{2}\frac{h_2}{2}\left[f(a) + f(b) + 2f(a + h_2)\right] + \frac{h_2}{2}\left[f\left(a + \frac{h_2}{2}\right) + f\left(a + 3\frac{h_2}{2}\right)\right]$$

$$= \frac{1}{2}R_{2,1} + \frac{1}{2}h_2\left[f\left(a + \frac{h_2}{2}\right) + f\left(a + 3\frac{h_2}{2}\right)\right]$$

. . . .

Proceeding in a similar way, we may prove that in general case

$$R_{k,1} = \frac{1}{2}R_{k-1,1} + \frac{1}{2}h_{k-1}\sum_{i=1}^{2^{k-2}} f\left[a + (i - 0.5)h_{k-1}\right] \tag{5.21}$$

The recursive formula (5.21) makes possible reducing the number of calculations necessary to compute the integral (5.8) with a prescribed accuracy. Further convergence improvement of approximations $R_{k,1}$, of the desired accurate value of the integral, can be achieved using the Richardson extrapolating procedure, described by the formula:

$$R_{i,j} = \frac{2^{2(j-1)}R_{i,j-1} - R_{i-1,j-1}}{2^{2(j-1)} - 1} \tag{5.22}$$

where $i = 1, 2, 3, \ldots, m$ and $j = 2, 3, \ldots, i$. This extrapolation gives approximate values $R_{i,j}$ of the integral, which can be represented in form of the following triangular table

$$
\begin{array}{llll}
R_{1,1} \\
R_{2,1} & R_{2,2} \\
R_{3,1} & R_{3,2} & R_{3,3} \\
R_{4,1} & R_{4,2} & R_{4,3} & R_{4,4} \\
\cdots\cdots\cdots\cdots\cdots\cdots\cdots\cdots \\
R_{m,1} & R_{m,2} & R_{m,3} & \cdots\cdots\cdots & R_{m,m}
\end{array}
\qquad(5.23)
$$

It was theoretically proved and confirmed by numerous experiments that values of $R_{i,j}$, lying on the main diagonal of the table, approach exact value of the integral faster than the approximations $R_{i,1}$ lying in the first column. As the criterion for ending the computations, the following condition is used most frequently.

$$
\left| R_{m,m} - R_{m-1,m-1} \right| \leq \varepsilon \qquad(5.24)
$$

where ε is a given sufficiently small positive number.

Example 5.3 In Table 5.1 we find several consecutive values of $R_{i,1}$, $R_{i,i}$, $R_{i,i} - R_{i,1}$ and $\left| R_{i,i} - R_{i-1,i-1} \right|$ computed by means of the Romberg method for the integral

$$
\int_0^1 \frac{4}{1-x^2}dx = \pi = 3.141592653589\ldots
$$

Table 5.1

i	$R_{i,1}$	$R_{i,i}$	$R_{i,i} - R_{i,1}$	$\left| R_{i,i} - R_{i-1,i-1} \right|$
1	3.000000000	3.000000000		
2	3.100000000	3.133333333	3.333×10^{-2}	1.333×10^{-1}
3	3.131176470	3.142117647	1.094×10^{-2}	8.784×10^{-3}
4	3.138988494	3.141585783	2.597×10^{-3}	5.318×10^{-4}
5	3.140941612	3.141592665	6.510×10^{-4}	6.881×10^{-6}
6	3.141429893	3.141592653	1.627×10^{-4}	1.163×10^{-8}
7	3.141551963	3.141592653	4.069×10^{-5}	4.852×10^{-11}
8	3.141582481	3.141592653	1.017×10^{-5}	7.110×10^{-14}

Results given above confirm the fact that the values $R_{i,i}$ lying on the main diagonal of the triangular table (5.16) approach accurate value of the integral faster than approximations $R_{i,1}$ placed in the first column.

5.2.3 The Simpson Method of Integration

One of the numerical integration methods, most frequently used for solving practical problems, is the Simpson method. In order to introduce this algorithm, let us divide an integration interval $[x_0, x_n]$ into $n/2$ equal sections, where n is an even number. The elementary segments (subintervals) obtained in this way

$$[x_0, x_2], [x_2, x_4], [x_4, x_6], \ldots, [x_{n-2}, x_n] \tag{5.25}$$

have equal lengths $2h = 2(b-a)/n$. On each elementary subinterval, the integrand $f(x)$ is interpolated by the Lagrange polynomial of second degree

$$L_2^{(i)}(x) \approx f(x) \text{ for } x_{i-1} \leq x \leq x_{i+1} \tag{5.26}$$

where $i = 1, 3, 5, \ldots, n-1$, see Fig. 5.7.

The interpolating polynomial (5.26), written according to the notation as in Fig. 5.7, has the form:

$$\begin{aligned}
L_2^{(i)}(x) &= \frac{(x - x_i)(x - x_{i+1})}{(x_{i-1} - x_i)(x_{i-1} - x_{i+1})} y_{i-1} + \frac{(x - x_{i-1})(x - x_{i+1})}{(x_i - x_{i-1})(x_i - x_{i+1})} y_i \\
&\quad + \frac{(x - x_{i-1})(x - x_i)}{(x_{i+1} - x_{i-1})(x_{i+1} - x_i)} y_{i+1} \\
&= \frac{y_{i-1}}{2h^2}(x - x_i)(x - x_i - h) - \frac{y_i}{h^2}(x - x_i + h)(x - x_i - h) \\
&\quad + \frac{y_{i+1}}{2h^2}(x - x_i + h)(x - x_i)
\end{aligned}$$

The value of elementary term S_i, see Fig. 5.7, is

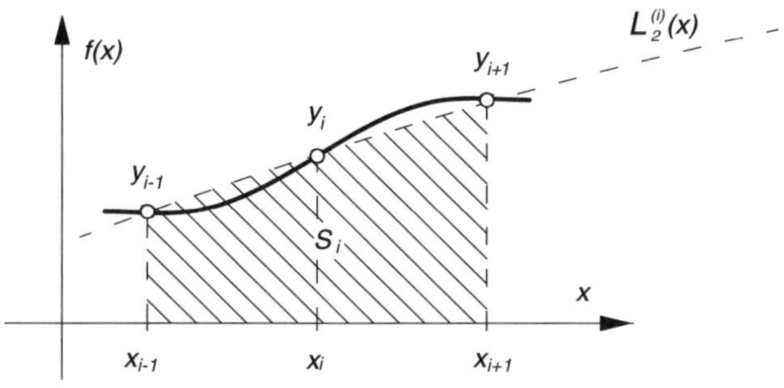

Fig. 5.7

$$S_i = \int\limits_{x_{i-1}}^{x_{i+1}} L_2^{(i)}(x)dx = \frac{y_{i-1}}{2h^2} \int\limits_{x_i-h}^{x_i+h} (x - x_i)(x - x_i - h)dx$$

$$+ \left(-\frac{y_i}{h^2}\right) \int\limits_{x_i-h}^{x_i+h} (x - x_i + h)(x - x_i - h)dx \qquad (5.27)$$

$$+ \frac{y_{i+1}}{2h^2} \int\limits_{x_i-h}^{x_i+h} (x - x_i + h)(x - x_i)dx$$

Calculating this integral we obtain:

$$S_i = \frac{h}{3}(y_{i-1} + 4 \cdot y_i + y_{i+1}) \qquad (5.28)$$

where $i = 1, 3, 5, \ldots, n - 1$. The sum of all terms S_i determined over the whole interval $[x_0, x_n]$ is equal to:

$$S = \frac{h}{3}(y_0 + 4y_1 + y_2 + y_2 + 4y_3 + y_4 + y_4 + 4y_5 + y_6 + \ldots\ldots$$
$$+ y_{n-4} + 4y_{n-3} + y_{n-2} + y_{n-2} + 4y_{n-1} + y_n) \qquad (5.29)$$

After grouping of terms, the sum (5.29) can be written in the following simpler form:

$$S = \frac{h}{3}[y_0 + 4(y_1 + y_3 + y_5 + \ldots + y_{n-1}) + 2(y_2 + y_4 + y_6 + \ldots + y_{n-2}) + y_n] \qquad (5.30)$$

known in the literature as the Simpson integral formula. Simplicity of its computing algorithm, accompanied by relatively good accuracy, constitutes very precious property of this algorithm. It was proved in the literature that the difference between accurate value of this integral and its approximation, given by the sum (5.30), is a second order quantity. It means that reducing twice the length of the subinterval $\Delta x_i \equiv h$, this difference decreases no less than $2^4 = 16$ times. The rule given in Sect. 5.2.1, concerning optimal partition of integration interval, saying that the sum of interpolation (method) errors and computation (computer processing) errors attains minimum, refers also to the Simpson method. Further accuracy improvement can be received by repeated calculation of the integral for different lengths of the subintervals (h) and appropriate "processing" of the results. As an example, let us consider the Aitken extrapolation procedure, in which computation of an integral is performed three times, i.e., the integration interval $[a, b]$ is divided into the subintervals with lengths h_1, h_2 and h_3, respectively, related according to the formula[7]:

$$\frac{h_2}{h_1} = \frac{h_3}{h_2} = q$$

Assume that we have three values of the integral obtained for different lengths h_1, h_2 and h_3 of subintervals, i.e.:

$$I_1 = I(h_1), \quad I_2 = I(h_2), \quad I_3 = I(h_3)$$

A more accurate value of the integral I is found according to the following relation:

$$I = I_1 - \frac{(I_1 - I_2)^2}{I_1 - 2I_2 + I_3}$$

Precision of the integral calculated in this way is of the order p, $R(h^p)$, where:

$$p = \frac{1}{\ln(q)} \ln \left(\frac{I_3 - I_2}{I_2 - I_1} \right)$$

and h is the longest subinterval among h_1, h_2 and h_3.

Example 5.4 As an illustration of the Simpson method algorithm and confirmation of conclusions given above, let us calculate the integral

$$\int_0^1 \frac{1}{1+x^2} dx$$

assuming that an integration interval was partitioned into $n = 10$ equal subintervals of the length $h = 0.1$. Discrete values of x_i and related to them values $y_i = f(x_i)$ of the integrand are given below.

$$
\begin{aligned}
x_0 &= 0.0, & y_0 &= f(x_0) = 1.000000000 \\
x_1 &= 0.1, & y_1 &= f(x_1) = 0.990099009 \\
x_2 &= 0.2, & y_2 &= f(x_2) = 0.961538461 \\
x_3 &= 0.3, & y_3 &= f(x_3) = 0.917431192 \\
x_4 &= 0.4, & y_4 &= f(x_4) = 0.862068965 \\
x_5 &= 0.5, & y_5 &= f(x_5) = 0.800000000 \\
x_6 &= 0.6, & y_6 &= f(x_6) = 0.735294117 \\
x_7 &= 0.7, & y_7 &= f(x_7) = 0.671140939 \\
x_8 &= 0.8, & y_8 &= f(x_8) = 0.609756097 \\
x_9 &= 0.9, & y_9 &= f(x_9) = 0.552486187 \\
x_{10} &= 1.0, & y_{10} &= f(x_{10}) = 0.500000000
\end{aligned}
$$

According to Simpson formula (5.30) we can write:

$$S = \frac{0.1}{3}[y_0 + 4(y_1 + y_3 + y_5 + y_7 + y_9) + 2(y_2 + y_4 + y_6 + y_8) + y_{10}] = 0.785398153$$

This integral can be also calculated using the analytical formula:

$$\int\limits_0^1 \frac{1}{1+x^2}\,dx = arctg(1) - arctg(0) = \frac{\pi}{4} \approx 0.785398163$$

Analytical and numerical results give the same answer up to eight significant digits, thus confirming our conclusion about relatively high accuracy of computation achieved.

5.2.4 The Newton–Cotes Method of Integration

Assume that discrete values $y_i = f(x_i)$ of integrand $f(x)$ are defined for

$$x_i = x_0 + i \cdot h \tag{5.31}$$

where $i = 0, 1, 2, 3, \ldots, n$, $h = (b-a)/n$, $x_0 \equiv a$ and $x_n \equiv b$. The main feature of the Newton–Cotes method is the interpolation of the integrand $y_i = f(x_i)$ defined above by the Lagrange polynomial of degree n

$$L_n(x) = \sum_{i=0}^{n} y_i \delta_i(x) \tag{5.32}$$

where the function:

$$\delta_i(x) = \frac{(x-x_0)(x-x_1)(x-x_2)\ldots(x-x_{i-1})(x-x_{i+1})\ldots(x-x_{n-1})(x-x_n)}{(x_i-x_0)(x_i-x_1)(x_i-x_2)\ldots(x_i-x_{i-1})(x_i-x_{i+1})\ldots(x_i-x_{n-1})(x_i-x_n)}$$

denotes a polynomial assigned to the term i of the series (5.32). Introducing the parameters

$$q = \frac{x-x_0}{h}, \quad Q_n = q(q-1)(q-2)(q-3)\ldots(q-n+1)(q-n) \tag{5.33}$$

polynomial $\delta_i(x)$ can be written in the form:

$$\delta_i(x) = \frac{hqh(q-1)h(q-2)\ldots h(q-i+1)h(q-i-1)\ldots h(q-n+1)h(q-n)}{ih(i-1)h(i-2)h\ldots(1)h(-1)h\ldots[-(n-i+1)]h[-(n-i)]h}$$

which after multiplying the numerator and the denominator by $(q-i)$, and some elementary simplifying transformations takes finally the form:

$$\delta_i(x) = \frac{h^n Q_n}{i!h^i(-1)^{n-i}h^{n-i}(n-i)!} \cdot \frac{1}{q-i} = (-1)^{n-i}\frac{1}{i!(n-i)!} \cdot \frac{Q_n}{q-i} \tag{5.34}$$

Thus, the interpolating polynomial (5.32) is equal to:

$$L_n(x) = \sum_{i=0}^{n} y_i \left[(-1)^{n-i} \frac{1}{i!(n-i)!} \cdot \frac{Q_n}{q-i} \right] \tag{5.35}$$

After replacing integrand $f(x)$ by polynomial (5.35) we obtain:

$$\int_a^b f(x)dx \approx \int_a^b L_n(x)dx = \sum_{i=0}^{n} y_i \left[(-1)^{n-i} \frac{1}{i!(n-i)!} \int_{x_0}^{x_n} \frac{Q_n}{q-i}dx \right] = \sum_{i=0}^{n} y_i A_i \tag{5.36}$$

Formula (5.36) is known in the literature as the Newton–Cotes quadrature. Coefficients A_i, where $i = 0, 1, 2, 3, \ldots, n$, of this quadrature are often represented in the form of a product:

$$A_i = (x_n - x_0)H_i = (b-a)H_i \tag{5.37}$$

where

$$H_i = \frac{1}{n}(-1)^{n-i} \frac{1}{i!(n-i)!} \int_0^n \frac{Q_n}{q-i}dq$$

is the i-Cotes coefficient. Relation (5.37) can be obtained by introducing new variables taking $(x = q \cdot h - x_0, dx = h \cdot dq, h = (b-a)/n)$ and putting the integration limits as in the expression (5.36) defining the coefficient A_i. From the relation (5.37), it follows that:

$$\sum_{i-0}^{n} H_i = 1, \quad H_i = H_{n-i} \quad \text{for } i = 0, 1, 2, 3, \ldots, n \tag{5.38}$$

Values of the Cotes coefficients calculated for $n = 1, 2, 3, 4$ and 5 are given in Table 5.2 [3].

Example 5.5 As an example illustrating the algorithm of Newton–Cotes method, consider the integral

Table 5.2

n	$d_n H_0, d_n H_n$	$d_n H_1, d_n H_{n-1}$	$d_n H_2, d_n H_{n-2}$	$d_n H_3, d_n H_{n-3}$	d_n
1	1				2
2	1	4			6
3	1	3			8
4	7	32	12		90
5	19	75	50		288
6	41	216	27	272	840

$$\int_0^1 \left(5x^4 + x\right)dx$$

which can be calculated analytically and equals 1.5. The integrand $f(x)$ is the polynomial of degree 4, and therefore an integration interval $[0, 1]$ should be divided into four equal parts ($h = 0.25$). According to this partition and Table 5.1 we obtain the values

$$
\begin{aligned}
x_0 &= 0.00, & y_0 &= f(x_0) = 0, & H_0 &= 7/90 \\
x_1 &= 0.25, & y_1 &= f(x_1) = 0.26953125, & H_1 &= 32/90 \\
x_2 &= 0.50, & y_2 &= f(x_2) = 0.8125, & H_2 &= 12/90 \\
x_3 &= 0.75, & y_3 &= f(x_3) = 2.33203125, & H_3 &= 32/90 \\
x_4 &= 1.00, & y_4 &= f(x_4) = 6, & H_4 &= 7/90
\end{aligned}
$$

which we introduce to Eqs. (5.37) and (5.36) obtaining

$$
\int_0^1 \left(5x^4 + x\right)dx = 1 \cdot \left[\frac{7}{90} \cdot 0 + \frac{32}{90} \cdot 0.26953125 + \frac{12}{90} \cdot 0.8125 \right.
$$
$$
\left. + \frac{32}{90} \cdot 2.33203125 + \frac{7}{90} \cdot 6 \right] = 1.5
$$

In this particular case, numerical result is the same as the accurate value of the integral obtained analytically. This is a direct consequence of the fact that this integrand function is a polynomial, and that the Lagrange interpolating polynomial is identically the same.

5.2.5 The Cubic Spline Function Quadrature

Acting in the same manner as shown in the previous section, we divide the whole integration interval $[a, b] \equiv [x_0, x_n]$ into n identical subintervals of the length

$$h = \frac{b - a}{n}$$

Assume that for a given set of $(n + 1)$ interpolation points

$$x_i = x_0 + i \cdot h, \quad i = 0, 1, 2, 3, \ldots, n$$

discrete values $y_i = f(x_i)$ of the integrand are known. When we calculate the definite integral

$$\int\limits_a^b f(x)dx \tag{5.39}$$

using the spline function method, the integrand $f(x)$ is interpolated over the whole interval $[x_0, x_n]$ by the spline function composed of n trinomials of the form:

$$q_i(x) = a_i + b_i(x - x_{i-1}) + c_i(x - x_{i-1})^2 + d_i(x - x_{i-1})^3 \tag{5.40}$$

where $x_{i-1} \leq x \leq x_i, i = 1, 2, 3, \ldots, n$ [5]. Values of the coefficients a_i, b_i, c_i and d_i of individual trinomials (5.40) are determined similarly as described in Sect. 4.1.5. Using general rules of integration calculus, the definite integral (5.39) can be transformed as follows:

$$\int\limits_a^b f(x)dx = \sum_{i=1}^n \int\limits_{x_{i-1}}^{x_i} f(x)dx \approx \sum_{i=1}^n \int\limits_{x_{i-1}}^{x_i} q_i(x)dx \tag{5.41}$$

Substituting the trinomial (5.40) into the expression (5.41) and performing elementary integration we get:

$$\int\limits_a^b f(x)dx = \sum_{i=1}^n \left(a_i \cdot h + \frac{1}{2}b_i \cdot h^2 + \frac{1}{3}c_i \cdot h^3 + \frac{1}{4}d_i \cdot h^4 \right) \tag{5.42}$$

Using the following equations

$$y_{i-1} = q_i(x = x_{i-1}) = a_i$$

$$y_i = q_i(x = x_i) = a_i + b_i h + c_i h^2 + d_i h^3$$

and following from the interpolation rule concerning the trinomials (5.40), the quadrature (5.42) can be written in the form:

$$\int\limits_a^b f(x)dx = \frac{1}{2}\sum_{i=1}^n h(y_{i-1} + y_i) - \frac{1}{12}\sum_{i=1}^n h^3(2c_i + 3d_i h) \tag{5.43}$$

having a simple physical interpretation. First sum of the equation (5.43) is nothing else but an approximate value of the integral computed by means of the method of trapezoids. The second sum is a correction term showing the difference between the results obtained using the spline function and the result given by the much simpler method of trapezoids.

5.2.6 The Gauss and Chebyshev Quadratures

It results from the considerations described in the previous sections that the definite integral of one variable can be expressed by the following general relation:

$$\int_a^b f(x)dx = \sum_{i=1}^n A_i f(x_i) \tag{5.44}$$

which in the literature is called the quadrature. Let us assume that the integrand function $f(x)$ is interpolated by the polynomial $W_n(x)$ of degree n over the interval $[a, b]$. At this assumption it is obvious that the values of coefficients A_i of the quadrature (5.44) depend on degree n and interpolation nodes x_i. The essence of the Gauss and Chebyshev quadratures presented below is that these nodes are not equally spaced over the interval $[a, b]$.

5.2.6.1 The Gauss Quadrature

Let us consider again the definite integral

$$\int_{-1}^1 f(t)dt = \sum_{i=1}^n A_i f(t_i) \tag{5.45}$$

defined over the interval $[-1, 1]$. Nodes $t_1, t_2, t_3, \ldots, t_n$ and coefficients $A_1, A_2, A_3, \ldots, A_n$ of the Gauss quadrature (5.45) should take such values for which this quadrature will be accurate for the integrand $f(t)$ in the form of a polynomial $f(t) \equiv W_m(t)$ of possibly highest degree m. The total number of nodes t_i and coefficients A_i of the quadrature (5.45) is equal to $2n$. The same number of independent coefficients has a polynomial $W_m(t)$ of degree $m = 2n - 1$. Thus the quadrature being evaluated should be accurate for polynomials

$$f_k(t) = t^k \tag{5.46}$$

where $k = 0, 1, 2, 3, \ldots, m = 2n - 1$. This requirement is expressed by the following equations:

$$\int_{-1}^1 t^k dt = \sum_{i=1}^n A_i f_k(t_i) = \sum_{i=1}^n A_i t_i^k \quad \text{for } k = 0, 1, 2, 3, \ldots, 2n - 1 \tag{5.47}$$

Let us assume that the integrand is a linear combination of polynomials (5.46), i.e., $f(t) = \sum_{k=0}^{2n-1} c_k t^k$, where c_k denotes kth real coefficient. It is not difficult to justify using Eq. (5.47) that the integral $\int\limits_{-1}^{1} f(t)dt$ can be written in the following form:

$$\int\limits_{-1}^{1} f(t)dt = \sum_{k=0}^{2n-1} c_k \int\limits_{-1}^{1} t^k dt = \sum_{k=0}^{2n-1} c_k \sum_{i=1}^{n} A_i t_i^k = \sum_{i=1}^{n} A_i \sum_{k=0}^{2n-1} c_k t_i^k = \sum_{i=1}^{n} A_i f(t_i)$$

(5.48)

identical as the quadrature (5.45). Thus, in order to evaluate the nodes $t_1, t_2, t_3, \ldots, t_n$ and coefficients $A_1, A_2, A_3, \ldots, A_n$ of the quadrature (5.48) it is necessary to solve the equation system (5.47) written as follows:

$$\int\limits_{-1}^{1} 1 dt = \sum_{i=1}^{n} A_i = 2$$

$$\int\limits_{-1}^{1} t dt = \sum_{i=1}^{n} A_i t_i = 0$$

$$\int\limits_{-1}^{1} t^2 dt = \sum_{i=1}^{n} A_i t_i^2 = \frac{2}{2+1}$$

$$\cdots\cdots\cdots\cdots\cdots\cdots\cdots\cdots\cdots\cdots$$

$$\int\limits_{-1}^{1} t^k dt = \sum_{i=1}^{n} A_i t_i^k = \frac{1-(-1)^{k+1}}{k+1}$$

$$\cdots\cdots\cdots\cdots\cdots\cdots\cdots\cdots\cdots\cdots$$

$$\int\limits_{-1}^{1} t^{2n-1} dt = \sum_{i=1}^{n} A_i t_i^{2n-1} = 0$$

(5.49)

Unfortunately, the equation system (5.49) is nonlinear with respect to nodes and for this reason rather difficult to solve. The approach presented below makes it possible much easier. In its first stage, the nodes $t_1, t_2, t_3, \ldots, t_n$ are evaluated in an indirect manner. For this purpose, the following property

$$\int\limits_{-1}^{1} t^k P_n(t)dt = 0 \quad \text{when} \quad k < n, n = 1, 2, 3, \ldots$$

of Legendre polynomials $P_n(t)$, see relations (4.84) and (4.85), is utilized. Let us assume that the function $f(t)$ has the form:

$$f(t) = t^k P_n(t)$$

(5.50)

where $k = 0, 1, 2, 3, \ldots, n - 1$. For $0 \leq k \leq n - 1$, the function (5.50) can be treated as a polynomial of degree at most $2n - 1$. For such integrand function the quadrature (5.45) should be accurate. The above requirement is expressed by the following equations:

$$\int_{-1}^{1} t^k P_n(t)dt = \sum_{i=1}^{n} A_i t_i^k P_n(t_i), \quad k = 0, 1, 2, 3, \ldots, n - 1 \tag{5.51}$$

Left side of equation (5.51) is equal to 0 due to the orthogonality of the Legendre polynomials $P_n(t)$ and polynomials t^k over the interval $[-1, 1]$ when $k < n$. According to this conclusion

$$\sum_{i=1}^{n} A_i t_i^k P_n(t_i) = 0, \quad k = 0, 1, 2, 3, \ldots, n - 1 \tag{5.52}$$

Thus, Eq. (5.52) is satisfied for any coefficients A_i if

$$P_n(t_i) = 0 \quad \text{for} \quad i = 1, 2, 3, \ldots, n \tag{5.53}$$

It may be concluded from condition (5.53) that for fixed value of degree n the nodes $t_1, t_2, t_3, \ldots, t_n$ of the quadrature (5.45) should be the same as roots of the Legendre polynomial of the same degree n. Values of nodes $t_1, t_2, t_3, \ldots, t_n$ determined in this way for $n = 2, 3, 4$ and 5 are given in Table 5.3. For these nodes, the equation system (5.49) becomes linear with respect to desired coefficients $A_1, A_2, A_3, \ldots, A_n$. Consequently, these coefficients can be evaluated by means of one of the direct methods described in Chap. 1. Naturally, the Gauss elimination method with the choice of the main element is the most suitable for this purpose. Values of coefficients $A_1, A_2, A_3, \ldots, A_n$ evaluated in this manner for $n = 2, 3, 4$ and 5 are also given in Table 5.3 [3].

Table 5.3

$n \rightarrow$	2	3	4	5
t_1	−0.577350269	−0.774596669	−0.861136311	−0.906179846
t_2	0.577350269	0.000000000	−0.339981043	−0.538469310
t_3		0.774596669	0.339981043	0.000000000
t_4			0.861136311	0.538469310
t_5				0.906179846
A_1	1.000000000	0.555555555	0.347854845	0.236926885
A_2	1.000000000	0.888888888	0.652145155	0.478628670
A_3		0.555555555	0.652145155	0.568888888
A_4			0.347854845	0.478628670
A_5				0.236926885

For example, when $n = 3$ the nodes of the Gauss–Legendre quadrature are: $t_1 = -\sqrt{3/5} = -0.774596669$, $t_2 = 0$ and $t_3 = \sqrt{3/5} = 0.774596669$. In this case, equations (5.49) take the form:

$$A_1 + A_2 + A_3 = 0$$
$$-\sqrt{3/5}A_1 + 0 \cdot A_2 + \sqrt{3/5}A_3 = 0$$
$$(3/5)A_1 + 0 \cdot A_2 + (3/5)A_3 = 2/3$$

The solution of this linear equation system is: $A_1 = A_3 = 5/9 \approx 0.555555555$ and $A_2 = 8/9 \approx 0.888888888$. Finally, the three-node Gauss–Legendre quadrature can be written as:

$$\int_{-1}^{1} f(t)dt = \frac{1}{9}\left[5 \cdot f(-\sqrt{3/5}) + 8 \cdot f(0) + 5 \cdot f(\sqrt{3/5})\right] \qquad (5.54)$$

Of course, the Gauss–Legendre quadrature can be generalized for definite integrals defined in arbitrary closed interval $[a, b]$ different from $[-1, 1]$. This can be done in a manner similar to that described in further part of this section, see transformation formulas (5.60) and (5.61). This problem is also illustrated by the Example 5.7.

Example 5.6 It is concluded in the literature that the five-node Gauss–Legendre quadrature ensures the relatively good accuracy of integration for the most applications. The results of calculations presented below confirm well the validity of the above conclusion. Thus, let us calculate the definite integral

$$\int_{-1}^{1} \frac{1}{x+2}dx = \ln(3) - \ln(1) = \ln(3)$$

by means of the five-node Gauss–Legendre quadrature. Using general formula (5.45) and values of appropriate ($n = 5$) nodes and coefficients given in Table 5.3 we can write:

$$\int_{-1}^{1} \frac{1}{x+2}dx \approx 0.236926885\left(\frac{1}{-0.906179846 + 2} + \frac{1}{0.906179846 + 2}\right)$$

$$+ 0.478628670\left(\frac{1}{-0.538469310 + 2} + \frac{1}{0.538469310 + 2}\right)$$

$$+ 0.568888888\left(\frac{1}{0 + 2}\right) = 1.098609241$$

In this example, the numerical approximation of the given integral differs from accurate value $\ln(3) \approx 1.098612289$ less than 3.1×10^{-6}.

5.2.6.2 The Chebyshev Quadrature

The Chebyshev method considered in this section represents the methods belonging to the third group (III), see Fig. 5.1, called in general the Gaussian methods. Short description of methods of this kind was given at the beginning of this chapter. The quadrature introduced by Chebyshev can be written in the following general form:

$$\int_{-1}^{1} f(t)dt = \sum_{i=1}^{n} A_i \cdot f(t_i) \tag{5.55}$$

where A_i, for $i = 1, 2, 3, \ldots, n$ are the fixed coefficients, [3]. The main idea of the Chebyshev method is the determination of such discrete values of t_i, for which:

- All coefficients A_i are equal,
- The quadrature (5.55) is accurate if the integrand $f(t)$ is the polynomial of the degree not greater than n.

First condition is satisfied for the coefficients $A_1 = A_2 = A_3 = \ldots = A_n = A$ determined from relation (5.55), assuming that $f(t) \equiv 1$ is the polynomial of degree 0 (zero). According to this relation:

$$\int_{-1}^{1} 1 \cdot dt = 2 = \sum_{i=1}^{n} A_i \cdot 1 = nA$$

Hence

$$A = A_1 = A_2 = A_3 = \ldots = A_n = \frac{2}{n} \tag{5.56}$$

After introducing coefficients (5.56) in Eq. (5.55) we obtain relation:

$$\int_{-1}^{1} f(t)dt = \frac{2}{n} \sum_{i=1}^{n} f(t_i) \tag{5.57}$$

known in the literature as the Chebyshev quadrature formula. This quadrature, according to the above assumption, should be accurate for integrands $f(t)$ which are polynomials: $t, t^2, t^3, t^4, \ldots, t^n$. This condition formulated for the polynomial of degree k, i.e., for $f(t) \equiv t^k$, can be written as:

$$\int_{-1}^{1} t^k dt = \frac{2}{n} \sum_{i=1}^{n} t_i^k = \frac{1}{k+1}[1 - (-1)^{k+1}] = \frac{2}{n}(t_1^k + t_2^k + t_3^k + \ldots + t_n^k) \tag{5.58}$$

Equation (5.58), formulated for $k = 1, 2, 3, \ldots, n$, respectively, constitutes the following nonlinear system:

$$t_1^1 + t_2^1 + t_3^1 + \ldots + t_n^1 = 0$$

$$t_1^2 + t_2^2 + t_3^2 + \ldots + t_n^2 = \frac{n}{3}$$

$$t_1^3 + t_2^3 + t_3^3 + \ldots + t_n^3 = 0$$

$$t_1^4 + t_2^4 + t_3^4 + \ldots + t_n^4 = \frac{n}{5} \tag{5.59}$$

$$\ldots \ldots \ldots \ldots \ldots \ldots \ldots \ldots$$

$$t_1^n + t_2^n + t_3^n + \ldots + t_n^n = \frac{n[1 - (-1)^{n+1}]}{2(n+1)}$$

Solutions of this system, evaluated for $n = 2, 3, 4, 5, 6$ and 7, are given in Table 5.4. We should emphasize the fact that the equation system (5.59) has no real solutions for $n = 8$ and $n \geq 10$. This fact constitutes some kind of limitation for the method presented above [3].

The Chebyshev quadrature formula can be generalized to definite integrals defined in an arbitrary closed interval $[a, b]$, different from $[-1, 1]$. Any definite integral

$$\int_a^b f(x)dx \tag{5.60}$$

can be transformed to the canonical form (5.55) using the following relation:

$$t(x) = \frac{2x}{b-a} - \frac{b+a}{b-a}, \quad dt = \frac{2}{b-a}dx$$

transforming the interval $[a, b]$, $(a \leq x \leq b)$ into the interval $[-1, 1]$, $(-1 \leq t \leq 1)$. Resulting formula has now the form:

$$\int_a^b f(x)dx = \frac{b-a}{2} \int_{-1}^1 f(t)dt = \frac{b-a}{2} \cdot \frac{2}{n} \sum_{i=1}^n f(t_i) = \frac{b-a}{n} \sum_{i=1}^n f(x_i) \tag{5.61}$$

Table 5.4

n	$t_1, t_n = -t_1$	$t_2, t_{n-1} = -t_2$	$t_3, t_{n-2} = -t_3$	$t_4, t_{n-3} = -t_4$
2	−0.577350			
3	−0.707107	0		
4	−0.794654	−0.187592		
5	−0.832498	−0.374541	0	
6	−0.866247	−0.422519	−0.266635	
7	−0.883862	−0.529657	−0.323912	0

where

$$x_i = \frac{b+a}{2} + \frac{b-a}{2} t_i \tag{5.62}$$

The values of roots t_i, where $i = 1, 2, 3, \ldots, n \leq 7$, are given in Table 5.4.

Example 5.7 As an illustration of the Chebyshev quadrature method, let us calculate the integral

$$\int_0^1 \frac{x}{1+x} dx$$

taking the value $n = 7$. The points (nodes) x_i, where $i = 1, 2, 3, \ldots, 7$, calculated according to (5.62), and related to them values $f(x_i)$ of the integrand are equal to:

$$x_1 = 0.5 + 0.5(-0.883862) = 0.0580690, \quad f(x_1) = 0.054882054$$
$$x_2 = 0.5 + 0.5(-0.529657) = 0.2351715, \quad f(x_2) = 0.190395827$$
$$x_3 = 0.5 + 0.5(-0.323912) = 0.3380440, \quad f(x_3) = 0.252640421$$
$$x_4 = 0.5 + 0.5(0) = 0.5 \qquad\qquad\qquad f(x_4) = 0.333333333$$
$$x_5 = 0.5 + 0.5(0.323912) = 0.6619560, \quad f(x_5) = 0.398299353$$
$$x_6 = 0.5 + 0.5(0.529657) = 0.7648285, \quad f(x_6) = 0.433372040$$
$$x_7 = 0.5 + 0.5(0.883862) = 0.9411931, \quad f(x_7) = 0.485048644$$

Introducing these values of the function $f(x_i)$, where $i = 1, 2, 3, \ldots, 7$, into the formula (5.61) we obtain:

$$\int_0^1 \frac{x}{1+x} dx \approx \frac{1-0}{7} \sum_{i=1}^{7} f(x_i) = \frac{1}{7} \cdot 2.147971672 = 0.306853096$$

The calculated value of this quadrature is very close to the value of the integral found analytically, that is:

$$\int_0^1 \frac{x}{1+x} dx = \int_0^1 1 \cdot dx - \int_0^1 \frac{1}{1+x} dx = 1 - 0 - [\ln(2) - \ln(1)] = 0.306852819$$

Correctness of our computations as well as the fact that precision of the method is sufficient for practical applications can be considered as confirmed.

5.3 Methods for Numerical Integration of Two Variable Functions

5.3.1 The Method of Small (Elementary) Cells

One of the simplest methods for numerical computation of double definite integral

$$I = \iint\limits_{G} f(x, y)dx\,dy \tag{5.63}$$

is the method of elementary cells. By a cell, see Fig. 5.8, we understand such small rectangular area D_{ij}:

$$x_i \leq x \leq x_i + \Delta x_i$$

$$y_j \leq y \leq y_j + \Delta y_j$$

for which the following relation can be formulated:

$$\iint\limits_{D_{ij}} f(x, y)dx\,dy \approx \Delta x_i \cdot \Delta y_j \cdot f\left(x_i + \frac{\Delta x_i}{2}, y_j + \frac{\Delta y_j}{2}\right) \tag{5.64}$$

The above formula, used for each elementary (small) area G, permits to substitute the integral (5.44) by its approximation written in form of the following double sum

$$\iint\limits_{G} f(x, y)dx\,dy \approx \sum_{i=1}^{m}\sum_{j=1}^{n} f\left(x_i + \frac{\Delta x_i}{2}, y_j + \frac{\Delta y_j}{2}\right)\Delta x_i \cdot \Delta y_j \tag{5.65}$$

In the literature we can find a proof of the fact that an approximation error introduced by the relation given above is a second order quantity with respect to Δx and Δy. When the area G is not a rectangle, in many cases it can be transformed into the rectangular one by means of changing the variables. To illustrate this procedure

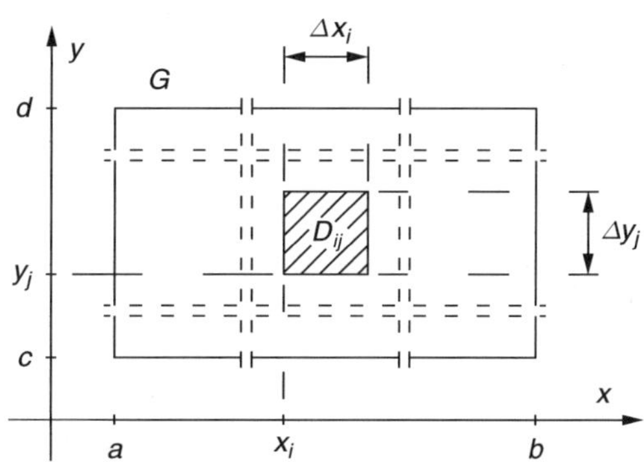

Fig. 5.8

Fig. 5.9

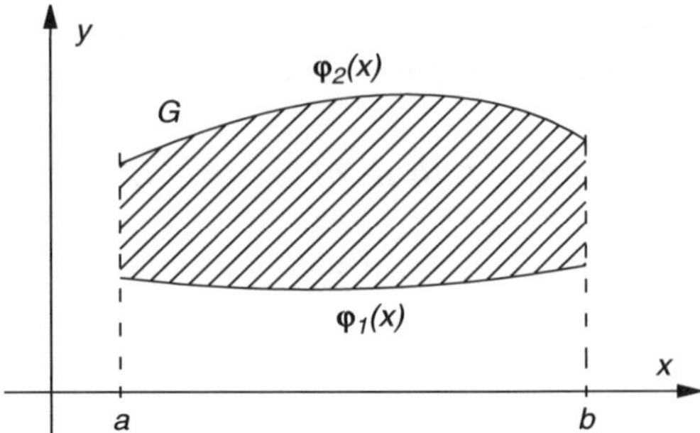

assume that an integration area G has the form of a curvilinear quadrangle shown in Fig 5.9.

Coordinates of each point of this area are defined by the relations:

$$a \leq x \leq b$$
$$\varphi_1(x) \leq y \leq \varphi_2(x) \tag{5.66}$$

This area can be transformed into the rectangular cell in the coordinate system (x, t) by means of the following new variable:

$$t = \frac{y - \varphi_1(x)}{\varphi_2(x) - \varphi_1(x)}, 0 \leq t \leq 1 \tag{5.67}$$

5.3.2 The Simpson Cubature Formula

Better approximation, as compared with the one obtained by means of the formula (5.64) can be obtained using the relation called often the Simpson type mechanic cubature [3, 8]. Let us begin with an assumption that we have rectangular integration area D_{ij} given by the relations:

$$x_i - h \leq x \leq x_i + h,$$
$$y_j - k \leq y \leq y_j + k \tag{5.68}$$

and having nine characteristic points, which are shown in Fig. 5.10.

Double integral defined on this area can be written in the following form:

$$\iint\limits_{D_i} f(x, y)dx\,dy = \int\limits_{x_i-h}^{x_i+h} \left[\int\limits_{y_j-k}^{y_j+k} f(x, y)dy \right] dx = \int\limits_{x_i-h}^{x_i+h} dx \int\limits_{y_j-k}^{y_j+k} f(x, y)dy \tag{5.69}$$

The integrand $f(x, y)$ appearing in the second integral can be interpreted as the function of one variable y, depending on the parameter x. Single integral of this

Fig. 5.10

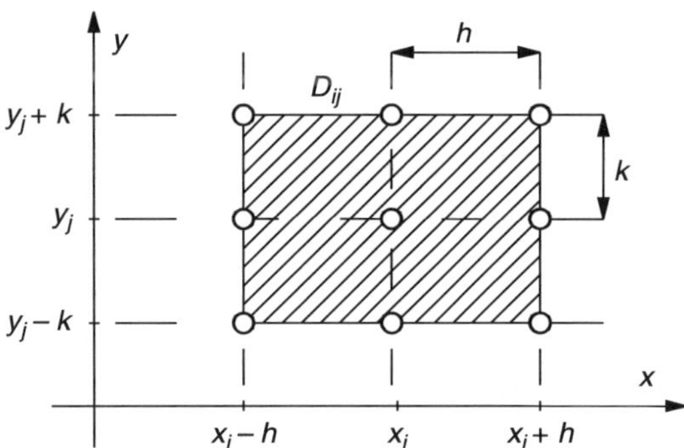

function, depending on the parameter, can be calculated using the quadrature Simpson formula (5.28). After introducing (5.28) to the formula (5.69) we obtain:

$$\oiint_{D_i} f(x, y)dx\, dy = \int_{x_i-h}^{x_i+h} dx \cdot \frac{k}{3}[f(x, y_j - k) + 4f(x, y_j) + f(x, y_j + k)]$$

$$= \frac{k}{3} \int_{x_i-h}^{x_i+h} f(x, y_j - k)dx + 4 \cdot \frac{k}{3} \int_{x_i-h}^{x_i+h} f(x, y_j)dx + \frac{k}{3} \int_{x_i-h}^{x_i+h} f(x, y_j + k)dx$$

$$(5.70)$$

Using for the second time the Simpson quadrature formula with respect to each of the single integrals given above, we obtain the following formula:

$$\oiint_{D_i} f(x, y)dx\, dy = \frac{k \cdot h}{9}[f(x_i - h, y_j - k) + 4f(x_i, y_j - k) + f(x_i + h, y_j - k)]$$

$$+ 4 \cdot \frac{k \cdot h}{9}[f(x_i - h, y_j) + 4f(x_i, y_j) + f(x_i + h, y_j)]$$

$$+ \frac{k \cdot h}{9}[f(x_i - h, y_j + k) + 4f(x_i, y_j + k) + f(x_i + h, y_j + k)]$$

$$(5.71)$$

known in the literature under the name of the Simpson cubature formula, or mechanical cubature presented often in the form:

$$\oiint_{D_i} f(x, y)dx\, dy = \frac{k \cdot h}{9}(A_{ij} + 4 \cdot B_{ij} + 16 \cdot C_{ij}) \qquad (5.72)$$

where

$$A_{ij} = f(x_i - h, y_j - k) + f(x_i + h, y_j - k) + f(x_i - h, y_j + k) + f(x_i + h, y_j + k)$$

$$B_{ij} = f(x_i, y_j - k) + f(x_i, y_j + k) + f(x_i - h, y_j) + f(x_i + h, y_j)$$
$$C_{ij} = f(x_i, y_j)$$

Example 5.8 Using the Simpson cubature formula (5.72) we calculate the integral

$$I = \iint\limits_{D_i} \frac{10}{xy} dx \, dy$$

defined on the surface of the rectangle D_{ij}: ($2 \leq x \leq 2.4, 4 \leq y \leq 4.6$). In this case $h = (2.4 - 2)/2 = 0.2, k = (4.6 - 4)/2 = 0.3, x_i = (2 + 2.4)/2 = 2.2,$ $y_i = (4 + 4.6)/2 = 4.3$. Values of the integrand needed for the computation, see Fig. 5.9, are given in Table 5.5.

The partial sums calculated according to Eq. (5.72) are equal to:

$$A_{ij} = 1.25 + 1.041666666 + 1.086956522 + 0.905797101 = 4.284420289$$

$$B_{ij} = 1.136363636 + 0.988142292 + 1.162790698 + 0.968992248 = 4.256288874$$

$$C_{ij} = 1.057082452$$

Finally, we obtain the following numerical approximation of the integral:

$$I_n = 0.2 \cdot 0.3/9(4.284420289 + 4 \cdot 4.256288874 + 16 \cdot 1.057082452) = 0.254819296$$

The accurate value of the same integral can be found analytically and is equal to:

$$I_a = 10 \cdot \ln(2.4/2) \cdot \ln(4.6/4) = 0.254816148$$

Next, we find relative error of this numerical approximation

$$\delta = \frac{0.254819296 - 0.254816148}{0.254816148} \approx 1.235 \cdot 10^{-5}$$

which is fully acceptable for the majority of applications. In case when dimensions of the integration area R are large, in order to assure sufficient computation accuracy the whole area should be divided into $n \times m$ sufficiently small identical rectangles D_{ij}, where $i = 1, 2, 3, \ldots, n, j = 1, 2, 3, \ldots, m$, and the sides are parallel to the axes x and y. Using the above procedure to all such rectangles we obtain the

Table 5.5

$y_j \downarrow /x_i \rightarrow$	2.0	2.2	2.4
4.0	1.250000000	1.136363636	1.041666666
4.3	1.162790698	1.057082452	0.968992248
4.6	1.086956522	0.988142292	0.905797101

Fig. 5.11

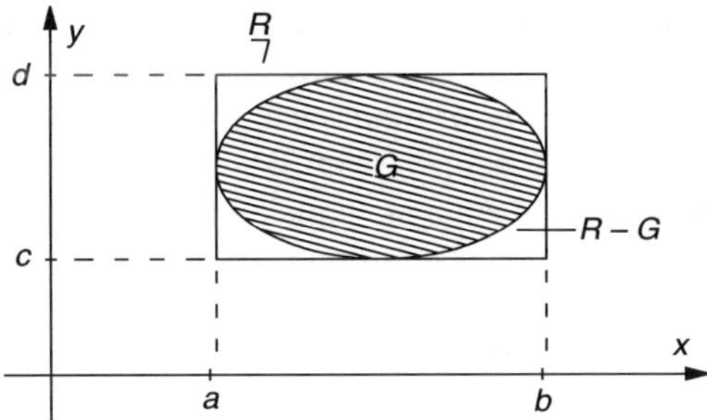

set of partial results, the sum of which constitutes the approximate value of the calculated integral. In the similar way, we proceed when the integration area G is the curvilinear quadrangle, as shown in Fig. 5.11.

This area should be approximated (with a surplus) by a set R composed of $n \times m$ rectangles D_{ij}, on which we define the following auxiliary integrand

$$
f^*(x, y) = \begin{vmatrix} f(x, y), & (x, y) \subset G \\ \\ 0, & (x, y) \not\subset G \end{vmatrix}
\tag{5.73}
$$

For such integrand we can write:

$$
\oiint_G f(x, y)dx\, dy = \oiint_R f^*(x, y)dx\, dy
\tag{5.74}
$$

The double integral standing on the right-side of Eq. (5.55) can be found using the procedure similar to that described above.

5.4 An Example of Applications

The basic goal of navigation is the determination of a geographical position (position) of an object moving in a three-dimensional (3D) space, in order to guide this object to the predetermined position in that space with a predetermined accuracy and at the right time. This problem is usually solvable using the autonomous methods, which utilize different board instruments. Historically, the first magnetic, mechanical and optical instruments of this kind were various types of compasses, magnetometers, logs, gyrocompasses, optical locators, sextants and infrared direction finders. Such equipments made possible determination of the geographic position of an object on the base of the Earth magnetic field measurements, position of the Sun, the Moon, as well as other celestial bodies. In the modern navigation systems, the satellites play a role of "reference points", whose position with respect to the Earth (more precisely with respect to the center of the Earth and the adopted map graticule) at any moment of universal time (UT) is known and monitored systematically.

A special three-dimensional ellipsoidal model of the Earth surface was developed and its WGS84 version (World Geodetic System, standard USA) is broadly used in the majority of multi-channel global positioning system (GPS) and differential global positioning system (DGPS,) receivers. All these systems and corresponding measurement equipment make possible the determination of the so-called *observed position*. Another position used for navigation purposes is the *counted position*. Its proper meaning defines position of an object determined on the base of its last observed position and of its trajectory, counted by means of the on-board measurement equipment. As an example, let us consider a procedure serving for determination the counted position of a submarine, which, at the emergency time, is located at point $P_0 \equiv (x_0, y_0, z_0 = 0)$, as shown in Fig. 5.12.

Assume now that this observed position was determined for the fixed moment of time $t = t_0$, on the basis of data received from Navy Navigation Satellite System (NNSS – TRANSIT) or GPS systems. At emergence time, measurement of the temperature and atmospheric pressure at the water surface is also performed. To simplify our further considerations, let us assume that the point P_0 is identical to the origin of a dextrose Cartesian coordinate system (x, y, z), in which the x-axis points to the north (N). Every one hour the actual position is recorded in the deck log. This position is being determined each time in the full submergence conditions based on the following data:

- last observed position,
- north direction indicated by the gyrocompass,

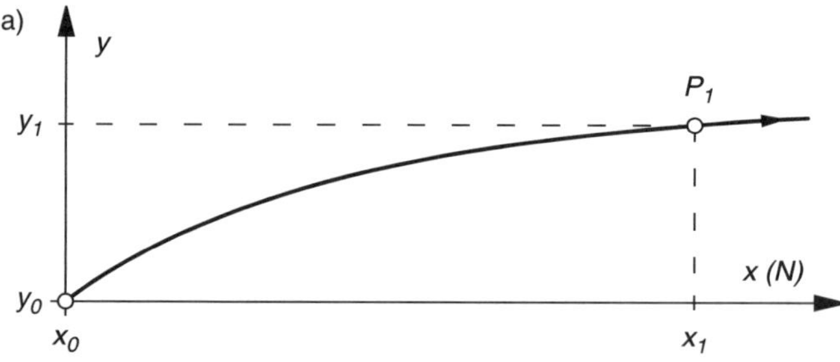

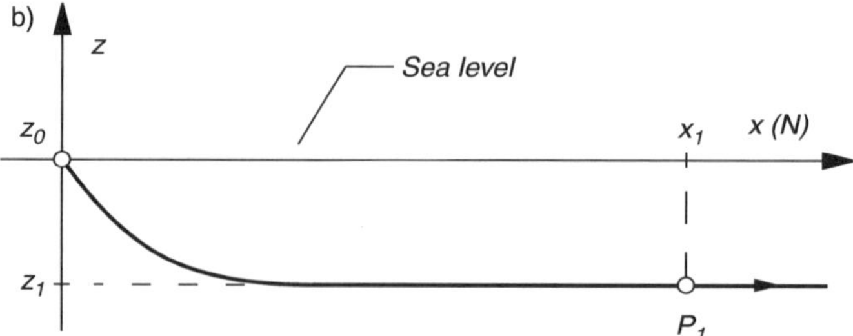

Fig. 5.12

- vertical direction indicated by the gravity force,
- indications of the vector speedometer,
- indications of the vector accelerometer, and
- indication of the board clock.

On the basis of indications of the on-board instruments mentioned above we form the sets of discrete values of the speed vector components, that is:

$$v_x(t_i) = v_x(t_0 + i \cdot \Delta t)$$

$$v_y(t_i) = v_y(t_0 + i \cdot \Delta t) \tag{5.75}$$

$$v_z(t_i) = v_z(t_0 + i \cdot \Delta t)$$

where $i = 1, 2, 3, \ldots, n$, and Δt is a small interval determining the sampling frequency. These values serve to determine (by interpolation) the functions $v_x(t)$, $v_y(t)$, $v_z(t)$, continuous in the time interval $t_0 \leq t \leq t_1 = t_0 + n \cdot \Delta t$. The coordinates (x_1, y_1, z_1) of the first counted position are equal to:

$$x_1 = x_0 + \int_{t_0}^{t_1} v_x(t)dt$$

$$y_1 = y_0 + \int_{t_0}^{t_1} v_y(t)dt \tag{5.76}$$

$$z_1 = z_0 + \int_{t_0}^{t_1} v_z(t)dt$$

$$t_1 = t_0 + n \cdot \Delta t$$

Integrals (5.57) are usually computed by means of one of the methods presented earlier in this chapter. In the full draught time only one of the desired coordinates can be experimentally verified. It is not difficult to guess that we are interested in the z_1-coordinate, because it defines the draught depth. The draught depth $h \equiv z_1$ with respect to the water surface can be found on the base of hydrostatic pressure measurement. Hydrostatic pressure, at a certain depth h, becomes greater than the pressure on the water surface, and the difference is equal to $p = g \cdot \rho \cdot h$, where g is the earth gravity known for this sea area, and ρ is a given (measured) water density. Measurement of this pressure is performed by means of mechanical instrument whose operation principle is similar to that of an aneroid, broadly used in meteorology. Each consecutive counted position is accompanied by the greater error than the previous one, and it is therefore necessary to correct this position as quickly as possible by determining the next observed position.

References

1. Abramowitz M. and I.A. Stegun, Handbook of mathematical functions. Dover, New York 1954
2. Forsythe G.E., Malcolm M.A. and C.B. Moler, Computer methods for mathematical computations. Prentice-Hall, Englewood Cliffs, NJ, 1977
3. Demidovitch B.P. and I. A. Maron, Fundamentals of numerical mathematics (in Russian). Publishing house "Science", Moscow, 1970
4. Mathews J.H., Numerical methods for mathematics, science and engineering. Prentice-Hall Intern. Inc., Englewood Cliffs, NJ, 1992
5. Shoup T.E., Applied numerical methods for the microcomputer. Prentice-Hall, Inc., Englewood Cliffs, NJ, 1984
6. Young D.M. and R.T. Gregory, A survey of numerical mathematics. Addison-Wesley Comp., London, 1973
7. Turchak L.I. and P.W. Plotnikov, Fundamentals of numerical methods (2nd edition in Russian). Publishing house "Physmathlit", Moscow, 2002
8. Tuma J.J., Handbook of numerical calculations in engineering. McGraw-Hill, New York, 1989

Chapter 6
Numerical Differentiation of One and Two Variable Functions

A moment of reflection on the usefulness of function derivatives in one or many variables for natural and technical sciences would certainly make us aware of the importance of problems considered in the present chapter. Although formulation of clear and exhaustive conclusions on this subject is not an easy task, it induces us to a reflection that introduction of derivatives made in great extent possible and accelerated the development of science and technology. It is not necessary to convince anybody that derivatives still play a fundamental role, because they offer mathematical, rigorous manner of describing dynamics of various changes taking place in isolated environments or in technical devices. For example, the instant speed $v(t)$ is the first derivative of function $L(t)$ describing relation of the path L to time t. The time derivative of instant speed is in turn a measure of acceleration. Acceleration is therefore, mathematically speaking, the second derivative of the path $L(t)$ with respect to time t. As another example, we analyze changes in time of the charge $Q(t)$ stored on the plates of a capacitor. The current flowing through this capacitor is proportional to the first derivative of function $Q(t)$. We may give numerous examples of this kind, because all processes which we meet in our environment are dynamic processes representing changes in function with respect to time. In this context the stationary process, understood as a process in which there are no changes, is a purely abstract process, nonexisting in real-world. Moreover, the very notion of time and its scale is difficult to define for such fictitious process. In order to specify our future considerations, let us assume that a real process is described by function $f(x)$, bounded and continuous over an interval $[a, b]$. The first derivative of the function $f(x)$, defined at an arbitrary point x_0 of this interval, is given by the following limit:

$$\frac{df(x)}{dx} = f'(x_0) = \lim_{\Delta x \to 0} \frac{f(x_0 + \Delta x) - f(x_0)}{\Delta x} \tag{6.1}$$

in which an increment Δx can be positive or negative. We therefore have to do with right-sided or left-sided limits, respectively. If both limits are equal then the function $f'(x)$ is continuous for $x = x_0$. Derivatives of many elementary functions such as polynomials, exponential, logarithmic, hyperbolic, trigonometric functions as well

S. Rosłoniec, *Fundamental Numerical Methods for Electrical Engineering,*
© Springer-Verlag Berlin Heidelberg 2008

as their various combinations can be in most cases described by similar kinds of functions, as we may see looking at some examples given below.

Function	First derivative

$$x^n \qquad\qquad \frac{d}{dx}x^n = nx^{n-1}$$

$$e^x \qquad\qquad \frac{d}{dx}e^x = e^x$$

$$\ln(x),\ x \neq 0 \qquad\qquad \frac{d}{dx}\ln(x) = \frac{1}{x}$$

$$ch(x) = \frac{1}{2}(e^x + e^{-x}) \qquad\qquad \frac{d}{dx}[ch(x)] = sh(x)$$

$$sh(x) = \frac{1}{2}(e^x - e^{-x}) \qquad\qquad \frac{d}{dx}[sh(x)] = ch(x)$$

$$th(x) = \frac{e^x - e^{-x}}{e^x + e^{-x}} \qquad\qquad \frac{d}{dx}[th(x)] = \frac{1}{ch^2(x)}$$

$$\cos(x) \qquad\qquad \frac{d}{dx}\cos(x) = -\sin(x)$$

$$\sin(x) \qquad\qquad \frac{d}{dx}\sin(x) = \cos(x)$$

$$tg(x) = \frac{\sin(x)}{\cos(x)},\ \cos(x) \neq 0 \qquad \frac{d}{dx}tg(x) = \frac{1}{\cos^2(x)}$$

$$(6.2)$$

Second derivatives of this functions can be determined according to the following general rule of differential calculus:

$$\frac{d^2}{dx^2}f(x) = \frac{d}{dx}\left[\frac{d}{dx}f(x)\right] \tag{6.3}$$

In most cases, the expressions found in this way are also combinations of elementary functions mentioned above. For example

$$\frac{d^2}{dx^2}ch(x) = \frac{d}{dx}sh(x) = \frac{d}{dx}\left[\frac{1}{2}(e^x - e^{-x})\right] = \frac{1}{2}[e^x - (-e^{-x})] = ch(x) \tag{6.4}$$

Some more compound functions can also be written in the form of elementary function series containing finite or infinite number of terms. A classical example of such functions, broadly used in electrodynamics and for description of the angle modulated signals, are the Bessel functions of the first kind [1, 2].

$$J_n(x) = \left[\frac{(x/2)^n}{n!} - \frac{(x/2)^{n+2}}{1!(n+1)!} + \frac{(x/2)^{n+4}}{2!(n+2)!} - \frac{(x/2)^{n+6}}{3!(n+3)!} + \cdots\right] \tag{6.5}$$

where $n = 0, 1, 2, 3\ldots$ and x is the real positive number. First derivatives of these functions are obtained by differentiating series (6.5) term after term, obtaining the recursive formula:

$$\frac{d}{dx}[J_n(x)] = \frac{1}{2}\left[\frac{n(x/2)^{n-1}}{n!} - \frac{(n+2)(x/2)^{n+1}}{1!(n+1)!} + \frac{(n+4)(x/2)^{n+3}}{2!(n+2)!} - \cdots\right]$$

$$= \frac{1}{2}\frac{(x/2)^{n-1}}{n!}\left[n - \frac{(n+2)(x/2)^2}{1!(n+1)} + \frac{(n+4)(x/2)^4}{2!(n+1)(n+2)} - \cdots\right] \quad (6.6)$$

Second derivatives of the Bessel functions (6.5) are in turn calculated using principle (6.3), that is by differentiating all terms of the series (6.6)

$$\frac{d^2}{dx^2}[J_n(x)] = \frac{1}{4}\left[\frac{n(n-1)(x/2)^{n-2}}{n!} - \frac{(n+2)(n+1)(x/2)^n}{1!(n+1)!}\right.$$

$$\left. + \frac{(n+4)(n+3)(x/2)^{n+2}}{2!(n+2)!} - \cdots\right] \quad (6.7)$$

The above considerations would help us to remind the basic principles of analytical calculating derivatives of one variable functions. They confirm implicitly an important conclusion saying that analytical method is the most appropriate form of calculation of derivatives, and that numerical methods, used to obtain approximate solutions should be used only in cases, when it cannot be avoided.

6.1 Approximating the Derivatives of One Variable Functions

Assume that we define the function $f(x)$, over an interval $[a, b] \equiv [x_0, x_n]$, for which we are not able to determine the derivatives analytically, i.e., in the manner described above. In such cases, it is necessary to determine the approximate values of these derivatives according to the formulas given below. For this end, we divide the interval $[a, b] \equiv [x_0, x_n]$ into n identical small subintervals of length $h = (b - a)/n$. Thus we have constructed a set of points.

$$x_i = x_0 + i \cdot h \quad (6.8)$$

where: $i = 0, 1, 2, 3\ldots$, see Fig. 6.1.

For each point of the set (6.8) we assign a value of the function

$$f_i = f(x_i) = f(x_0 + i \cdot h) \quad (6.9)$$

First derivative of the function $f(x)$, defined at an arbitrary point x_i of the set (6.8), can be approximated by means of the following quotients of finite differences

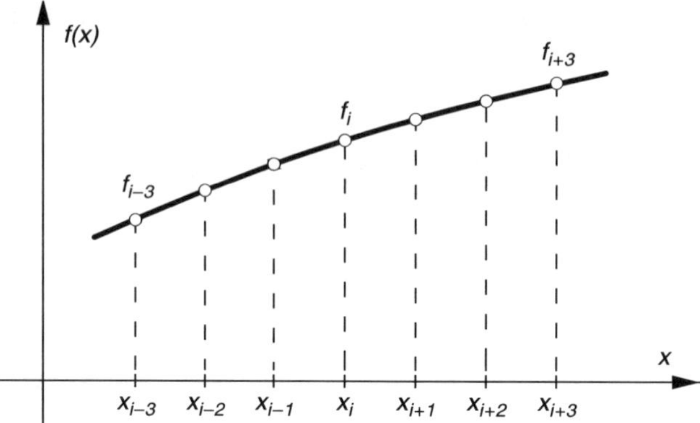

Fig. 6.1

– Left-sided approximation

$$f_l' \approx \frac{f(x_i - h) - f(x_i)}{-h} = \frac{f_{i-1} - f_i}{-h} \tag{6.10}$$

– Right-sided approximation

$$f_p' \approx \frac{f(x_i + h) - f(x_i)}{h} = \frac{f_{i+1} - f_i}{h} \tag{6.11}$$

– Two-sided (central) approximation

$$f_c' \approx \frac{f(x_i + h) - f(x_i - h)}{2h} = \frac{f_{i+1} - f_{i-1}}{2h} \tag{6.12}$$

Magnitude of the error emerging during numerical differentiation of the function $f(x)$, according to the formula (6.12), can be evaluated developing this function in the Taylor series for $\Delta x = h$ and $\Delta x = -h$. In case when $\Delta x = h$, we obtain the formula:

$$f(x_i + h) = f(x_i) + f^{(1)}(x_i)h + \frac{1}{2!}f^{(2)}(x_i)h^2 + \sum_{k=3}^{\infty} \frac{1}{k!}f^{(k)}(x_i)h^k \tag{6.13}$$

where $f^{(k)}(x_i)$ is the derivative of order k, calculated for $x = x_i$. The series (6.13) can be written in the form:

$$f(x_i + h) = f(x_i) + f^{(1)}(x_i)h + \frac{1}{2!}f^{(2)}(x_i)h^2 + O^{(+)}(h^3) \tag{6.14}$$

where

$$O^{(+)}(h^3) = \sum_{k=3}^{\infty} \frac{1}{k!} f^{(k)}(x_i) h^k = \frac{h^3}{3!} f^{(3)}(c_1), \quad c_1 \in [x_i, x_i + h]$$

is the truncation error [3, 4]. Similarly, for $\Delta x = -h$

$$f(x_i - h) = f(x_i) - f^{(1)}(x_i)h + \frac{1}{2!} f^{(2)}(x_i)h^2 + O^{(-)}(h^3) \qquad (6.15)$$

where

$$O^{(-)}(h^3) = \sum_{k=3}^{\infty} (-1)^k \frac{1}{k!} f^{(k)}(x_i) h^k = \frac{h^3}{3!} f^{(3)}(c_2), \quad c_2 \in [x_i, x_i - h]$$

is also the truncation error. Taking the difference of the series (6.14) and (6.15), we obtain the following equation:

$$f(x_i + h) - f(x_i - h) = 2f^{(1)}(x_i)h + \left[O^{(+)}(h^3) - O^{(-)}(h^3) \right]$$

which can be written in the form:

$$f^{(1)}(x_i) = \frac{f(x_i + h) - f(x_i - h)}{2h} - \frac{1}{2} \left[O^{(+)}(h^2) - O^{(-)}(h^2) \right] \qquad (6.16)$$

The comparison of expressions (6.12) and (6.16) shows that the first derivative calculated according to the formula (6.12) is loaded by an error of the second order $O(h^2)$. Similarly, we can easily prove that the approximation error following from the application of Eqs. (6.10) and (6.11) is the first order quantity $O(h)$. In order to obtain a formula serving for approximate computing of second derivative of the function $f(x)$ we add side-by-side the series (6.14) and (6.15). In consequence, we obtain the following expression:

$$f(x_i + h) + f(x_i - h) = 2f(x_i) + f^{(2)}(x_i)h^2 + O^{(+)}(h^4) + O^{(-)}(h^4) \qquad (6.17)$$

where:

$$O^{(+)}(h^4) = \sum_{k=4}^{\infty} \frac{1}{k!} f^{(k)}(x_i) h^k = \frac{h^4}{4!} f^{(4)}(c_1), \quad c_1 \in [x_i, x_i + h]$$

$$O^{(-)}(h^4) = \sum_{k=4}^{\infty} (-1)^k \frac{1}{k!} f^{(k)}(x_i) h^k = \frac{h^4}{4!} f^{(4)}(c_2), \quad c_2 \in [x_i, x_i - h]$$

are truncation errors. Equation (6.17) can be represented in an equivalent form:

$$f^{(2)}(x_i) = \frac{f(x_i + h) - 2f(x_i) + f(x_i - h)}{h^2} + \left[O^{(+)}(h^2) + O^{(-)}(h^2)\right]$$

which after neglecting the truncation errors is the desired difference formula:

$$f^{(2)}(x_i) \approx \frac{f(x_i + h) - 2f(x_i) + f(x_i - h)}{h^2} = \frac{f_{i-1} - 2f_i + f_{i+1}}{h^2} \qquad (6.18)$$

Approximation error of second derivative of the function $f(x)$, emerging when using the formula (6.18), is the second order quantity $O(h^2)$. The expressions (6.12) and (6.18), which we have just found, are called in the English language literature the central difference formulas of the second order. First and second derivatives of the considered function $f(x)$ can be calculated more accurately by means of the following, however more complicated expressions:

$$f^{(1)}(x_i) \approx \frac{-f_{i+2} + 8f_{i+1} - 8f_{i-1} + f_{i-2}}{12h} \qquad (6.19)$$

$$f^{(2)}(x_i) \approx \frac{-f_{i+2} + 16f_{i+1} - 30f_i + 16f_{i-1} - f_{i-2}}{12h^2} \qquad (6.20)$$

called central difference formulas of the fourth order. In order to develop the first of them and to evaluate the corresponding approximation error, let us develop the functions $f(x_i + h)$, $f(x_i - h)$, $f(x_i + 2h)$, $f(x_i - 2h)$ in the Taylor series, taking into account only the first six terms. First two functions are approximated by the series:

$$f(x_i + h) \approx f(x_i) + f^{(1)}(x_i)h + \frac{1}{2!}f^{(2)}(x_i)h^2 + \frac{1}{3!}f^{(3)}(x_i)h^3 + \frac{1}{4!}f^{(4)}(x_i)h^4$$
$$+ \frac{1}{5!}f^{(5)}(x_i)h^5$$

$$f(x_i - h) \approx f(x_i) - f^{(1)}(x_i)h + \frac{1}{2!}f^{(2)}(x_i)h^2 - \frac{1}{3!}f^{(3)}(x_i)h^3 + \frac{1}{4!}f^{(4)}(x_i)h^4$$
$$- \frac{1}{5!}f^{(5)}(x_i)h^5$$

which make possible to write an expression:

$$8[f(x_i + h) - f(x_i - h)] \approx 16f^{(1)}(x_i)h + \frac{16}{3!}f^{(3)}(x_i)h^3 + \frac{16}{5!}f^{(5)}(x_i)h^5 \qquad (6.21)$$

Proceeding in the similar way, one can prove that:

$$f(x_i + 2h) - f(x_i - 2h) \approx 4f^{(1)}(x_i)h + \frac{16}{3!}f^{(3)}(x_i)h^3 + \frac{64}{5!}f^{(5)}(x_i)h^5 \qquad (6.22)$$

Subtracting expressions (6.21) and (6.22) side-by-side we obtain:

$$8[f(x_i + h) - f(x_i - h)] - [f(x_i + 2h) - f(x_i - 2h)] \approx 12 f^{(1)}(x_i) h$$

$$- 0.4 f^{(5)}(x_i) h^5 = 12 f^{(1)}(x_i) h + O(h^5) \tag{6.23}$$

After rearranging and dividing both sides by the term h, expression (6.23) takes the form:

$$f^{(1)}(x_i) \approx \frac{8 f(x_i + h) - 8 f(x_i - h) - f(x_i + 2h) + f(x_i - 2h)}{12h} + O(h^4) \tag{6.24}$$

which, neglecting the truncation error $O(h^4)$, is identical to the formula (6.19). Similarly, we develop the Eq. (6.20) and formulas for calculating third order derivatives, given in Tables 6.1 and 6.2, [4].

Example 6.1 Using Eqs. (6.18) and (6.20), approximate values of the second derivative for function $f(x) = e^x - x^2/2$, where $x = 0.5$, were determined. Calculations were performed for four different values of $\Delta x \equiv h$ equal to $0.2, 0.1, 0.01$ and 0.001, respectively. These obtained results, see the second and fourth columns of Table 6.3, were compared with exact value $f^{(2)}(0.5) = e^{0.5} - 1 = 0.648721271$.

As it follows from our above discussion, one of the means for obtaining precision increase when using differential formulas in derivative computing, is the increase of the number n of function values $f_i = f(x_i)$ used in these formulas. Yet, with

Table 6.1

Central difference formulas of the second order, $O(h^2)$

$$f^{(1)}(x_i) \approx \frac{f(x_i + h) - f(x_i - h)}{2h} = \frac{f_{i+1} - f_{i-1}}{2h}$$

$$f^{(2)}(x_i) \approx \frac{f(x_i + h) - 2f(x_i) + f(x_i - h)}{h^2} = \frac{f_{i+1} - 2f_i + f_{i-1}}{h^2}$$

$$f^{(3)}(x_i) \approx \frac{f(x_i + 2h) - 2f(x_i + h) + 2f(x_i - h) - f(x_i - 2h)}{2h^3}$$

$$= \frac{f_{i+2} - 2f_{i+1} + 2f_{i-1} - f_{i-2}}{2h^3}$$

Table 6.2

Central difference formulas of the fourth order, $O(h^4)$

$$f^{(1)}(x_i) \approx \frac{8f(x_i + h) - 8f(x_i - h) - f(x_i + 2h) + f(x_i - 2h)}{12h}$$

$$= \frac{-f_{i+2} + 8f_{i+1} - 8f_{i-1} + f_{i-2}}{12h}$$

$$f^{(2)}(x_i) \approx \frac{-f(x_i + 2h) + 16f(x_i + h) - 30f(x_i) + 16f(x_i - h) - f(x_i - 2h)}{12h^2}$$

$$= \frac{-f_{i+2} + 16f_{i+1} - 30f_i + 16f_{i-1} - f_{i-2}}{12h^2}$$

$$f^{(3)}(x_i) \approx \frac{-f(x_i + 3h) + 8f(x_i + 2h) - 13f(x_i + h) + 13f(x_i - h) - 8f(x_i - 2h) + f(x_i - 3h)}{8h^3}$$

$$= \frac{-f_{i+3} + 8f_{i+2} - 13f_{i+1} + 13f_{i-1} - 8f_{i-2} + f_{i-3}}{8h^3}$$

Table 6.3

Step h	Formula (6.18) $O(h^2)$	Error of the formula (6.18)	Formula (6.20) $O(h^4)$	Error of the formula (6.20)
0.2	0.654224341	5.503×10^{-3}	0.648691855	-2.941×10^{-5}
0.1	0.650095663	1.374×10^{-3}	0.648719437	-1.834×10^{-6}
0.01	0.648735010	1.374×10^{-5}	0.648721270	$\approx 1 \times 10^{-9}$
0.001	0.648721408	1.370×10^{-7}	0.648721270	$\approx 1 \times 10^{-9}$

an increase in the number n, these expressions become more complicated. It leads directly to the increase of the amount of computations and of the processing error involved. In practical cases, we take most frequently $n \leq 4$, but more accurate approximation of the derivative may be achieved using Runge or Romberg procedures [4, 5]. In order to explain the essence of the Runge procedure, assume that the derivative $f(x)$ is approximated by means of a differential expression $f(x, h)$ where $h \equiv \Delta x$ is a fixed calculation step. Let R be the approximation error, for which the principal term can be written in the form $h^p \phi(x)$, that is:

$$R = h^p \phi(x) + 0(h^{p+1}) \tag{6.25}$$

where p is the precision order of the given differential formula. With this assumption we can write:

$$f(x) = f(x, h) + h^p \phi(x) + 0(h^{p+1}) \tag{6.26}$$

Derivative (6.26) written for the different step $h_1 = k \cdot h$ has the form:

$$\begin{aligned}
f(x) &= f(x, k \cdot h) + (k \cdot h)^p \phi(x) + 0[(k \cdot h)^{p+1}] \\
&= f(x, k \cdot h) + k^p \cdot h^p \phi(x) + k^{p+1} 0(h^{p+1})
\end{aligned} \tag{6.27}$$

Subtracting expressions (6.26) and (6.27) side-by–side, we obtain principal term of approximation error:

$$h^p \phi(x) = \frac{f(x, h) - f(x, k \cdot h)}{k^p - 1} - \frac{1 - k^{p+1}}{1 - k^p} 0(h^{p+1}) \tag{6.28}$$

After introducing Eq. (6.28) into (6.26) we obtain the following formula:

$$f(x) = f(x, h) + \frac{f(x, h) - f(x, k \cdot h)}{k^p - 1} + 0'(h^{p+1}) \tag{6.29}$$

known in the literature as the Runge extrapolation formula [5]. It makes possible more accurate calculation of the derivative based on the calculation results obtained by means of chosen differential formulas for two different steps, namely h and $k \cdot h$. The effective order of approximation precision obtained in this way is greater $(p+1)$ than precision order (p) of the differential formula used above by one. Another example of similarly "constructed" two-step extrapolation procedure is the Richardson

Table 6.4

x	1.8	1.9	2.0	2.1	2.2
$F(x)$	5.832	6.859	8.000	9.261	10.648

procedure [6]. Among other procedures, for which the number of steps is greater than two ($h_1, h_2, h_3, .., h_q$), the most popular is the Romberg procedure.

Example 6.2 In Table 6.4, some discrete values of function $F(x) = x^3$ are given, for which the derivative $f(x) = 3x^2$ for $x = 2$ takes the value $f(2) = 12.00$.

Central approximations for two values of the derivative $f(x, h)$, calculated according to the formula (6.12), where $x = 2, h = 0.1$ and $h_1 = 2 \cdot h = 0.2$ are equal to:

$$f(2, 0.1) = \frac{9.261 - 6.859}{2 \cdot 0.1} = 12.01, \quad f(2, 0.2) = \frac{10.648 - 5.832}{2 \cdot 0.2} = 12.04$$

Differential formula (6.12) is the second order expression ($p = 2$). Introducing these numbers into the Runge formula (6.29) we get the following, more accurate approximation of the derivative

$$f(2) = 12.01 + \frac{12.01 - 12.04}{2^2 - 1} = 12.00$$

which in this case is equal to the exact value obtained analytically.

6.2 Calculating the Derivatives of One Variable Function by Differentiation of the Corresponding Interpolating Polynomial

Essence of the algorithms serving for determination of one variable function derivative, introduced in the present section, may be found simply when reading its title. First, the function $f(x)$ considered above, as shown in Eqs. (6.8) and (6.9), is interpolated by one of the polynomials introduced in Chap. 4 or by their linear combination, which has the form of a spline function.

6.2.1 Differentiation of the Newton–Gregory Polynomial and Cubic Spline Functions

As first example of interpolating function used for calculation of derivatives, let us consider the Newton–Gregory polynomial described in Sect. 4.1.3. This polynomial, for $x_i < x < x_{i+1}$ has the following form:

$$N(x_i + t \cdot h) = f_i + t\Delta f_i + \frac{t(t-1)}{2!}\Delta^2 f_i + \frac{t(t-1)(t-2)}{3!}\Delta^3 f_i + \ldots$$
$$\ldots\ldots\ldots + \frac{t(t-1)(t-2)\ldots(t-n+1)}{n!}\Delta^n f_i \tag{6.30}$$

where $t = (x - x_i)/h$ and $\Delta f_i, \Delta^2 f_i, \Delta^3 f_i, \Delta^4 f_i, \ldots$ are finite differences of the degree i. Polynomial (6.30) is known in the literature as the first Newton–Gregory forward interpolation polynomial. As a rule, it is used to calculate values of the function at points lying in the left-half of the interpolation interval $[x_0, x_n]$. This circumstance can be justified in the following way. Finite differences $\Delta^m f_i$ can be found using the values $f_i, f_{i+1}, f_{i+2}, f_{i+3}, \ldots, f_{i+m}$, with $i + m \le n$. For i close to n, finite differences of higher orders are not calculated. For example, if $i = n - 3$, only the differences $\Delta f_i, \Delta^2 f_i, \Delta^3 f_i$ are present in the polynomial (6.30). If our task is to determine values of the function at points belonging to the right-half of the interpolation interval $[x_0, x_n]$, it is recommended to use the polynomial

$$N(x_n + th) = f_n + t\Delta f_{n-1} + \frac{t(t+1)}{2!}\Delta^2 f_{n-2} + \frac{t(t+1)(t+2)}{3!}\Delta^3 f_{n-3} + \ldots$$
$$\ldots\ldots\ldots + \frac{t(t+1)(t+2)\ldots(t+n-1)}{n!}\Delta^n f_0 \tag{6.31}$$

defined for $t = (x - x_n)/h \le 0$ [7]. This form of a polynomial is called second Newton–Gregory backward interpolation polynomial. Differentiating polynomial (6.30) with respect to the variable x lying in the subinterval $[x_i, x_{i+1}]$, we obtain approximate value of the first derivative of the interpolated function $f(x)$:

$$f^{(1)}(x) \approx \frac{dN(x)}{dx} = \frac{1}{h} \cdot \frac{dN}{dt} \approx \frac{1}{h}\left(\Delta f_i + \frac{2t-1}{2!}\Delta^2 f_i + \frac{3t^2 - 6t + 2}{3!}\Delta^3 f_i\right.$$
$$+ \frac{4t^3 - 18t^2 + 22t - 6}{4!}\Delta^4 f_i + \frac{5t^4 - 40t^3 + 105t^2 - 100t + 24}{5!}\Delta^5 f_5 + \ldots\Big) \tag{6.32}$$

Second derivative of the interpolated function $f(x)$ is:

$$f^{(2)}(x) \approx \frac{1}{h^2}(\Delta^2 f_i + \frac{6t-6}{3!}\Delta^3 f_i + \frac{12t^2 - 36t + 22}{4!}\Delta^4 f_i$$
$$+ \frac{20t^3 - 120t^2 + 210t - 100}{5!}\Delta^5 f_5 + \ldots) \tag{6.33}$$

Similar relations can be obtained differentiating the interpolation polynomial (6.31) with respect to x.

Example 6.3 For the function given in the first and second columns of Table 6.5, calculate the approximate values of first and second derivatives at $x = 0.05$. From

Table 6.5

x_i	$f(x_i)$	Δf_i	$\Delta^2 f_i$	$\Delta^3 f_i$	$\Delta^4 f_i$	$\Delta^5 f_i$
0.0	1.000000					
		0.205171				
0.1	1.205171		0.011060			
		0.216231		1.163×10^{-3}		
0.2	1.421402		0.012224		1.223×10^{-4}	
		0.228456		1.285×10^{-3}		1.287×10^{-5}
0.3	1.649858		0.013509		1.352×10^{-4}	
		0.241965		1.421×10^{-3}		
0.4	1.891824		0.014930			
		0.256896				
0.5	2.148721					

Table 6.5, we find that $h = 0.1$. Therefore, $t = (x - x_0)/h = (0.05 - 0.00)/0.1 = 0.5$. Using Eqs. (6.32), (6.33) and Table 6.5 we can write:

$$f^{(1)}(0.05) \approx \frac{1}{0.1}\Big[0.205171 + \frac{2 \cdot 0.5 - 1}{2}0.011060$$
$$+ \frac{3 \cdot (0.5)^2 - 6 \cdot 0.5 + 2}{6}1.163 \cdot 10^{-3}$$
$$+ \frac{4 \cdot (0.5)^3 - 18 \cdot (0.5)^2 + 22 \cdot 0.5 - 6}{24}1.223 \cdot 10^{-4}$$
$$+ \frac{5 \cdot (0.5)^4 - 40 \cdot (0.5)^3 + 105 \cdot (0.5)^2 - 100 \cdot 0.5 + 24}{120}1.287 \cdot 10^{-5}$$
$$+ \dots\Big] = 2.051271$$

$$f^{(2)}(0.05) \approx \frac{1}{(0.1)^2}\Big[0.011060 + \frac{6 \cdot 0.5 - 6}{6}1.163 \cdot 10^{-3}$$
$$+ \frac{12 \cdot (0.5)^2 - 36 \cdot 0.5 + 22}{24}1.223 \cdot 10^{-4}$$
$$+ \frac{20 \cdot (0.5)^3 - 120 \cdot (0.5)^2 + 210 \cdot 0.5 - 100}{120}1.287 \cdot 10^{-5}$$
$$+ \dots\Big] = 1.051175$$

Example 6.4 Recalculate the first and second derivatives of the function analyzed in Example 6.3 at point $x = 0.15$. In this case, we obtain: $t = (x - x_1)/h = (0.15 - 0.10)/0.1 = 0.5$

$$f^{(1)}(0.15) \approx \frac{1}{0.1}\Big[0.216231 + \frac{2 \cdot 0.5 - 1}{2}0.012224$$
$$+ \frac{3 \cdot (0.5)^2 - 6 \cdot 0.5 + 2}{6}1.285 \cdot 10^{-3}$$

$$\left. + \frac{4 \cdot (0.5)^3 - 18 \cdot (0.5)^2 + 22 \cdot 0.5 - 6}{24} 1.352 \cdot 10^{-4} \right.$$

$$+ \frac{5 \cdot (0.5)^4 - 40 \cdot (0.5)^3 + 105 \cdot (0.5)^2 - 100 \cdot 0.5 + 24}{120} 0 \right] = 2.161831$$

$$f^{(2)}(0.15) \approx \frac{1}{(0.1)^2}\left[0.012224 + \frac{6 \cdot 0.5 - 6}{6} 1.285 \cdot 10^{-3} \right.$$

$$\left. + \frac{12 \cdot (0.5)^2 - 36 \cdot 0.5 + 22}{24} 1.352 \cdot 10^{-4} + 0 + \dots \right] = 1.162093$$

The derivatives calculated in the Examples 6.3 and 6.4 are very close to corresponding values evaluated analytically. It has been possible because values of $f(x_i)$ given in Table 6.6 are only discrete values of function $f(x) = \exp(x) + x$. This calculation examples confirm the conclusion established in the literature that the method discussed above reveals good precision, sufficient for practical applications, and has unsophisticated computation algorithm. These unquestionable advantages were very important in the past "pre-computer" times. We must however remember that the degree n of the interpolation polynomial, regardless of its form (power, Lagrange or Newton–Gregory), increases with the number of nodes, and the derivatives calculated on the basis of the polynomial of high degree can be charged with considerable errors. This remark only to a small extent refers to the spline function, see Sect. 4.4. Each polynomial of this function has the form:

$$q_i(x) = k_{i0} + k_{i1}x + k_{i2}x^2 + k_{i3}x^3 \tag{6.34}$$

where: $i = 1, 2, 3$, $x_{i-1} \leq x \leq x_i$, and $k_{i0}, k_{i1}, k_{i2}, k_{i3}$ are fixed coefficients. Determination of derivatives by means of the trinomials (6.34) is an elementary operation, which need not be explained further. Differentiation of the polynomial (6.35) given below, serving for the same purpose, is a little bit more difficult task [3, 5]:

$$\begin{aligned} q_i\left[t(x)\right] &= t \cdot f_i + \bar{t} \cdot f_{i-1} + \Delta x_i \left[(k_{i-1} - d_i) \cdot t \cdot \bar{t}^2 - (k_i - d_i) \cdot t^2 \cdot \bar{t}\right] \\ &= t \cdot f_i + (1 - t)f_{i-1} + \Delta x_i \left[(k_{i-1} - d_i)(t^3 - 2t^2 + t)\right] \\ &\quad - \Delta x_i \left[(k_i - d_i)(t^2 - t^3)\right] \end{aligned} \tag{6.35}$$

Table 6.6

i	0	1	2	3
x_i	1.0	3.0	5.0	7.0
$f(x_i)$	2.0	3.5	3.8	3.0

where

$$\Delta x_i \equiv h = x_i - x_{i-1}, \quad \Delta f_i = f_i - f_{i-1}, \quad d_i = \frac{\Delta f_i}{\Delta x_i}, \quad t = \frac{x - x_{i-1}}{\Delta x_i}, \quad \bar{t} = 1 - t.$$

First derivative of the trinomial (6.35) with respect to the variable x is:

$$\frac{dq_i[t(x)]}{dx} = \frac{dq_i(t)}{dt} \cdot \frac{dt}{dx} = \frac{1}{\Delta x_i} \cdot \frac{dq_i(t)}{dt}$$
$$= \frac{1}{\Delta x_i}(f_i - f_{i-1}) + (k_{i-1} - d_i)(3t^2 - 4t + 1) - (k_i - d_i)(2t - 3t^2)$$

$$(6.36)$$

Differentiating the expression (6.36) with respect to the variable x, we obtain the formula used for calculating the second derivative, namely:

$$\frac{d^2 q_i[t(x)]}{dx^2} = \frac{2}{(\Delta x_i)^2}[(k_{i-1} - d_i)(3t - 2) + (k_i - d_i)(3t - 1)] \qquad (6.37)$$

Example 6.5 In Table 6.6 some discrete values of the function $f(x)$ interpolated by a spline function $Q(x)$ composed of three trinomials of the type (6.35) are given.

Coefficients k_i, where $i = 0, 1, 2$ and 3, appearing in these trinomials are equal to $k_0 = 0.855555$, $k_1 = 0.538889$, $k_2 = -0.311111$ and $k_3 = -0.044444$. Chosen values of the function $Q(x) \approx f(x)$ and its derivatives $dQ(x)/dx$ and $d^2 Q(x)/dx^2$ are given in Table 6.7.

Table 6.7

x	$Q(x) \approx f(x)$	$dQ(x)/dx$	$d^2 Q(x)/dx^2$
1.0	2.0	0.855555	$< 10^{-6}$
1.5	2.424479	0.835763	−0.039583
2.0	2.829166	0.776389	−0.079166
2.5	3.194270	0.677430	−0.118750
2.9	3.444554	0.569764	−0.150416
3.0	3.5	0.538889	−0.158333
3.1	3.552287	0.506680	−0.163750
3.5	3.727604	0.367013	−0.185416
4.0	3.862500	0.168055	−0.212500
4.5	3.891145	−0.057986	−0.239558
4.9	3.828462	−0.258319	−0.261249
5.0	3.8	−0.311111	−0.266666
5.1	3.766333	−0.361111	−0.233333
5.5	3.591666	−0.494444	−0.100000
6.0	3.333333	−0.511111	0.066666
6.5	3.108333	−0.361111	0.233336
7.0	3.0	−0.044444	0.400000

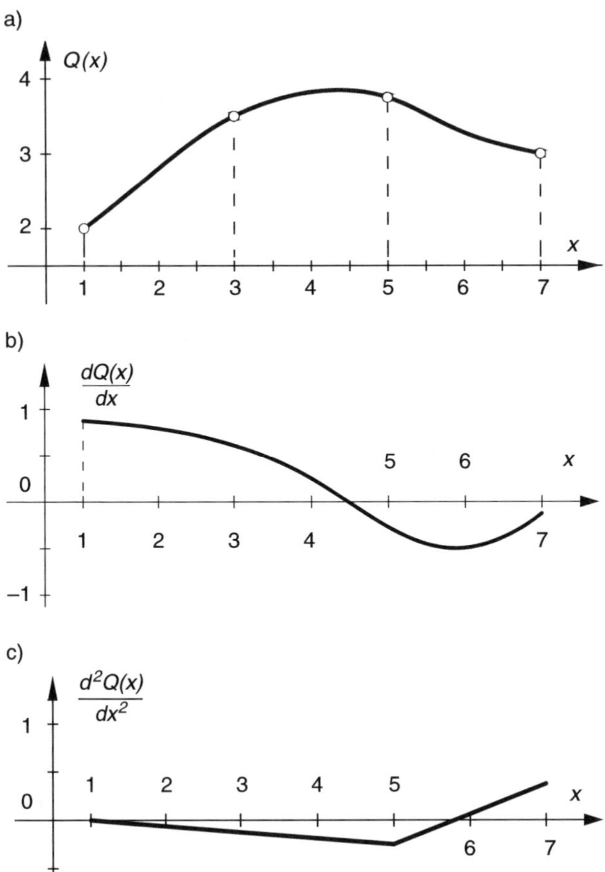

Fig. 6.2

Remaining results of the calculations have been used to draw up curves of these functions, which are shown in Fig. 6.2.

From analysis of the derivatives shown in Fig. 6.2(b, c) it follows that they are continuous in the internal interpolation points, the fact which confirms correctness of our calculation and illustrates implicitly properties of the applied interpolation method.

6.3 Formulas for Numerical Differentiation of Two Variable Functions

Let us consider a function with two variables $f(x, y)$, for which chosen discrete values $f(x_i, y_j)$, defined for $x_i = x_0 + i \cdot h_1$, $y_j = y_0 + j \cdot h_2$, where $i = 0, 1, 2, 3, \ldots, n$ and $j = 0, 1, 2, 3, \ldots, m$, are shown in Fig. 6.3.

Fig. 6.3

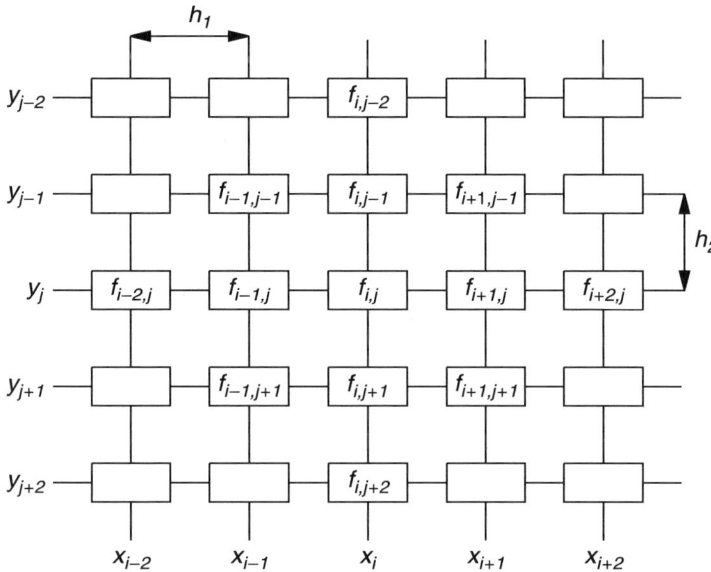

First partial derivatives of this function can be approximated using the quotients of finite differences calculated for sufficiently small values of steps h_1 and h_2.

$$\frac{\partial f(x, y)}{\partial x} \approx \frac{f(x + h_1, y) - f(x, y)}{h_1}$$
$$\frac{\partial f(x, y)}{\partial y} \approx \frac{f(x, y + h_2) - f(x, y)}{h_2} \tag{6.38}$$

At an arbitrary point (x_i, y_j) derivatives (6.38) can be expressed by the discrete values of function $f(x, y)$ specified in Fig. 6.3.

$$\left(\frac{\partial f}{\partial x}\right)_{ij} \approx \frac{f_{i+1,j} - f_{ij}}{h_1}$$
$$\left(\frac{\partial f}{\partial y}\right)_{ij} \approx \frac{f_{i,j+1} - f_{ij}}{h_2} \tag{6.39}$$

As we can see, relations (6.39) describe right-sided approximation of the desired derivatives. In the same manner as in case of the functions with one variable, see relations (6.10), (6.11) and (6.12), we can also find the left-sided and central approximations of these derivatives. In order to determine the central approximation, characterized by augmented accuracy, we develop the function $f(x, y)$ in the binomial Taylor series:

$$f(x + \Delta x, y + \Delta y)$$

$$= f(x, y) + \frac{\partial f}{\partial x}\Delta x + \frac{\partial f}{\partial y}\Delta y$$

$$+ \frac{1}{2!}\left(\frac{\partial^2 f}{\partial x^2}\Delta x^2 + 2\frac{\partial^2 f}{\partial x \partial y}\Delta x \Delta y + \frac{\partial^2 f}{\partial y^2}\Delta y^2\right)$$

$$+ \frac{1}{3!}\left(\frac{\partial^3 f}{\partial x^3}\Delta x^3 + 3\frac{\partial^3 f}{\partial x^2 \partial y}\Delta x^2 \Delta y + 3\frac{\partial^3 f}{\partial x \partial y^2}\Delta x \Delta y^2 + \frac{\partial^3 f}{\partial y^3}\Delta y^3\right) + \ldots$$

$$(6.40)$$

For $x = x_i$, $\Delta x = h_1$, $y = y_j$, $\Delta y = 0$, series (6.40) can be written as:

$$f(x_i + h_1, y_j) = f_{i+1,j} = f_{i,j} + \left(\frac{\partial f}{\partial x}\right)_{ij} h_1 + \frac{1}{2!}\left(\frac{\partial^2 f}{\partial x^2}\right)_{ij} h_1^2 + \frac{1}{3!}\left(\frac{\partial^3 f}{\partial x^3}\right)_{ij} h_1^3 + \ldots$$

$$(6.41)$$

Similarly, for $x = x_i$, $\Delta x = -h_1$, $y = y_j$, $\Delta y = 0$ series (6.40) takes the form:

$$f(x_i - h_1, y_j) = f_{i-1,j} = f_{i,j} - \left(\frac{\partial f}{\partial x}\right)_{ij} h_1 + \frac{1}{2!}\left(\frac{\partial^2 f}{\partial x^2}\right)_{ij} h_1^2 - \frac{1}{3!}\left(\frac{\partial^3 f}{\partial x^3}\right)_{ij} h_1^3 + \ldots$$

$$(6.42)$$

Subtracting expressions (6.41) and (6.42) side-by-side we can obtain the relation:

$$f_{i+1,j} - f_{i-1,j} = 2h_1 \left(\frac{\partial f}{\partial x}\right)_{ij} + O(h_1{}^3)$$

from which we obtain the central second order formula serving for calculating the first partial derivative with respect to variable x:

$$\left(\frac{\partial f}{\partial x}\right)_{ij} \approx \frac{f_{i+1,j} - f_{i-1,j}}{2h_1} - O(h_1{}^2) \approx \frac{f_{i+1,j} - f_{i-1,j}}{2h_1} \qquad (6.43)$$

Adding up side-by-side both series (6.41) and (6.42), we obtain the relation:

$$f_{i+1,j} + f_{i-1,j} = 2f_{ij} + \frac{2}{2!}h_1{}^2\left(\frac{\partial^2 f}{\partial x^2}\right)_{ij} + O(h_1{}^4)$$

which can be written in the following equivalent form:

$$\left(\frac{\partial^2 f}{\partial x^2}\right)_{ij} \approx \frac{f_{i+1,j} - 2f_{ij} + f_{i-1,j}}{h_1{}^2} - O(h_1{}^2) \approx \frac{f_{i+1,j} - 2f_{ij} + f_{i-1,j}}{h_1{}^2} \qquad (6.44)$$

The formula we have just obtained is called the second order central difference formula for calculating the second partial derivative with respect to variable x. In the similar way, the difference formulas serving for calculating the approximate values of first and second derivatives with respect to variable y can be determined. Consequently,

$$\left(\frac{\partial f}{\partial y}\right)_{ij} \approx \frac{f_{i,j+1} - f_{i,j-1}}{2h_2}, \quad \left(\frac{\partial^2 f}{\partial x^2}\right)_{ij} \approx \frac{f_{i,j+1} - 2f_{ij} + f_{i,j-1}}{h_2{}^2} \tag{6.45}$$

Expressions (6.45), similarly as their equivalents (6.43) and (6.44), are called central difference formulas of the second order. A function of two variables can have in general $2^2 = 4$ different second derivatives, including two mixed derivatives. If the mixed derivatives are continuous, then according to the Schwartz theorem they are also equal. Functions which can be expanded in the Fourier series satisfy this assumption, and therefore approximate values of these derivatives can be calculated according to the following central difference formula of the second order:

$$\left(\frac{\partial^2 f}{\partial x \partial y}\right)_{ij} = \left(\frac{\partial^2 f}{\partial y \partial x}\right)_{ij} \approx \frac{f_{i+1,j+1} - f_{i+1,j-1} - f_{i-1,j+1} + f_{i-1,j-1}}{4h_1 h_2} \tag{6.46}$$

which can be derived similarly as the previous one presented above. Values of the function $f(x, y)$, shown in Fig. 6.3, correspond to different expansions of the series (6.40). Based on these values and related Fourier series expansions, more accurate formulas for calculating partial derivatives can also be developed. In Table 6.8, some examples of several relations obtained in this way are given. They proved to be particularly useful to solve the Laplace equation in a two-dimensional space as well as for some optimization strategies.

Example 6.6 Table 6.9 contains nine discrete values of the function $f(x, y) = 1/(x^2 + 2y^2)$, which were determined in close neighborhood of the point ($x = 1, y = 1$).

Approximate values of partial derivatives of this function calculated at point ($x = 1, y = 1$) by means of formulas (6.43), (6.44), (6.45) and (6.46) are given in the second column of Table 6.10.

These approximate values are very close to their accurate ones, given in the third column. All results presented in Table 6.10 confirm well correctness of difference formulas under discussion, as well as their usefulness for engineering calculations.

Table 6.8

Central difference formulas of the fourth order $O(h_1^4, h_2^4)$
$\left(\dfrac{\partial f}{\partial x}\right)_{ij} \approx \dfrac{f_{i+1,j+1} - f_{i-1,j+1} + f_{i+1,j-1} - f_{i-1,j-1}}{4h_1}$
$\left(\dfrac{\partial f}{\partial y}\right)_{ij} \approx \dfrac{f_{i+1,j+1} - f_{i+1,j-1} + f_{i-1,j+1} - f_{i-1,j-1}}{4h_2}$
$\dfrac{\partial^2 f}{\partial x^2} \approx \dfrac{-f_{i+2,j} + 16f_{i+1,j} - 30f_{ij} + 16f_{i-1,j} - f_{i-2,j}}{12h_1{}^2}$
$\dfrac{\partial^2 f}{\partial y^2} \approx \dfrac{-f_{i,j+2} + 16f_{i,j+1} - 30f_{ij} + 16f_{i,j-1} - f_{i,j-2}}{12h_2{}^2}$

Table 6.9

$y \downarrow /x \rightarrow$	0.95	1.00	1.05
0.95	0.369344413	0.356506238	0.343938091
1.00	0.344530577	0.333333333	0.322320709
1.05	0.321802091	0.312012480	0.302343159

Table 6.10

Derivative	Approximate value	Accurate value	Relative error
$\dfrac{\partial f}{\partial x}$	−0.222098	−0.222222	-5.580×10^{-4}
$\dfrac{\partial f}{\partial y}$	−0.444937	−0.444444	1.109×10^{-3}
$\dfrac{\partial^2 f}{\partial x \partial y}$	0.594739	0.592592	3.623×10^{-3}
$\dfrac{\partial^2 f}{\partial x^2}$	0.073848	0.074074	-3.051×10^{-3}
$\dfrac{\partial^2 f}{\partial y^2}$	0.740820	0.740740	1.080×10^{-4}

6.4 An Example of the Two-Dimensional Optimization Problem and its Solution by Using the Gradient Minimization Technique

One of the popular plays in times of my childhood was blowing bubbles with the soap solution. A piece of straw, small plate filled with water and a little bit of gray soap was all that needed to conjure up beautiful, lazily moving spherical envelopes. The blown soap bubble is nothing else but a closed, very thin water layer surrounding a portion of heated air. Due to the surface tension, this envelope adopts spontaneously the spherical shape, by which its surface attains the minimum. In such simple, but at the same time in a suggestive manner the nature proves that the sphere is such optimal geometric solid, for which the ratio of the total lateral area to the volume is the smallest possible; that is it attains the minimum. In other words, the process of shaping of this soap bubble is the optimization process, in the sense of the criterion defined above. This observation draws us to the conclusion that all containers used for gas substances, designed according to this criterion should have spherical shape. For such form, the quantity of stuff used for their manufacturing would be smallest. Each departure from this principle should be well justified – the fact that each designer should always have in his mind. Following the rule just explained, let us design a parallelepiped tin container having the volume of $V = 1\,\mathrm{m}^3$ under condition that one of its geometrical dimensions, see Fig. 6.4, cannot be less than $1.3\,\mathrm{m}$.

It is known from the elementary mathematics that volume of this container is:

$$V = x \cdot y \cdot z = 1 m^3$$

Fig. 6.4

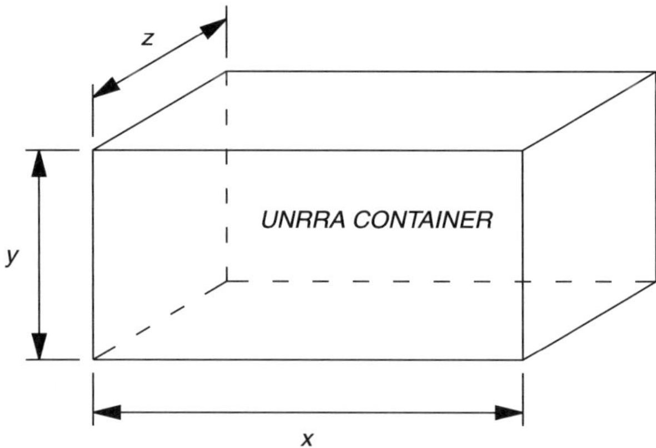

and that one of its geometrical dimensions, for example z, can be expressed by two remaining, i.e.:

$$z = \frac{V}{x \cdot y} = \frac{1}{x \cdot y} \tag{6.47}$$

The quantity, which should attain minimum, is the total lateral surface of the container $S(x, y, z) = 2(xy + yz + xz)$, which after introducing relation (6.47) can be calculated using the formula:

$$S(x, y) = 2\left(xy + \frac{1}{x} + \frac{1}{y}\right) \tag{6.48}$$

While performing minimization process for the surface (6.48) we cannot forget about the given design constraint saying that one of the dimensions, for example x, should not be less than 1.3 m. This limitation can be taken into account during the minimization (optimization) process by adding to the relation (6.48) the following easily analytically differentiable term:

$$P(x) = \exp\left[t\left(1 - \frac{x}{1.3}\right)\right] \tag{6.49}$$

where t is a positive fixed parameter with an appropriate value.

Two typical curves of the function (6.49) calculated for $t = 100$ and $t = 200$ are shown in Fig. 6.5.

For sufficiently large values of parameter t (for example, $t > 300$) the penalty component (6.49) is small in comparison with the expected value of the function (6.48), if only x belongs to the area of acceptable solutions [8, 9]. If a current solution is located close to the boundary of the acceptable area, or outside this boundary, then the penalty component (6.49) becomes large in comparison with the value of the minimized function (6.48). In this case, the penalty term, more precisely its

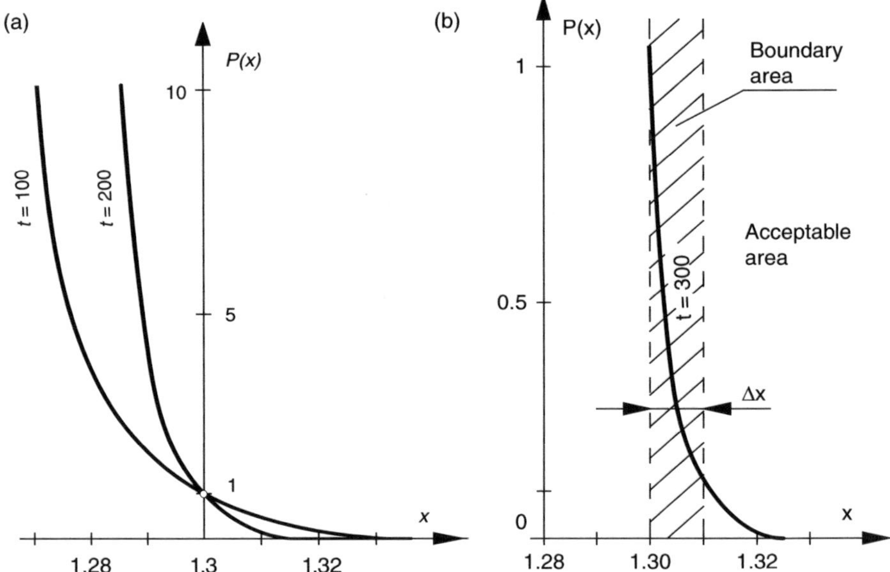

Fig. 6.5

gradient, "pushes" the point corresponding to this solution towards the acceptable area. As our current problem is concerned, the area of acceptable solution is
bounded by the inequality $x \geq 1.3 + \Delta x$, where Δx is the width of a relatively
narrow boundary area, see Fig. 6.5(b). Similarly, we can take into account the limitations for the variable y, which should always be positive.

$$Q(y) = \exp(-t \cdot y) \tag{6.50}$$

After adding the relations (6.49) and (6.50) to (6.48), we obtain the following
extended objective function:

$$F(x, y, t) = 2\left(xy + \frac{1}{x} + \frac{1}{y}\right) + \exp\left[t\left(1 - \frac{x}{1.3}\right)\right] + \exp(-t \cdot y) \tag{6.51}$$

which can be minimized by means of one of the gradient optimization methods.
It has been confirmed experimentally that the relatively simple steepest descent
method is suitable for this purpose [9, 10]. Thus, let us consider its algorithm step-
by-step for didactic reasons. Before we discuss the merits of this algorithm, let us
remind the basic concept of the gradient defined for many variable functions, and
explain why it is used in iterative optimization strategies. The gradient of a function
$f(x_1, x_2, \ldots, x_n) \equiv f(\mathbf{x})$ is the vector, whose coordinates are partial derivatives of
this function determined with respect to particular variables, i.e.:

$$\nabla f(\mathbf{x}) \equiv \left(\frac{\partial f}{\partial x_1}, \frac{\partial f}{\partial x_2}, \ldots, \frac{\partial f}{\partial x_n} \right) \tag{6.52}$$

It is not difficult to prove that the gradient (6.52) is a vector pointing at the steepest growth direction of the function $f(x_1, x_2, \ldots, x_n) \equiv f(\mathbf{x})$, in close neighborhood of the $n-$dimensional point $(x_1, x_2, \ldots, x_n) \equiv (\mathbf{x})$. The vector $-\nabla f(\mathbf{x})$ would of course point the direction of steepest descent, and this property is used for searching minimum of the function being optimized [9, 10]. In case of the objective function (6.51), the gradient will have two components:

$$\frac{\partial F}{\partial x} = 2y - \frac{2}{x^2} - \frac{t}{1.3} \exp\left[t \left(1 - \frac{x}{1.3} \right) \right]$$
$$\frac{\partial F}{\partial y} = 2x - \frac{2}{y^2} - t \cdot \exp(-t \cdot y) \tag{6.53}$$

Equations (6.53) serve to determine the unitary vector pointing at minimization direction $\mathbf{d} = (d_x, d_y)$, having the following components:

$$d_x = \frac{-1}{|\nabla F(x, y)|} \cdot \frac{\partial F}{\partial x}$$
$$d_y = \frac{-1}{|\nabla F(x, y)|} \cdot \frac{\partial F}{\partial y} \tag{6.54}$$

where

$$|\nabla F(x, y)| = \sqrt{\left(\frac{\partial F}{\partial x} \right)^2 + \left(\frac{\partial F}{\partial y} \right)^2}.$$

As defined in (6.54), absolute value of the vector $\mathbf{d} = (d_x, d_y)$ is equal to 1. The steepest descent method is an iterative method; that is the result obtained in the i iteration is subsequently used as the "starting point" for computation executed during next iteration. Assume now that we know the coordinate of the point (x_i, y_i) belonging to the admissible region. At this point we determine the minimization vector $\mathbf{d}^{(i)}$ using relations (6.53) and (6.54). Position of the new point (x_{i+1}, y_{i+1}) is established, making the search along the line

$$x = x_i + j \cdot h \cdot d_x^{(i)}$$
$$y = y_i + j \cdot h \cdot d_y^{(i)} \tag{6.55}$$

where h is the step, and j is the parameter taking consecutive values of $1, 2, 3, \ldots$ For each subsequent value of the parameter j, the objective function (6.51) is calculated and compared with the value obtained previously; that is for $j-1$. If the value of the objective function is decreasing, then the process of searching for its minimal value along the line (6.55) is continued. Component values (6.55), for which the objective function begins to increase are taken as the desired coordinates of the optimal point (x_{i+1}, y_{i+1}). For this point, we determine a new vector $\mathbf{d}^{(i+1)}$ and continue the computing in the new iteration $(i+2)$ using the same algorithm. The following condition is most frequently used as the criterion of ending the computation process.

$$|\nabla F(x, y)| = \sqrt{\left(\frac{\partial F}{\partial x}\right)^2 + \left(\frac{\partial F}{\partial y}\right)^2} \le \varepsilon \tag{6.56}$$

The parameter ε is an arbitrary small positive number. In Table 6.11 we have summarized some results obtained by means of the algorithm explained above for the following data: $[x_0, y_0] \equiv [1.5, 1]$, $h = 0.0001$, $t = 1000$ and $\varepsilon = 0.15$.

In Table 6.12 we have shown for comparison similar results obtained with an assumption that all dimensions of our container can take arbitrary values. For optimization the problem is formulated in such a way that only the function (6.48) is minimized.

Comparing the subsequent values of x_i given in the second column of Table 6.11 we deduce that the solution of our design problem lies very close to the boundary of the acceptable region ($x = 1.3$ m). Assuming that the penalty term is absent and would not modify the original goal function, we would obtain the following optimization result: $x = 1.3$, $y = 0.877058019$. Total surface area of such container would be equal to $S = 6.099163239$ m^2. Using these results we can find relative error of the approximation obtained above, that is:

$$\delta = \frac{6.104739 - 6.099163239}{6.099163239} \approx 9.142 \cdot 10^{-4}$$

Table 6.11

Iteration "i"	x_i, m	y_i, m	S, m^2	$|\nabla F(x, y)|$
0	1.500000	1.000000	6.333333	1.494847
1	1.304662	0.824196	6.110166	20.835037
2	1.310161	0.824284	6.112767	0.366847
...	...	...	...	...
50	1.309949	0.863675	6.105203	0.131874
...	...	...	...	...
100	1.309677	0.873755	6.104739	0.131963

Table 6.12

Iteration "i"	x_i, m	y_i, m	S, m^2	$|\nabla S(x, y)|$
0	1.500000	1.000000	6.333333	1.494847
1	1.277011	0.799310	6.109770	0.686100
2	1.127286	1.031168	6.038563	0.615014
...	...	...	...	...
5	0.999044	0.985557	6.000453	6.918×10^{-2}
...	...	...	...	...
8	1.002596	1.002465	6.000039	$< 1 \times 10^{-2}$

This error is rather small and this fact confirms practical usefulness of the method explained above. Surface area of an equivalent spherical container having the same volume $V = 1$ m^3, equals 4.835976 m^2 and is approximately 1.262 times less than the total area of the cubicoidal container designed above.

References

1. Abramowitz M. and I.A. Stegun, Handbook of mathematical functions. Dover, New York, 1954
2. Kong J.A., Electromagnetic wave theory. John Wiley and Sons, New York, 1983
3. Forsythe G.E., Malcolm M.A. and C.B. Moler, Computer methods for mathematical computations. Prentice-Hall, Englewood Cliffs, NJ, 1977
4. Mathews J.H., Numerical methods for mathematics, science and engineering. Prentice-Hall Intern. Inc., Englewood Cliffs, NJ, 1992
5. Shoup T.E., Applied numerical methods for the microcomputer. Prentice-Hall Inc., Englewood Cliffs, NJ, 1984
6. Mathews J.H., Numerical methods for mathematics, science and engineering. Prentice-Hall, Inc., Englewood Cliffs, NJ, 1987
7. Turchak L.I. and P.W. Plotnikov, Fundamentals of numerical methods (2nd edition in Russian) Publishing house "Physmathlit", Moscow, 2002
8. Bazaraa M.S., Sherali H.D. and C.M. Shetty, Nonlinear programming. Theory and applications. John Wiley and Sons, New York, 1993
9. Himmelblau D.M., Applied nonlinear programming. McGraw-Hill, New York, 1972
10. Fletcher R., A review of methods for unconstrained optimization. Academic Press, New York, 1969

Chapter 7
Methods for Numerical Integration of Ordinary Differential Equations

The equations containing one or more derivatives are called differential equations. Depending on the number of independent variables and corresponding number of derivatives these equations are divided into:

- ordinary differential equations formulated for functions of one variable and
- partial differential equations formulated for functions of many variables.

Subject of our considerations in the present chapter is the ordinary differential equations, with some additional requirements for the function in question and some of its derivatives. If these requirements, given for the function and eventually for some of their derivatives, are defined only for one value of the independent variable, they are called initial conditions. In such cases, the problem of evaluating the function satisfying given differential equation with an initial condition is called the initial value problem, proposed originally by Cauchy [1, 2]. In case when these requirements are defined not for one, but for more values of the independent variables, they are called boundary conditions. Correspondingly, the problem of determining a function satisfying given differential equations and the boundary conditions is known as the boundary problem. In case of the initial value problem, time plays often the role of independent variable. Classical example of such problem is a mathematical description of free motion of the infinitesimally small material body, suspended on a long, infinitesimally thin, inextensible thread. Initial condition for this problem is defined by position and speed of this body in a chosen moment of time, say $t_0 = 0$. For the pendulum considered in this example, one can also formulate boundary conditions defining the trajectory of this body in a given time interval.

7.1 The Initial Value Problem and Related Solution Methods

Let us consider the initial value problem consists in evaluation of a function $y(x)$ that satisfies the following equation:

$$\frac{dy(x)}{dx} = f\,[x, y(x)] \tag{7.1}$$

S. Rosłoniec, *Fundamental Numerical Methods for Electrical Engineering*,
© Springer-Verlag Berlin Heidelberg 2008

with initial condition $y_0 = y(x_0)$. Numerical solution of this equation consists in evaluation of the set of discrete values $y_n = y(x_n) = y(x_0 + n \cdot h)$ of the unknown function, where $n = 1, 2, 3, \ldots$, and $h = \Delta x$ is an adopted integration step. Several methods for finding y_n are known. In the literature they are often classified as:

– one-step (self-starting) methods, and
– multi-step methods, called briefly the predictor–corrector methods.

As compared with equivalent one-step methods, the multi-step methods ensure better numerical efficiency. In other words, they make possible in obtaining more accurate approximation with less necessary computations. Unfortunately, they are not self-starting and several initial points must be given in advance. It means that knowledge of some first (approximate or accurate) values of the desired function is necessary. These values, called commonly initial sections, are determined most often using one-step methods of the same order, as for example the Runge–Kutta (RK 4) method described in Sect. 7.2.3.

7.2 The One-Step Methods

7.2.1 The Euler Method and its Modified Version

The simplest representative of the one-step methods is the Euler method, discussed below on the basis of Fig. 7.1.

Process of finding the consecutive values $y_n = y(x_n) = y(x_0 + n \cdot h)$ for $n = 1, 2, 3, \ldots$, begins from the starting point $P_0 \equiv (x_0, y_0)$, at which

$$\frac{dy(x)}{dx} = f\,[x_0, y(x_0)] = f\,[x_0, y_0] \tag{7.2}$$

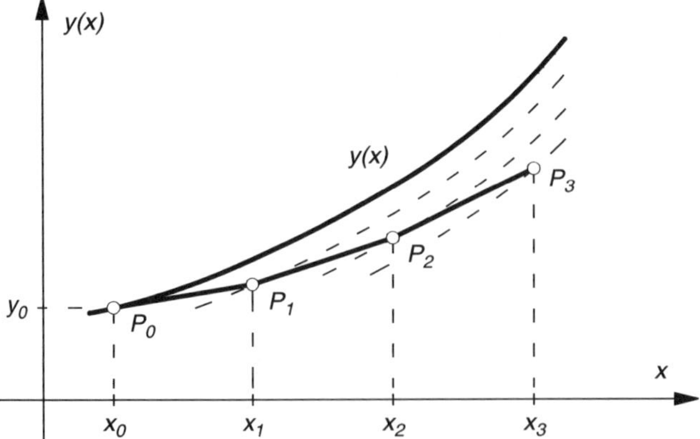

Fig. 7.1

In other words, the value of the function $f(x_0, y_0)$ is equal to the tangent of the angle, at which the tangent to the curve $y(x)$ satisfying Eq. (7.2) is inclined with respect to the x-axis. Therefore, first computed value of the desired function is:

$$y_1 = y(x_0 + h) = y_0 + h \cdot f[x_0, y(x_0)]$$

Thus we have obtained the point $P_1 \equiv (x_1, y_1)$, which can be treated as the starting point in the process of finding y_2, related to the point $P_2 = (x_2, y_2)$. Repeating this procedure several times, the set of discrete values y_n of the function approximating desired solution $y(x)$ is evaluated. Approximation accuracy of the function $y(x)$, obtained from the set of discrete values $y_n = y(x_n)$, calculated using the Euler method is rather small. The modified version of this method is therefore most frequently applied in practice. The point $P_{n+1} \equiv (x_{n+1}, y_{n+1})$, where $n = 0, 1, 2, 3, \ldots$, computed using standard Euler method, is placed at the intersection of the line, tangent to the integral curve at the point $P_n \equiv (x_n, y_n)$, and a line parallel to the y-axis, satisfying the abscissa $x_{n+1} = x_n + h$. Considerable increase of accuracy may be obtained when the slope coefficient of the tangent line is calculated not at the point $P_n \equiv (x_n, y_n)$, but at the new point Q_n having the coordinate $x = x_n + h/2$ and lying on the integral curve, see Fig. 7.2.

Unfortunately, determination of the coordinate y at the point Q_n is not possible, and therefore this point is replaced (approximated) in the algorithm of the modified Euler method by another point R_n whose coordinates are:

$$x_n + \frac{h}{2}, \quad y_n + \frac{h}{2} f(x_n, y_n)$$

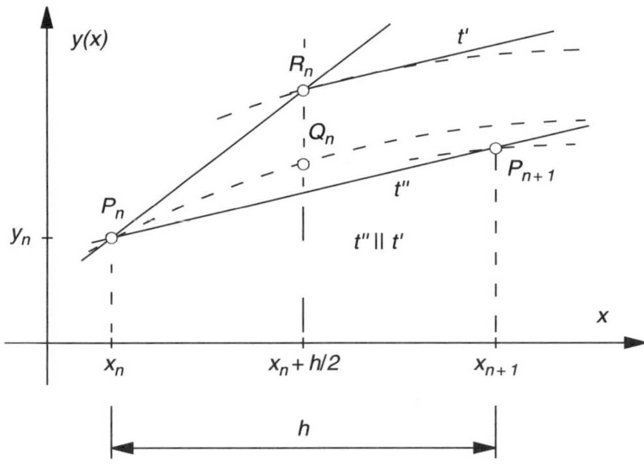

Fig. 7.2

Slope coefficient of the line tangent to the integral curve at point R_n is equal to:

$$f\left[x_n + \frac{h}{2}, y_n + \frac{h}{2}f(x_n, y_n)\right] \tag{7.3}$$

The line having the slope (7.3) and passing by the point $P_n \equiv (x_n, y_n)$, obtained by intersection with a line $x_{n+1} = x_n + h$ and parallel to the y-axis, determines a new point having the coordinate:

$$y_{n+1} = y_n + h \cdot f\left[x_n + \frac{h}{2}, y_n + \frac{h}{2}f(x_n, y_n)\right]$$

The coordinate y_{n+1} at this new point is treated as the next discrete value of the desired solution. After introduction of the notation

$$k_1 = hf(x_n, y_n), \quad k_2 = hf\left(x_n + \frac{h}{2}, y_n + \frac{k_1}{2}\right)$$

the value of y_{n+1} can be written as:

$$y_{n+1} = y_n + k_2 \tag{7.4}$$

Another version of the Euler method is the Heun method discussed in Sect. 7.2.2.

7.2.2 The Heun Method

Let us assume that the approximate value $y_n = y(x_n)$ of the desired function $y(x)$ is known, see Fig. 7.3.

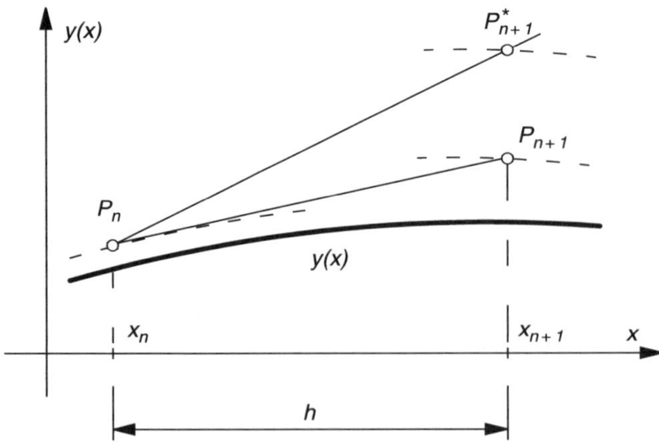

Fig. 7.3

In case of the Heun method, an auxiliary coordinate is calculated first

$$y_{n+1}^* = y_n + hf(x_n, y_n) \tag{7.5}$$

and used next to determine the quantity

$$f(x_{n+1}, y_{n+1}^*)$$

expressing the slope coefficient of the tangent to the curve described by Eq. (7.2) and passing through the point $P_{n+1}^* \equiv (x_{n+1}, y_{n+1}^*)$ being first approximation of the desired solution. The point, which gives a much better approximation, $P_{n+1} \equiv (x_{n+1}, y_{n+1})$, has the coordinate y_{n+1} calculated from the formula:

$$y_{n+1} = y_n + \frac{1}{2}h\left[f(x_n, y_n) + f(x_{n+1}, y_{n+1}^*)\right] \tag{7.6}$$

It is not difficult to see that the slope of the line passing through the points P_n and P_{n+1} is the arithmetical mean of the slopes of tangents at points $P_n \equiv (x_n, y_n)$ and $P_{n+1}^* \equiv (x_{n+1}, y_{n+1}^*)$. The point P_{n+1} determined in this way plays the role of starting point to the computation performed during next iteration $(n + 2)$. This algorithm can be described by the following formulas:

$$k_1 = hf(x_n, y_n), \qquad k_2 = hf(x_n + h, y_n + k_1)$$
$$y_{n+1} = y_n + \frac{1}{2}(k_1 + k_2) \tag{7.7}$$

where $n = 1, 2, 3, \ldots$ In comparison with the Euler method presented earlier, the Heun method ensures greater accuracy and better numerical stability. To determine the order of accuracy we develop the function $y(x)$ in the Taylor series, in close neighborhood of the point x_n

$$y(x_n + h) = y(x_n) + hy'(x_n) + \frac{1}{2}h^2 y''(x_n) + \ldots \tag{7.8}$$

Assume also that only first three terms of the series (7.8) will be taken into account. Second derivative $y''(x_n)$ visible in the third term of this series can be approximated as follows:

$$y''(x_n) \approx \frac{\Delta y'}{\Delta x} = \frac{y'(x_n + h) - y'(x_n)}{h}$$

Hence, the sum of three initial terms of the series (7.8) is equal to:

$$y(x_n + h) = y(x_n) + hy'(x_n) + \frac{1}{2}h^2\left[\frac{y'(x_n + h) - y'(x_n)}{h}\right]$$
$$= y(x_n) + \frac{1}{2}h\left[y'(x_n + h) + y'(x_n)\right] \tag{7.9}$$

and is identical, as in the formula (7.6). According to our previous assumption, about neglecting the terms of the series containing the multiplier h^n, for $n \geq 3$, the Heun methods may be classified to the group of methods of the second order. In other words, it guarantees accuracy comparable to the Taylor series approximation (7.8), in which all initial terms are present, including the term with second derivative. Acting in much the same way, it is possible to prove that the modified Euler method discussed earlier has the same (second) order accuracy. In the extrapolating formula (7.9), expressing the essence of the Heun method, second derivative is represented by the term containing two values of the first derivative defined for the left and right end of the subinterval $[x_n, x_{n+1}]$. In order to express third derivative in terms of the finite differences, knowledge of the second derivative at two different points is necessary. Hence, there is a necessity of defining the slope of the desired curve at one additional point, which lies inside a subinterval $[x_n, x_{n+1}]$. Reasoning in a similar way as above, we come to the conclusion that in order to determine higher derivatives we must compute the slopes of the desired function at many points inside a subinterval $[x_n, x_{n+1}]$. This last conclusion becomes starting point for elaborating the whole group of one-step methods, such as the Runge–Kutta method.

7.2.3 The Runge–Kutta Method (RK 4)

The Runge–Kutta method of the fourth order, denoted in the literature as RK 4, is an important representative of one-step methods. In this iteration method, the nth approximate value $y_n = y(x_n)$ of the evaluated function forms a basis for calculation of the next $(n+1)$ approximate value $y_{n+1} = y(x_n+h)$, where h denotes the adopted integration step. The calculations are performed according to formula:

$$y_{n+1} = y_n + \frac{1}{6}(k_1 + 2k_2 + 2k_3 + k_4) \tag{7.10}$$

where:

$$k_1 = hf(x_n, y_n), \qquad\qquad k_2 = hf\left(x_n + \frac{h}{2}, y_n + \frac{k_1}{2}\right)$$

$$k_3 = hf\left(x_n + \frac{h}{2}, y_n + \frac{k_2}{2}\right), \qquad k_4 = hf(x_n + h, y_n + k_3)$$

The derivation of these recursive formulas is behind the scope of the present handbook. It can be found in more advanced books on numerical analysis; for example in [3]. The Runge–Kutta RK 4 method is one of the most popular and broadly used methods in the field of engineering. According to the opinion established in the literature, it guarantees the accuracy sufficient in most applications (of the fourth order in the sense of the Taylor series accuracy) and is sufficiently stable. Another advantage of this method is also the simplicity of computation algorithm, see formula (7.10), for which the starting point is defined by initial condition. The computation formulas given above can be generalized for the case of a system of k first-order differential equations. The system of two ($k = 2$) first-order equations is especially useful in the field of engineering, and for that reason it is written below together with appropriate initial conditions.

$$\frac{dy_1(x)}{dx} = f_1\left[x, y_1(x), y_2(x)\right], \quad y_1(x_0) = y_{1,0}$$
$$\frac{dy_2(x)}{dx} = f_2\left[x, y_1(x), y_2(x)\right], \quad y_2(x_0) = y_{2,0} \tag{7.11}$$

Let us assume that the values of the functions $y_{1,n} = y_1(x_n)$ and $y_{2,n} = y_2(x_n)$, calculated during the n iterations are known. Subsequent discrete values $y_{1,n+1} = y_1(x_{n+1})$ and $y_{2,n+1} = y_2(x_{n+1})$ are determined using the following recursive formulas:

$$y_{1,n+1} = y_{1,n} + \frac{1}{6}(k_1 + 2k_2 + 2k_3 + k_4)$$
$$y_{2,n+1} = y_{2,n} + \frac{1}{6}(l_1 + 2l_2 + 2l_3 + l_4) \tag{7.12}$$

where

$$k_1 = hf_1(x_n, y_{1,n}, y_{2,n}), \qquad\qquad l_1 = hf_2(x_n, y_{1,n}, y_{2,n})$$
$$k_2 = hf_1\left(x_n + \frac{h}{2}, y_{1,n} + \frac{k_1}{2}, y_{2,n} + \frac{l_1}{2}\right), \quad l_2 = hf_2\left(x_n + \frac{h}{2}, y_{1,n} + \frac{k_1}{2}, y_{2,n} + \frac{l_1}{2}\right)$$
$$k_3 = hf_1\left(x_n + \frac{h}{2}, y_{1,n} + \frac{k_2}{2}, y_{2,n} + \frac{l_2}{2}\right), \quad l_3 = hf_2\left(x_n + \frac{h}{2}, y_{1,n} + \frac{k_2}{2}, y_{2,n} + \frac{l_2}{2}\right)$$
$$k_4 = hf_1(x_n + h, y_{1,n} + k_3, y_{2,n} + l_3), \quad l_4 = hf_2(x_n + h, y_{1,n} + k_3, y_{2,n} + l_3)$$

Example 7.1 In order to illustrate the algorithm of the Runge–Kutta RK 4 method, let us determine a function $y(t)$ satisfying the following differential equation:

$$\frac{d^2y(t)}{dt^2} - 4\frac{dy(t)}{dt} + 3y(t) = 0$$

with an initial condition:

$$y(t = 0) = 0, \quad \frac{dy(t = 0)}{dt} = -2.$$

By using a substitution

$$y(t) \equiv y_1(t), \quad \frac{dy(t)}{dt} \equiv y_2(t)$$

the second order differential equation given above can be replaced by the equivalent system of two differential equations of the first order, namely:

$$\frac{dy_1(t)}{dt} = y_2(t)$$

$$\frac{dy_2(t)}{dt} = 4y_2(t) - 3y_1(t)$$

Consequently, the initial conditions are: $y_1(t = 0) = 0$ and $y_2(t = 0) = -2$. Calculations of discrete values of the function $y_1(t) \equiv y(t)$ have been performed according to (7.12), for the integration step $\Delta t \equiv h = 0.001$. Some resaults obtained in this way are given in the second column of Table 7.1.

The differential equation under integration has an analytic solution $y(t) = e^t - e^{3t}$, which has been used to find comparative results given in the third column.

7.2.4 The Runge–Kutta–Fehlberg Method (RKF 45)

The methods of numerical solution of the differential equations can be characterized by constant step h in the whole given integration range. Magnitude of this step should be chosen in such a way that the sum of the method and machine (processing) errors is as small as possible. One of the simplest ways to guarantee sufficient accuracy of the solution obtained is to solve the same problem for two different integration steps, most frequently h and $h/2$. Discrete values of the two solutions obtained in this way are next compared at points x_n corresponding to the larger step of the two. If the differences between two solutions compared are not sufficiently small, the whole computation process should be repeated for the step reduced two times. Multiple solution of the same equation system, up to the moment when the two consecutive solutions are sufficiently close, generates the necessity of performing many directly useless computations.

Another, more efficient manner serving to guarantee the sufficiently accurate solution, is the integration of differential equations using variable step adjusted automatically at each point x_n of the independent variable. An example of the method in which this concept is used is the Runge–Kutta–Fehlberg method, denoted in the

Table 7.1

t	$y(t)$ Numerical solution	$y(t)$ Analytical solution
0.000	0.000 000 000	0.000 000 000
0.001	−0.002 004 004	−0.002 004 027
. . .	. . .	. . .
0.100	−0.244 687 903	−0.244 687 795
. . .	. . .	. . .
1.000	−17.367 257 827	−17.367 252 349

literature by the symbol RKF 45. The relatively simple selection procedure for the integration step h can be made based on the two calculated approximate solutions. The preselected admissible error control tolerance, denoted in the majority of numerical formulas by the symbol *Tol*, is the parameter of this procedure. In order to obtain possibly accurate description of the RKF45 algorithm assume that the following differential equation is given

$$\frac{dy(x)}{dx} = f[x, y(x)] \tag{7.13}$$

with an initial condition $y_0 = y(x_0)$. According to [2] for each point $[x_n, y_n = y(x_n)]$, where $n = 0, 1, 2, 3, \ldots$, the following parameters are calculated:

$$k_1 = h \cdot f(x_n, y_n)$$

$$k_2 = h \cdot f\left(x_n + \frac{1}{4}h, y_n + \frac{1}{4}k_1\right)$$

$$k_3 = h \cdot f\left(x_n + \frac{3}{8}h, y_n + \frac{3}{32}k_1 + \frac{9}{32}k_2\right)$$

$$k_4 = h \cdot f\left(x_n + \frac{12}{13}h, y_n + \frac{1932}{2197}k_1 - \frac{7200}{2197}k_2 + \frac{7296}{2197}k_3\right) \tag{7.14}$$

$$k_5 = h \cdot f\left(x_n + h, y_n + \frac{439}{216}k_1 - 8k_2 + \frac{3680}{513}k_3 - \frac{845}{4104}k_4\right)$$

$$k_6 = h \cdot f\left(x_n + \frac{1}{2}h, y_n - \frac{8}{27}k_1 + 2k_2 - \frac{3544}{2565}k_3 + \frac{1859}{4104}k_4 - \frac{11}{40}k_5\right)$$

The symbol h denotes the optimal step determined for previous value of the independent variable, namely for x_{n-1} with $n > 1$. When $n = 1$, the step h is determined *a priori*. Parameters k_1, k_3, k_4, k_5 and k_6 are then used to calculate first the approximate value of y_{n+1}, according to the following fourth order formula

$$y_{n+1} = y_n + \frac{25}{216}k_1 + \frac{1408}{2565}k_3 + \frac{2197}{4104}k_4 - \frac{1}{5}k_5 \tag{7.15}$$

Secondly, more accurate value of the desired solution, denoted by z_{n+1}, is calculated according to the fifth order formula, namely:

$$z_{n+1} = y_n + \frac{16}{135}k_1 + \frac{6656}{12825}k_3 + \frac{28561}{56430}k_4 - \frac{9}{50}k_5 + \frac{2}{55}k_6 \tag{7.16}$$

Optimum step size $s_{n+1} \cdot h$ for this case is obtained multiplying the step size h by the correction coefficient

$$s_{n+1} = \left(\frac{Tol \cdot h}{2|z_{n+1} - y_{n+1}|}\right)^{1/4} \approx 0.84 \left(\frac{Tol \cdot h}{|z_{n+1} - y_{n+1}|}\right)^{1/4} \tag{7.17}$$

where *Tol* denotes a given tolerance error defining approximation accuracy of the desired solution $y_n = y(x_n)$. Knowing the optimum integration step value equal to $s_{n+1} \cdot h$, the function value $y_{n+1} = y(x_{n+1})$ is then calculated by means of Eqs. (7.14) and (7.15) given above. The value of $y_{n+1} = y(x_{n+1})$ and step $h \equiv s_{n+1} \cdot h$ play the role of starting values for calculating the next point $[x_{n+2}, y(x_{n+2})]$ of the desired solution. In the procedure of finding the integration step described by Eq. (7.17) we use absolute value of the difference between approximations (7.15) and (7.16). In case when it is less than a given sufficiently small positive number ε, that is when $|z_{n+1} - y_{n+1}| \leq \varepsilon$, this integration step should be incremented by a reasonably limited value, for example, less than two times. In Example 7.2 given below, we took $\varepsilon = 1 \times 10^{-10}$ and the rule that preserving the inequality $|z_{n+1} - y_{n+1}| \leq \varepsilon$, the integration step becomes increased by $\sqrt{2}$ times.

Example 7.2 Let us assume that the following differential equation is given

$$\frac{dy(x)}{dx} = 2 + \frac{1}{2}y^2(x)$$

with an initial condition $y(x=0)=0$. Taking initial integration step equal to $h=0.1$ and tolerance error $Tol=1\times10^{-7}$, this equation has been solved by means of the RKF 45 method over the range [0, 1.5]. Obtained results are presented in the second and third columns of Table 7.2. The corresponding exact results $y(x)=2 \cdot \tan(x)$, found analytically, are given in the fourth column. Absolute differences (found on the base of these solutions) given in the fifth column prove good quality of performed numerical calculations. For tutorial reasons, the initial value problem discussed in this example has been additionally solved using the RK 4 method. The results obtained for fixed step $h=0.1$ are shown in Table 7.3. All results presented above confirm fully the opinion known from the literature that the RKF 45 method, as compared with the RK 4, can be treated as more accurate and less sensitive with respect to the given (initial) size of the integration step.

Table 7.2

n	x_n	y_n	$2\tan(x)$	$|y_n - 2\tan(x)|$
1	0.095 216 216	0.191 010 017	0.191 010 013	4.363×10^{-9}
2	0.185 939 677	0.376 225 189	0.376 225 203	1.404×10^{-8}
3	0.267 491 821	0.548 119 411	0.548 119 366	4.501×10^{-8}
4	0.341 017 815	0.709 764 917	0.709 764 897	1.983×10^{-8}
5	0.408 932 224	0.866 724 642	0.866 724 669	2.778×10^{-8}
...	...	...	...	...
49	1.313 911 477	7.613 576 896	7.613 573 551	3.345×10^{-6}
50	1.319 331 445	7.785 044 767	7.785 041 333	3.435×10^{-6}
51	1.323 546 921	7.923 491 828	7.923 488 616	3.211×10^{-6}
...	...	...	...	...
559	1.500 295 103	28.321 312 327	28.321 310 043	2.284×10^{-6}
560	1.500 461 817	28.388 665 044	28.388 641 357	2.368×10^{-5}
561	1.500 528 679	28.415 767 236	28.415 746 688	2.051×10^{-5}

Table 7.3

n	x_n	y_n	$2\tan(x)$	$|y_n - 2\tan(x)|$
1	0.10	0.200 669 181	0.200 669 348	1.671×10^{-7}
2	0.20	0.405 419 762	0.405 420 065	3.022×10^{-7}
3	0.30	0.618 672 088	0.618 672 430	3.424×10^{-7}
4	0.40	0.845 585 997	0.845 586 478	4.808×10^{-7}
5	0.50	1.092 604 632	1.092 604 994	3.619×10^{-7}
...	...	...	...	...
10	1.00	3.114 812 975	3.114 815 473	2.498×10^{-6}
11	1.10	3.929 493 277	3.929 519 891	2.661×10^{-5}
12	1.20	5.144 143 807	5.144 304 275	1.604×10^{-4}
13	1.30	7.203 127 497	7.204 206 943	$1.079 \cdot 10^{-3}$
14	1.40	11.583 951 747	11.595 766 067	1.181×10^{-2}
15	1.50	27.673 314 419	28.202 840 805	5.291×10^{-1}

7.3 The Multi-step Predictor–Corrector Methods

In case of all one-step methods discussed in the previous section, the value $y_{n+1} = y(x_{n+1})$ of the determined function is calculated on the basis of only one value $y_n = y(x_n)$, computed during the previous iteration. In the multi-step methods we use for this end not only the value $y_n = y(x_n)$ but also $y_{n-k+1} = y(x_{n-k+1})$, $y_{n-k+2} = y(x_{n-k+2})$, $y_{n-k+3} = y(x_{n-k+3})$, ..., $y_n = y(x_n)$, where the number of steps $k = 1, 2, 3, \ldots$ determines also order of the method. In order to determine integral expression constituting theoretical base of all multi-step methods, let us consider the following first-order differential equation:

$$\frac{dy(x)}{dx} = f[x, y(x)] \tag{7.18}$$

in which we integrate both sides over an interval from x_n to x_{n+1}. Integrating the left-side of the equation (7.18) we obtain:

$$\int_{x_n}^{x_{n+1}} \frac{dy(x)}{dx} dx = y(x_{n+1}) - y(x_n) \approx y_{n+1} - y_n \tag{7.19}$$

In case of the $k-$step method, the following discrete values of the function constituting the right-side of the equation (7.18) are known:

$$f_{n-k+1} = f(x_{n-k+1}, y_{n-k+1})$$

$$f_{n-k+2} = f(x_{n-k+2}, y_{n-k+2})$$

$$f_{n-k+3} = f(x_{n-k+3}, y_{n-k+3}) \tag{7.20}$$

$$\ldots\ldots\ldots\ldots\ldots\ldots\ldots\ldots$$

$$f_n = f(x_n, y_n)$$

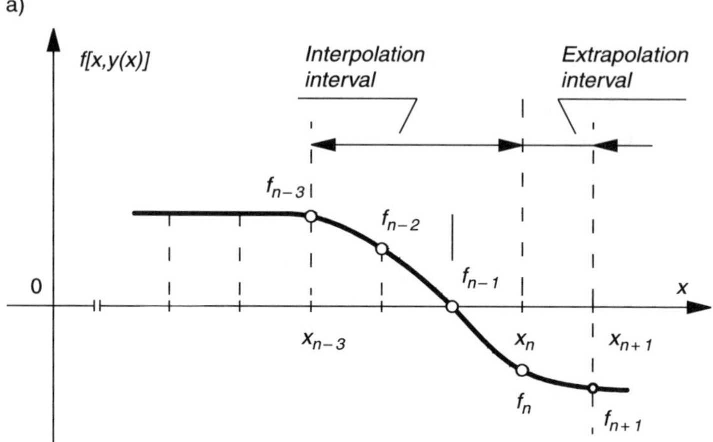

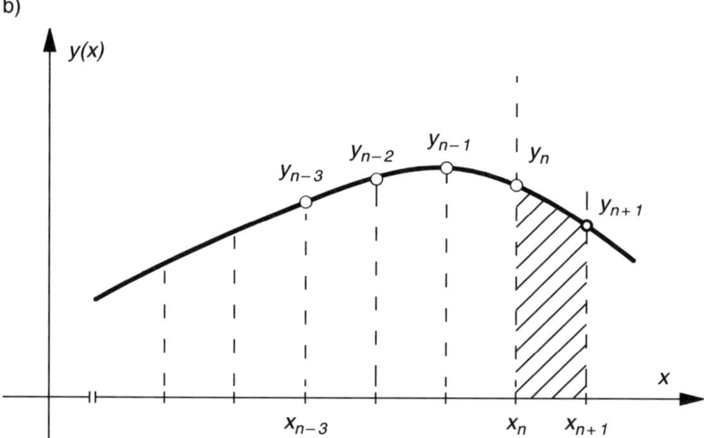

Fig. 7.4

These discrete values are represented by the corresponding points specified in Fig. 7.4.

In order to integrate the discrete function, constituting right-side of equation (7.18), it should be first interpolated or approximated over an interval $[x_{n-k+1}, x_n]$. On the basis of values (7.20), the interpolation polynomial $P_{k-1}(x)$ of degree $(k-1)$ is formed most frequently for this purpose. Next, this polynomial is used afterwards to extrapolate (predict) the function $f[x, y(x)]$ over the interval $[x_n, x_{n+1}]$, yelding:

$$\int_{x_n}^{x_{n+1}} f[x, y(x)]dx \approx \int_{x_n}^{x_{n+1}} P_{k-1}(x)dx \tag{7.21}$$

Comparison of integrals (7.19) and (7.21) gives the following general expression:

$$y_{n+1} = y_n + \int_{x_n}^{x_{n+1}} P_{k-1}(x)dx \tag{7.22}$$

making the theoretical basis for the group of multi-step methods, called in general the Adams methods. In case when $k = 1$, see relation (7.22), we deal with the simplest Adams method, which is identical to the one-step Euler method, discussed in the previous section. In practice, the four-step method ($k = 4$), assuring accuracy comparable to that obtained when using the Runge–Kutta fourth order method (RK 4), is commonly used. In this case it is convenient to use the third degree Newton–Gregory interpolating polynomial (see section 4.1.3) expanded with respect to x_n.

$$\begin{aligned} P_{4-1}(x) \equiv N_3(x) = a_0 &+ a_1(x - x_n) \\ &+ a_2(x - x_n)(x - x_{n-1}) \\ &+ a_3(x - x_n)(x - x_{n-1})(x - x_{n-2}) \end{aligned} \tag{7.23}$$

Assume that numerical integration of Eq. (7.18) is performed with a constant step h. Hence we obtain $x_{n-1} = x_n - h$, $x_{n-2} = x_n - 2h$ and $x_{n-3} = x_n - 3h$. Polynomial (7.23) takes at points x_n, $x_{n-1} = x_n - h$, $x_{n-2} = x_n - 2h$ and $x_{n-3} = x_n - 3h$ the following values:

$$P_3(x_n) = f_n = a_0$$

$$P_3(x_{n-1}) = f_{n-1} = a_0 + a_1(-h)$$

$$P_3(x_{n-2}) = f_{n-2} = a_0 + a_1(-2h) + a_2(-2h)(-h)$$

$$P_3(x_{n-3}) = y_{n-3} = a_0 + a_1(-3h) + a_2(-3h)(-2h) + a_3(-3h)(-2h)(-h)$$

These values make possible in evaluating the polynomial coefficients a_0, a_1, a_2 and a_3 from the formulas:

$$\begin{aligned} a_0 &= f_n \\ a_1 &= \frac{f_n - f_{n-1}}{h} = \frac{\Delta f_n}{h} \\ a_2 &= \frac{f_n - 2f_{n-1} + f_{n-2}}{2h^2} = \frac{\Delta^2 f_n}{2h^2} \\ a_3 &= \frac{f_n - 3f_{n-1} + 3f_{n-2} - f_{n-3}}{6h^3} = \frac{\Delta^3 f_n}{6h^3} \end{aligned} \tag{7.24}$$

Hence, by using relations (7.24) and introducing a new variable $t = x - x_n$, the interpolating polynomial can be written as:

$$P_3(t) = f_n + \frac{\Delta f_n}{h} \cdot t + \frac{\Delta^2 f_n}{2h^2} \cdot t(t+h) + \frac{\Delta^3 f_n}{6h^3} \cdot t(t+h)(t+2h)$$

$$= f_n + \frac{\Delta f_n}{h} \cdot t + \frac{\Delta^2 f_n}{2h^2} \cdot (t^2 + th) + \frac{\Delta^3 f_n}{6h^3} \cdot (t^3 + 3ht^2 + 2h^2 t)$$

$$(7.25)$$

Changes of variable x in an interval $x_n \le x \le x_{n+1} = x_n + h$ correspond to the variations of variable t in an interval $0 \le t \le h$. It implies that polynomial (7.25) should be integrated over the interval $[0, h]$. This integration yelds the formula:

$$\int_0^h P_3(t)dt = \int_0^h \left[f_n + \frac{\Delta f_n}{h} \cdot t + \frac{\Delta^2 f_n}{2h^2} \cdot (t^2 + th) + \frac{\Delta^3 f_n}{6h^3} \cdot (t^3 + 3ht^2 + 2h^2 t) \right] dt$$

$$= f_n h + \frac{h}{2}\Delta f_n + \frac{5h}{12}\Delta^2 f_n + \frac{3h}{8}\Delta^3 f_n$$

$$(7.26)$$

which defines the increment Δy_n. After introduction of the integral (7.26) into (7.22) we obtain the following extrapolating formula:

$$y_{n+1} = y_n + h\left(f_n + \frac{1}{2}\Delta f_n + \frac{5}{12}\Delta^2 f_n + \frac{3}{8}\Delta^3 f_n \right) \qquad (7.27)$$

for the four-step Adams method [3]. The finite differences appearing in this last formula Δf_n, $\Delta^2 f_n$ and $\Delta^3 f_n$ are related to the values of f_n, f_{n-1}, f_{n-2} and f_{n-3} in the following manner, see also relations (7.24)

$$\Delta f_n = f_n - f_{n-1}$$
$$\Delta^2 f_n = f_n - 2f_{n-1} + f_{n-2} \qquad (7.28)$$
$$\Delta^3 f_n = f_n - 3f_{n-1} + 3f_{n-2} - f_{n-3}$$

Substituting the above expressions into formula (7.27) we obtain finally

$$y_{n+1} = y_n + h\left[f_n + \frac{1}{2}(f_n - f_{n-1}) + \frac{5}{12}(f_n - 2f_{n-1} + f_{n-2}) \right.$$

$$\left. + \frac{3}{8}(f_n - 3f_{n-1} + 3f_{n-2} - f_{n-3}) \right]$$

$$= y_n + \frac{h}{24}(55f_n - 59f_{n-1} + 37f_{n-2} - 9f_{n-3}) \qquad (7.29)$$

This formula, equivalent to (7.27), is called the explicit extrapolating formula of the four-step Adams, or Adams–Bashforth method. It was elaborated in 1855 by Adams on request of the famous British artilleryman Bashforth. Adams elaborated this method using the Lagrange polynomial of third degree to interpolate the function $f[x, y(x)]$. The equivalence of relations (7.27) and (7.29) proved above is not accidental. It relults directly from a fact that Newton–Gregory and Lagrange polynomials used to interpolate the function $f[x, y(x)]$ are identical. The identity of these interpolating polynomials can be confirmed in turn by means of the corresponding Weierstrass theorem.

7.3.1 The Adams–Bashforth–Moulthon Method

The values f_n, f_{n-1}, f_{n-2} and f_{n-3} specified in Fig. 7.4(a) serve, according to formula (7.29), to determine predicted approximate value $y_{n+1} = y(x_{n+1})$ of the desired function. At the same time, the value of the function $f_{n+1} = f[x_{n+1}, y_{n+1}]$ is computed. The values $y_{n+1} = y(x_{n+1})$ and $f_{n+1} = f[x_{n+1}, y_{n+1}]$ found in this manner make possible generalization of the Adams–Bashforth method discussed above by adding the correction stage in which consecutive, more accurate approximations of $y_{n+1} = y(x_{n+1})$ are evaluated, namely:

$$y_{n+1}^{(1)}, y_{n+1}^{(2)}, y_{n+1}^{(3)}, y_{n+1}^{(4)}, \dots \tag{7.30}$$

In order to explain the essence of this stage, we find the Lagrange third order polynomial interpolating the function $f[x, y(x)]$ at points (x_{n-2}, f_{n-2}), (x_{n-1}, f_{n-1}), (x_n, f_n) and at the newly determined point (x_{n+1}, f_{n+1}). In case when, $x_{n-2} = x_n - 2h$, $x_{n-1} = x_n - h$ and $x_{n+1} = x_n + h$ the interpolating polynomial takes the form:

$$\begin{aligned}
L_3(x) = \; & f_{n-2} \frac{(x - x_n + h)(x - x_n)(x - x_n - h)}{-6h^3} \\
& + f_{n-1} \frac{(x - x_n + 2h)(x - x_n)(x - x_n - h)}{2h^3} \\
& + f_n \frac{(x - x_n + 2h)(x - x_n + h)(x - x_n - h)}{-2h^3} \\
& + f_{n+1} \frac{(x - x_n + 2h)(x - x_n + h)(x - x_n)}{6h^3}
\end{aligned} \tag{7.31}$$

After introduction of an auxiliary variable $t = x - x_n$, this polynomial can be written in the following more convenient form for further integration:

$$\begin{aligned}
L_3(t) = \; & f_{n-2} \cdot \frac{t^3 - th^2}{-6h^3} + f_{n-1} \cdot \frac{t^3 + ht^2 - 2h^2 t}{2h^3} \\
& + f_n \cdot \frac{t^3 + 2ht^2 - h^2 t - 2h^3}{-2h^3} + f_{n+1} \cdot \frac{t^3 + 3ht^2 + 2h^2 t}{6h^3}
\end{aligned} \tag{7.32}$$

More accurate values $y_{n+1} = y(x_{n+1})$ can be found by means of a formula, similar to expression (7.22), in which the polynomial $P_{k-1}(x)$ is repalced by another polynomial, namely by (7.31). According to the correction rule discussed above we can write the following general formula to evaluate the consecutive approximations of $y_{n+1} = y(x_{n+1})$:

$$y_{n+1}^{(i+1)} = y_n + \int_0^h L_3(t)dt = y_n + \frac{f_{n-2}}{-6h^3}\left(\frac{1}{4}h^4 - \frac{1}{2}h^2h^2\right)$$

$$+ \frac{f_{n-1}}{2h^3}\left(\frac{1}{4}h^4 + \frac{1}{3}h \cdot h^3 - h^2h^2\right)$$

$$+ \frac{f_n}{-2h^3}\left(\frac{1}{4}h^4 + \frac{2}{3}h \cdot h^3 - \frac{1}{2}h^2h^2 - 2h^3h\right)$$

$$+ \frac{f_{n+1}}{6h^3}\left(\frac{1}{4}h^4 + h \cdot h^3 + h^2h^2\right)$$

$$= y_n + \frac{h}{24}\left[f_{n-2} - 5f_{n-1} + 19f_n + 9f_{n+1}\left(y_{n+1}^{(i)}\right)\right] \tag{7.33}$$

where $i = 0, 1, 2, 3, \ldots$ The formula (7.33) is called the implicit Adams–Moulthon correction formula. The word implicit means here that computation of the consecutive $(i + 1)$ approximation of y_{n+1} is found using the value f_{n+1} dependent on the approximation y_{n+1} determined in the previous iteration i. The correction process is continued iteratively until the following condition is satisfied:

$$\left|y_{n+1}^{(i)} - y_{n+1}^{(i+1)}\right| \le \varepsilon \tag{7.34}$$

where ε is a positive, arbitrary small number. When condition (7.34) is satisfied the value $y_{n+1}^{(i+1)}$ is accepted as y_{n+1}. Naturally, such evaluated value y_{n+1} is used in the next step $(n + 2)$, aiming at finding the value $y_{n+2} = y(x_{n+2})$ by means of similar two-stage procedure, where $n = 0, 1, 2, 3, \ldots$ Predictor (7.29) and corrector (7.33) constitute theoretical basis for two-stage prediction and correction method, called commonly the Adams–Boshforth–Moulthon method.

7.3.2 The Milne–Simpson Method

Another popular predictor–corrector method is the Milne–Simpson method. The predictor (extrapolation formula) of this method can be determined on the base of the general relation:

$$y_{n+1} = y_{n-3} + \int_{x_{n-3}}^{x_{n+1}} L_3(x)dx \tag{7.35}$$

where $L_3(x)$ is the Lagrange polynomial of third degree, interpolating the function $f[x, y(x)]$, standing on the right-side of Eq. (7.18). The points (nodes) of this interpolation are (x_{n-3}, f_{n-3}), (x_{n-2}, f_{n-2}), (x_{n-1}, f_{n-1}) and (x_n, f_n). In the special case when $x_{n-3} = x_n - 3h$, $x_{n-2} = x_n - 2h$ and $x_{n-1} = x_n - h$, the interpolating polynomial has the following form:

$$L_3(x) = f_{n-3} \frac{(x - x_n + 2h)(x - x_n + h)(x - x_n)}{-6h^3}$$
$$+ f_{n-2} \frac{(x - x_n + 3h)(x - x_n + h)(x - x_n)}{2h^3}$$
$$+ f_{n-1} \frac{(x - x_n + 3h)(x - x_n + 2h)(x - x_n)}{-2h^3} \qquad (7.36)$$
$$+ f_n \frac{(x - x_n + 3h)(x - x_n + 2h)(x - x_n + h)}{6h^3}$$

After substitution of an auxiliary variable $t = x - x_n$ into polynomial (7.36) it transforms itself to:

$$L_3(t) = f_{n-3} \cdot \frac{1}{-6h^3} \cdot (t^3 + 3ht^2 + 2h^2 t)$$
$$+ f_{n-2} \cdot \frac{1}{2h^3} \cdot (t^3 + 4ht^2 + 3h^2 t)$$
$$+ f_{n-1} \cdot \frac{1}{-2h^3} \cdot (t^3 + 5ht^2 + 6h^2 t) \qquad (7.37)$$
$$+ f_n \cdot \frac{1}{6h^3} \cdot (t^3 + 6ht^2 + 11h^2 t + 6h^3)$$

According to formula (7.35), the polynomial (7.36) should be integrated over the interval $[x_n - 3h, x_n + h]$. In case of using an equivalent polynomial (7.37), integration is performed from $-3h$ to h. The process of integration is described by

$$\int_{-3h}^{h} L_3(t)dt = \frac{f_{n-3}}{-6h^3}\left(\frac{1}{4}h^4 - \frac{81}{4}h^4 + h \cdot h^3 + h27h^3 + h^2 h^2 - h^2 9h^2\right)$$
$$+ \frac{f_{n-2}}{2h^3}\left(\frac{1}{4}h^4 - \frac{81}{4}h^4 + \frac{4}{3}h \cdot h^3 + \frac{4}{3}h27h^3 + \frac{3}{2}h^2 h^2 - \frac{3}{2}h^2 9h^2\right)$$
$$+ \frac{f_{n-1}}{-2h^3}\left(\frac{1}{4}h^4 - \frac{81}{4}h^4 + \frac{5}{3}h \cdot h^3 + \frac{5}{3}h27h^3 + 3h^2 h^2 - 3h^2 9h^2\right)$$
$$+ \frac{f_n}{6h^3}\left(\frac{1}{4}h^4 - \frac{81}{4}h^4 + 2h \cdot h^3 + 2h27h^3\right.$$
$$\left. + \frac{11}{2}h^2 h^2 - \frac{11}{2}h^2 9h^2 + 6h^3 h + 6h^3 h\right)$$
$$= \frac{4h}{3}(0 \cdot f_{n-3} + 2f_{n-2} - f_{n-1} + 2f_n)$$

Finally, we get:

$$y_{n+1} = y_{n-3} + \int_{-3h}^{h} L_3(t)dt = y_{n-3} + \frac{4h}{3}(2f_n - f_{n-1} + 2f_{n-2}) \qquad (7.38)$$

This formula serves to determine first approximation of y_{n+1} and $f_{n+1} = f[x_{n+1}, y_{n+1}]$. The corrector can be determined in a similar way. For this end, we evaluate for the second time the Lagrange polynomial of third degree, interpolating the function $f[x, y(x)]$ at points (x_{n-2}, f_{n-2}), (x_{n-1}, f_{n-1}), (x_n, f_n) and at the newly found point (node) (x_{n+1}, f_{n+1}):

$$
\begin{aligned}
L_3(x) = {} & f_{n-2} \frac{(x - x_n + h)(x - x_n)(x - x_n - h)}{-6h^3} \\
& + f_{n-1} \frac{(x - x_n + 2h)(x - x_n)(x - x_n - h)}{2h^3} \\
& + f_n \frac{(x - x_n + 2h)(x - x_n + h)(x - x_n - h)}{-2h^3} \\
& + f_{n+1} \frac{(x - x_n + 2h)(x - x_n + h)(x - x_n)}{6h^3}
\end{aligned}
\tag{7.39}
$$

Next, the polynomial (7.39) is integrated over the range $[x_{n-1}, x_{n+1}]$. The result of this integration is the following relation:

$$
\begin{aligned}
\int_{x_{n-1}}^{x_{n+1}} L_3(x)\,dx = {} & \int_{x_n-h}^{x_n+h} L_3(x)\,dx \\
= {} & \frac{f_{n-2}}{-6h^3} \left(\frac{1}{4}h^4 - \frac{1}{4}h^4 - \frac{1}{2}h^2 h^2 + \frac{1}{2}h^2 h^2 \right) \\
& + \frac{f_{n-1}}{2h^3} \left(\frac{1}{4}h^4 - \frac{1}{4}h^4 + \frac{1}{3}h\,h^3 + \frac{1}{3}h\,h^3 - h^2 h^2 + h^2 h^2 \right) \\
& + \frac{f_n}{-2h^3} \left(\frac{1}{4}h^4 - \frac{1}{4}h^4 + \frac{2}{3}h\,h^3 + \frac{2}{3}h\,h^3 - \frac{1}{2}h^2 h^2 \right. \\
& \left. + \frac{1}{2}h^2 h^2 - 2h^3 h - 2h^3 h \right) \\
& + \frac{f_{n+1}}{6h^3} \left(\frac{1}{4}h^4 - \frac{1}{4}h^4 + h \cdot h^3 + h \cdot h^3 + h^2 h^2 - h^2 h^2 \right) \\
= {} & \frac{h}{3}(0 \cdot f_{n-2} + f_{n-1} + 4f_n + f_{n+1}) \\
= {} & \frac{h}{3}(f_{n+1} + 4f_n + f_{n-1})
\end{aligned}
\tag{7.40}
$$

that is similar to the Simpson formula given by (5.11). Thus the correction procedure is continued iteratively, according to the formula:

$$
y_{n+1}^{(i+1)} = y_{n-1} + \int_{x_{n-1}}^{x_{n+1}} L_3(x)\,dx = y_{n-1} + \frac{h}{3}\left[f_{n+1}\left(x_{n+1}, y_{n+1}^{(i)}\right) + 4f_n + f_{n-1} \right]
\tag{7.41}
$$

This process is continued until the following condition is satisfied:

$$\left| y_{n+1}^{(i)} - y_{n+1}^{(i+1)} \right| \le \varepsilon \tag{7.42}$$

where ε is a positive, arbitrary small number. When the condition (7.42) is satisfied we take $y_{n+1} \equiv y_{n+1}^{(i+1)}$ and pass to the next step $(n+2)$ in order to find, by means of similar two-stage technique, the values $y_{n+2} = y(x_{n+2})$, for $n = 0, 1, 2, 3, \ldots$. One modification of the Milne–Simpson method consists in adding the following term:

$$\Delta m_{n+1} = \frac{28}{29}(y_n - p_n) \tag{7.43}$$

(modifier) to the predictor formula (7.38). Consequently, the better approximation of the predictor is calculated recursively from the formulas:

$$p_{n+1} = y_{n-3} + \frac{4h}{3}(2f_n - f_{n-1} + 2f_{n-2})$$

$$m_{n+1} = p_{n+1} + \frac{28}{29}(y_n - p_n)$$

$$f_{n+1} = f(x_{n+1}, m_{n+1}) \tag{7.44}$$

$$y_{n+1} = y_{n-1} + \frac{h}{3}(f_{n-1} + 4f_n + f_{n+1})$$

Naturally, the corrected value (corrector) of the desired solution y_{n+1} is evaluated in the same manner, i.e. by using approach expressed by formulas (7.41) and (7.42) [3].

7.3.3 The Hamming Method

A predictor stage of the Hamming method is the same as the predictor stage of the Milne–Simpson method discussed above. It means that the first approximation of the desired solution is calculated according to the formula (7.38), i.e.:

$$y_{n+1} \approx p_{n+1} = y_{n-3} + \int_{-3h}^{h} L_3(t)dt = y_{n-3} + \frac{4h}{3}(2f_n - f_{n-1} + 2f_{n-2}) \tag{7.45}$$

where $n = 3, 4, 5, \ldots$ This first approximation is next used in the process of finding consecutive approximations of the desired solution, which are calculated iteratively according to the following corrector formula [4]:

$$y_{n+1}^{(i+1)} = \frac{-y_{n-2} + 9y_n}{8} + \frac{3h}{8}\left[-f_{n-1} + 2f_n + f_{n+1}\left(x_{n+1}, y_{n+1}^{(i)}\right)\right] \qquad (7.46)$$

The value of the function f_{n+1} included in this formula is calculated on the basis of $y_{n+1}^{(i)}$ evaluated during the previous iteration i. The iterative corrector process is interrupted, when the difference between two consecutive approximations, i.e., $\left|y_{n+1}^{(i)} - y_{n+1}^{(i+1)}\right|$, is smaller than the assumed admissible error ε.

Example 7.3 The subject of considerations in this example is an initial value problem formulated for the following differential equation:

$$\frac{dy(x)}{dx} = x^2 + 2x - y(x)$$

with the condition $y(x = 0) = 1$. This equation was solved by using the Hamming method for $0 \leq x \leq 3$, $dx \equiv h = 0.01$ and $\varepsilon = 10^{-12}$. Initial section (y_0, y_1, y_2, y_3), of the desired solution was determined by using the Runge–Kutta method RK 4. Some of predicted and corrected results obtained over the integration range $[0, 3]$ are presented in the second and third columns of Table 7.4, respectively.

The exact solution (obtained analytically) of the initial value problem under discussion is $y(x) = e^{-x} + x^2$. It makes possible to evaluate the maximum absolute value of the approximation error. It was verified that for $0 \leq x \leq 3$ such error does not exceed the value of 7×10^{-9}. Undoubtedly, the presented results confirm well the general opinion that the Hamming method is quite accurate, stable and easy to program. Thus, it is suitable for the most engineering purposes.

Table 7.4

n	x_n	p_n	y_n
0	0.00		1.000 000 000
1	0.01		0.990 149 833
2	0.02		0.980 598 673
3	0.03		0.971 345 533
	Start	Predicted value	Corrected value
4	0.04	0.962 389 439 213	0.962 389 438 629
5	0.05	0.953 729 423 796	0.953 729 423 952
6	0.06	0.945 364 533 329	0.945 364 533 037
7	0.07	0.937 293 819 408	0.937 293 819 357
8	0.08	0.929 516 345 914	0.929 516 345 839
9	0.09	0.922 031 184 774	0.922 031 184 725
10	0.10	0.914 837 417 539	0.914 837 417 492
...	...	...	...
300	3.00	9.049 787 068 288	9.049 787 068 285

7.4 Examples of Using the RK 4 Method for Integration of Differential Equations Formulated for Some Electrical Rectifier Devices

7.4.1 The Unsymmetrical Voltage Doubler

Figure 7.5 presents the electrical scheme of an unsymmetrical voltage doubler investigated in the present section.

The electronic circuit of this type, often called the Villard's doubler, generates a quasi constant output voltage $u_R(t) = u_2(t)$ with relatively small ripples. The maximum value of this output valtage is close to the doubled amplitude of the alternating control voltage $u_s(t)$. In the analysis presented below, node voltages $u_1(t)$ and $u_2(t)$ have been used as the state variables in the following equations:

$$i_{c1}(t) = C_1 \frac{du_s(t)}{dt} - C_1 \frac{du_1}{dt}$$

$$i_{d1}(t) = I_s \left[\exp\left[\frac{-u_1(t)}{V_T} \right] - 1 \right], \quad I_s = 10^{-8},\,\text{A}, \quad V_T = 0.026,\,\text{V}$$

$$i_{d2}(t) = I_s \left[\exp\left[\frac{u_1(t) - u_2(t)}{V_T} \right] - 1 \right] \tag{7.47}$$

$$i_{c2}(t) = C_2 \frac{du_2(t)}{dt}$$

$$i_R = \frac{1}{R} u_2(t)$$

According to Kirchhoff's law, sums of the currents at nodes 1 and 2, see Fig. 7.5, have to be equal to zero. This law is satisfied when:

$$i_{c1} + i_{d1} - i_{d2} = C_1 \frac{du_s(t)}{dt} - C_1 \frac{du_1(t)}{dt} + I_s \left[\exp\left[\frac{-u_1(t)}{V_T} \right] - 1 \right]$$

$$- I_s \left[\exp\left[\frac{u_1(t) - u_2(t)}{V_T} \right] - 1 \right] = 0$$

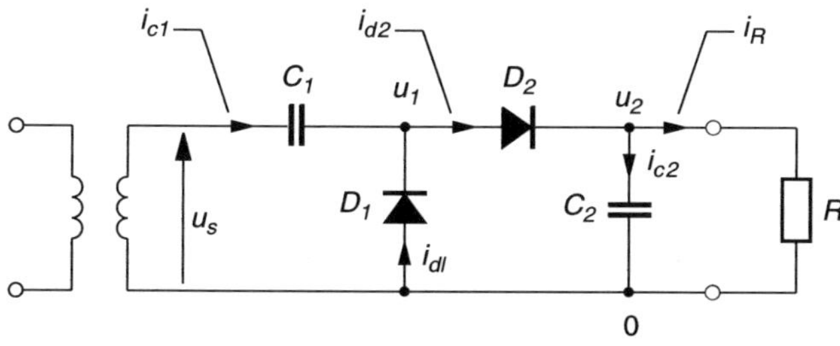

Fig. 7.5

$$i_{d2} - i_{c2} - i_R = I_s \left[\exp\left[\frac{u_1(t) - u_2(t)}{V_T} \right] - 1 \right] - C_2 \frac{du_2(t)}{dt} - \frac{1}{R} u_2(t) = 0$$

The above equations can be replaced by the following system of two first-order differential equations:

$$\frac{dx(t)}{dt} = \frac{du_s(t)}{dt} + \frac{I_s}{C_1} \left[\exp\left[\frac{-x(t)}{V_T} \right] - 1 \right] - \frac{I_s}{C_1} \left[\exp\left[\frac{x(t) - y(t)}{V_T} \right] - 1 \right]$$

$$= f_1[t, x(t), y(t)]$$

$$\frac{dy(t)}{dt} = \frac{I_s}{C_2} \left[\exp\left[\frac{x(t) - y(t)}{V_T} \right] - 1 \right] - \frac{1}{RC_2} y(t) = f_2[t, x(t), y(t)] \qquad (7.48)$$

where $x(t) \equiv u_1(t)$ and $y(t) \equiv u_2(t)$. Equation system (7.48) with the initial conditions $x(t_0) = x_0$ and $y(t_0) = y_0$ constitute the initial value problem. For solving this problem the Runge–Kutta method RK 4 have been used, see Sect. 7.2.3. Thus, the following formulas have been implemented in the computer program P7.5 written for this purpose:

$$x_{n+1} = x_n + \frac{1}{6}(k_1 + 2k_2 + 2k_3 + k_4)$$
$$\qquad (7.49)$$
$$y_{n+1} = y_n + \frac{1}{6}(l_1 + 2l_2 + 2l_3 + l_4)$$

where

$$k_1 = \Delta t \cdot f_1\,(t_n, x_n, y_n)$$
$$l_1 = \Delta t \cdot f_2\,(t_n, x_n, y_n)$$
$$k_2 = \Delta t \cdot f_1\left(t_n + \frac{\Delta t}{2}, x_n + \frac{k_1}{2}, y_n + \frac{l_1}{2} \right)$$
$$l_2 = \Delta t \cdot f_2\left(t_n + \frac{\Delta t}{2}, x_n + \frac{k_1}{2}, y_n + \frac{l_1}{2} \right)$$
$$k_3 = \Delta t \cdot f_1\left(t_n + \frac{\Delta t}{2}, x_n + \frac{k_2}{2}, y_n + \frac{l_2}{2} \right)$$
$$l_3 = \Delta t \cdot f_2\left(t_n + \frac{\Delta t}{2}, x_n + \frac{k_2}{2}, y_n + \frac{l_2}{2} \right)$$
$$k_4 = \Delta t \cdot f_1\,(t_n + \Delta t, x_n + k_3, y_n + l_3)$$
$$l_4 = \Delta t \cdot f_2\,(t_n + \Delta t, x_n + k_3, y_n + l_3)$$

The calculations of $u_1(t) \equiv x(t)$ and $u_2(t) \equiv y(t)$ have been performed for the following data: $u_s(t) = 1(t) \cdot 5 \cdot \sin(2\pi \cdot 50 \cdot t)$, V, $u_1(t = 0) = x_0 = 0$, $u_2(t = 0) = y_0 = 0$, $R = 10000\,\Omega$, $C_1 = 0.001\,\text{F}$, $C_2 = 0.001\,\text{F}$, $\Delta t = 0.000001\,\text{s}$, where $1(t)$ is the unit step function. Some of the most interesting results (transient state and a fragment of the steady-state) are given in Tables 7.5 and 7.6 and illustrated in Figs. 7.6 and 7.7, respectively.

Table 7.5 (Transient state)

t, s	$u_s(t), V$	$u_1(t), V$	$u_2(t), V$
0.000	0.000000	0.000000	0.000000
0.001	1.545085	1.008232	0.356839
0.002	2.938926	1.703043	1.235775
0.003	4.045085	2.251987	1.792836
0.004	4.755282	2.599078	2.155744
0.005	5.000000	2.698879	2.300436
0.006	4.755282	2.445129	2.309237
0.007	4.045085	1.734932	2.309006
0.008	2.938926	0.628773	2.308775
0.009	1.545085	−0.489326	2.308544
0.010	0.000000	−0.490678	2.308313
. . .	. . .	. . .	. . .
0.015	−5.000000	−0.416423	2.307159
0.016	−4.755282	−0.153684	2.306929
0.017	−4.045085	0.556514	2.306698
0.018	−2.938926	1.662673	2.306467
0.019	−1.545085	2.917029	2.445720
0.020	0.000000	3.690106	3.217445
0.021	1.545085	4.461840	3.990436
0.022	2.938926	5.156479	4.689203
0.023	4.045085	5.705252	5.246091
0.024	4.755282	6.052173	5.608822
0.025	5.000000	6.151825	5.753318

Table 7.6 (Steady-state)

t, s	$u_s(t), V$	$u_1(t), V$	$u_2(t), V$
0.200	0.000000	4.608779	9.179598
0.201	1.545085	6.153864	9.178680
0.202	2.938926	7.547709	9.177763
0.203	4.045085	8.653863	9.176845
0.204	4.755282	9.364059	9.175928
0.205	5.000000	9.588190	9.195594
0.206	4.755282	9.335794	9.202352
0.207	4.045085	8.625595	9.201432
0.208	2.938926	7.519436	9.200512
0.209	1.545085	6.125594	9.199592
0.210	0.000000	4.580508	9.198672
0.211	−1.545085	3.035424	9.197752
0.212	−2.938926	1.641583	9.196833
0.213	−4.045085	0.535425	9.195913
0.214	−4.755282	−0.174770	9.194993
0.215	−5.000000	−0.399888	9.194074
0.216	−4.755282	−0.144122	9.193154
0.217	−4.045085	0.566076	9.192235
0.218	−2.938926	1.672236	9.191316
0.219	−1.545085	3.066078	9.190397
0.220	0.000000	4.611163	9.189478

a)

$u_s(t), V$
4
2
0
-2
-4
-6
0
10
20
30
t, ms

b)

$u_1(t), V$
4
2
0
-2
0
10
20
30
t, ms

c)

$u_2(t), V$
4
2
0
0
10
20
30
t, ms

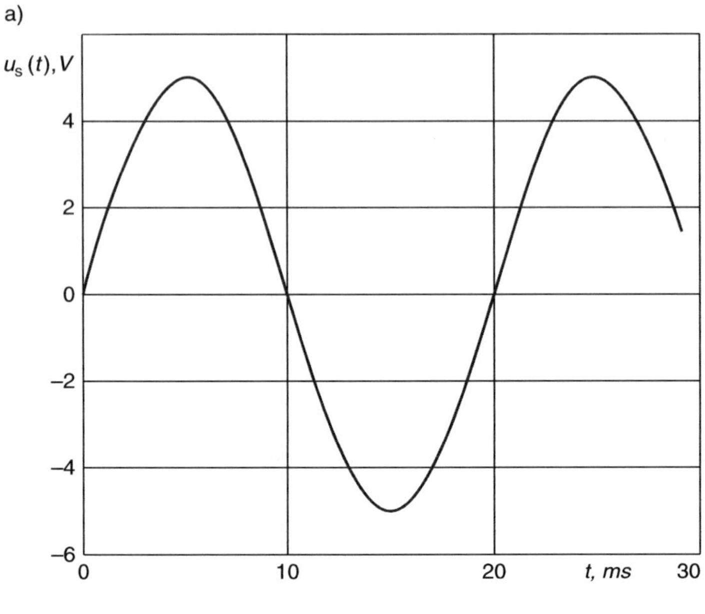

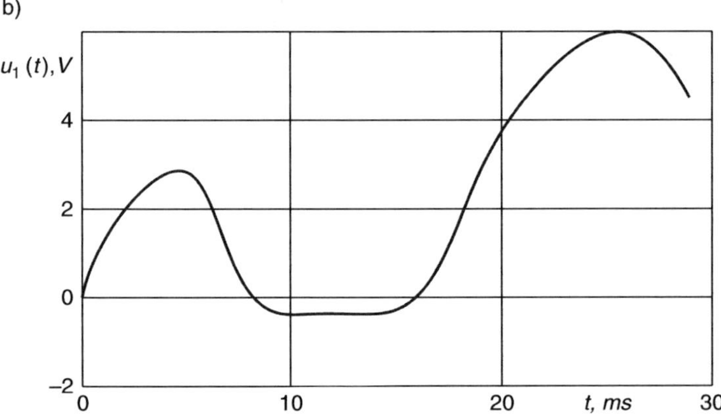

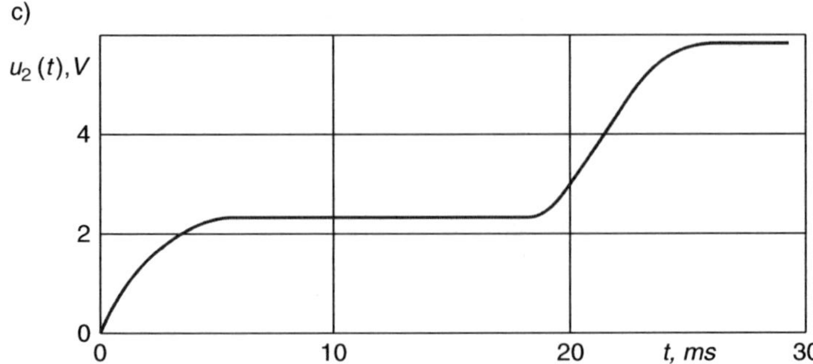

Fig. 7.6

a)

$u_s(t), V$
4
2
0
-2
-4
-6
200
210
220
t, ms
230

b)

V
8
u_2
u_1
6
~4.60
4
2
0
-2
200
210
220
t, ms
230

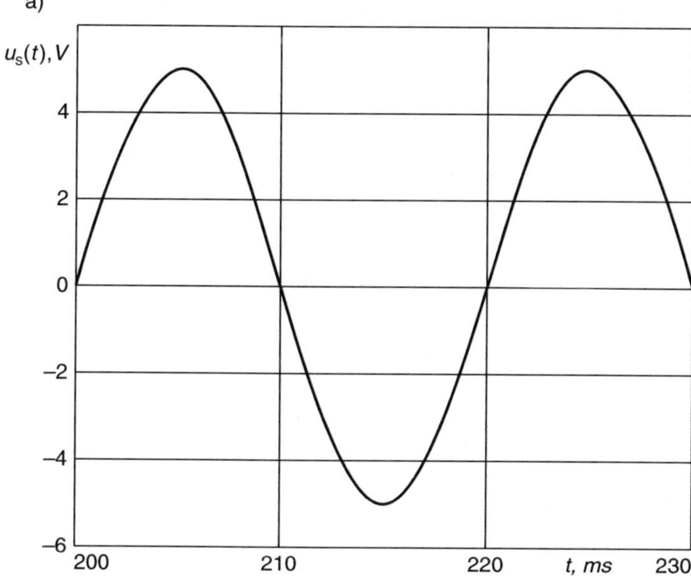

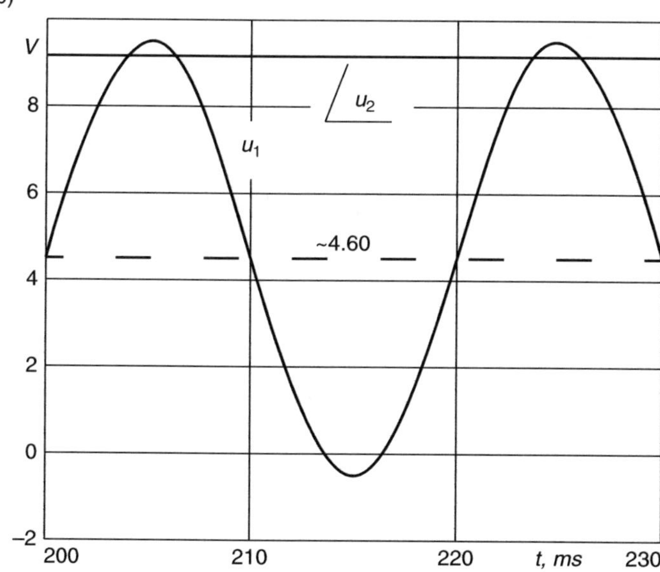

Fig. 7.7

All the results presented above are fully consistent with the corresponding results obtained by means of the PSpice simulation program intented for analysis of electrical and electronic circuits [5].

7.4.2 The Full-Wave Rectifier Integrated with the Three-Element Low-Pass Filter

The electrical scheme of the full-wave rectifier with a three-element low-pass filter is shown in Fig. 7.8.

Let us assume that the circuit is supplied by alternating voltage, which after transformation is equal to $u_s(t) = 10\sin(2\pi \cdot 50 \cdot t)$, V. The time-domain analysis of this circuit consists in determining the functions of voltage $u_1(t)$, current $i_L(t)$ and voltage $u_2(t)$, which are treated as the state variables. The instant values of currents and voltages in the individual branches, see Fig. 7.8, are related with state variables mentioned above by the following differential equations:

$$i_{c1}(t) = C_1\frac{du_1(t)}{dt}, \qquad i_R(t) = \frac{u_1(t)}{R}$$

$$i_L(t) = i_{c1}(t) + i_R(t)$$

$$u_L(t) = u_2(t) - u_1(t) = L\frac{di_L(t)}{dt} \tag{7.50}$$

$$i_{c2}(t) = C_2\frac{du_2(t)}{dt}, \qquad i_p(t) = I_s\left[\exp\left[\frac{|u_s(t)| - u_2^{(t)}}{2 \cdot V_T}\right] - 1\right]$$

$$i_p(t) = i_{c2}(t) + i_L(t)$$

where

$$i(t) = I_s\left[\exp\left[\frac{u_d(t)}{V_T}\right] - 1\right], \qquad I_s = 10^{-8}\,A, \qquad V_T = 0.026\,V$$

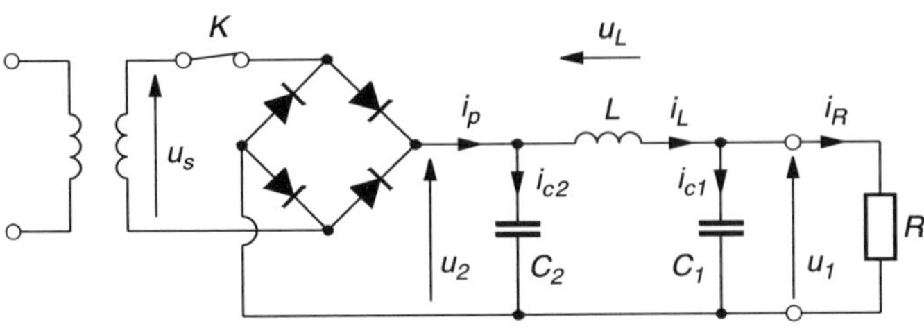

Fig. 7.8

is the current–voltage characteristic of one diode of the Gretz bridge [6]. After some rearrangements, Eq. (7.50) can be written in the form of the equivalent equation system:

$$\frac{du_1(t)}{dt} = \frac{1}{C_1} \cdot i_L(t) - \frac{1}{C_1 R} \cdot u_1(t)$$

$$\frac{di_L(t)}{dt} = \frac{1}{L} \cdot u_2(t) - \frac{1}{L} \cdot u_1(t) \tag{7.51}$$

$$\frac{du_2(t)}{dt} = \frac{1}{C_2} \cdot I_s \left[\exp \left[\frac{|u_s(t)| - u_2(t)}{2 \cdot V_T} \right] - 1 \right] - \frac{1}{C_2} \cdot i_L(t)$$

In the last equation of this system, only absolute values of the control voltage $|u_s(t)|$ are used. This is the consequence of rectifying properties and symmetry of the Gretz bridge being used. By assuming the notation

$$u_1(t) \equiv x(t), \quad i_L(t) \equiv y(t), \quad u_2(t) \equiv z(t)$$

the equation system (7.51) can be written as follows:

$$\frac{dx(t)}{dt} = \frac{1}{C_1} \cdot y(t) - \frac{1}{C_1 R} \cdot x(t) = f_1[t, x(t), y(t), z(t)]$$

$$\frac{dy(t)}{dt} = \frac{1}{L} \cdot z(t) - \frac{1}{L} \cdot x(t) = f_2[t, x(t), y(t), z(t)] \tag{7.52}$$

$$\frac{dz(t)}{dt} = \frac{1}{C_2} \cdot I_s \left[\exp \left[\frac{|u_s(t)| - z(t)}{2 \cdot V_T} \right] - 1 \right] - \frac{1}{C_2} \cdot y(t) = f_3[t, x(t), y(t), z(t)]$$

The above three differential equations together with the given initial conditions: $t_0, x_0 = x(t_0), y_0 = y(t_0), z_0 = z(t_0)$, and the time dependent control voltage $u_s(t)$ constitute the initial value problem. Also, in this case the Runge–Kutta method RK 4 has been used for solving this problem. Consequently, the following computational formulas have been implemented in the computer program P7.8 written for this purpose.

$$x_{n+1} = x_n + \frac{1}{6}(k_1 + 2k_2 + 2k_3 + k_4)$$

$$y_{n+1} = y_n + \frac{1}{6}(l_1 + 2l_2 + 2l_3 + l_4) \tag{7.53}$$

$$z_{n+1} = z_n + \frac{1}{6}(m_1 + 2m_2 + 2m_3 + m_4)$$

where

$$k_1 = \Delta t \cdot f_1 (t_n, x_n, y_n, z_n)$$

$$l_1 = \Delta t \cdot f_2 (t_n, x_n, y_n, z_n)$$

$$m_1 = \Delta t \cdot f_3 (t_n, x_n, y_n, z_n)$$

$$k_2 = \Delta t \cdot f_1 \left(t_n + \frac{\Delta t}{2}, x_n + \frac{k_1}{2}, y_n + \frac{l_1}{2}, z_n + \frac{m_1}{2} \right)$$

$$l_2 = \Delta t \cdot f_2 \left(t_n + \frac{\Delta t}{2}, x_n + \frac{k_1}{2}, y_n + \frac{l_1}{2}, z_n + \frac{m_1}{2} \right)$$

$$m_2 = \Delta t \cdot f_3 \left(t_n + \frac{\Delta t}{2}, x_n + \frac{k_1}{2}, y_n + \frac{l_1}{2}, z_n + \frac{m_1}{2} \right)$$

$$k_3 = \Delta t \cdot f_1 \left(t_n + \frac{\Delta t}{2}, x_n + \frac{k_2}{2}, y_n + \frac{l_2}{2}, z_n + \frac{m_2}{2} \right)$$

$$l_3 = \Delta t \cdot f_2 \left(t_n + \frac{\Delta t}{2}, x_n + \frac{k_2}{2}, y_n + \frac{l_2}{2}, z_n + \frac{m_2}{2} \right)$$

$$m_3 = \Delta t \cdot f_3 \left(t_n + \frac{\Delta t}{2}, x_n + \frac{k_2}{2}, y_n + \frac{l_2}{2}, z_n + \frac{m_2}{2} \right)$$

$$k_4 = \Delta t \cdot f_1 (t_n + \Delta t, x_n + k_3, y_n + l_3, z_n + m_3)$$

$$l_4 = \Delta t \cdot f_2 (t_n + \Delta t, x_n + k_3, y_n + l_3, z_n + m_3)$$

$$m_4 = \Delta t \cdot f_{31} (t_n + \Delta t, x_n + k_3, y_n + l_3, z_n + m_3)$$

The computer program P7.8 mentioned above makes it possible calculating the discrete values of functions $u_1(t)$, $i_L(t)$ and $u_2(t)$. The calculations have been carried out for the following data:

$$u_1(t = 0) = 0, \quad i_L(t = 0) = 0, \quad u_2(t = 0) = 0, \quad u_s(t) = 10 \sin(2\pi \cdot 50 \cdot t) \cdot 1(t), \quad V$$

$$R = 50, \ \Omega, \ C_1 = C_2 = 1000, \ \mu F, \ L = 0.1, \ H, \ \Delta t = 0.00001$$

where $1(t)$ is the unit step function. This unit step function is implemented by closing the key K at $t = 0$. Most interesting results (transient state and a fragment of the steady-state) obtained in this example are shown in Tables 7.7 and 7.8 and illustrated in Figs. 7.9 and 7.10.

The voltage functions $u_1(t)$ and $u_2(t)$ presented in Fig. 7.11 have been evaluated additionally over the steady-state for the large loading resistance $R = 200 \ \Omega$, i.e., for the smaller output current.

Also of interest is evaluating the voltage functions $u_1(t)$ and $u_2(t)$ over the turn-off range. It can be done by assuming that at any moment t_0 of the steady-state, the control voltage $u_s(t)$ rapidly decays, i.e., $u_s(t) = 0$ when $t \geq t_0$. Of course,

Table 7.7 (Transient state)

t, s	$u_1(t), V$	$i_L(t), A$	$u_2(t), V$
0.000	0.000000	0.000000	0.000000
0.001	0.001599	0.007073	2.075171
0.002	0.023661	0.042071	4.870406
0.003	0.092803	0.101948	7.097102
0.004	0.229249	0.179322	8.543801
0.005	0.444541	0.264967	9.093901
0.006	0.740428	0.349458	8.900919
0.007	1.111088	0.427389	8.511956
0.008	1.547555	0.497010	8.049006
0.009	2.039629	0.557010	7.521143
0.010	2.576131	0.606309	6.938546
0.011	3.145120	0.644011	6.312390
0.012	3.734136	0.669482	5.654603
0.013	4.332514	0.690456	7.083565
0.014	4.945724	0.722844	8.524475
0.015	5.581145	0.758900	9.052700
0.016	6.238178	0.789163	8.685337
0.017	6.905755	0.806423	7.898242
0.018	7.569908	0.808976	7.089315
0.019	8.216146	0.796887	6.285174
0.020	8.830489	0.770527	5.500300
0.021	9.399738	0.730544	4.748663
0.022	9.911818	0.678878	4.858998
0.023	10.365789	0.637741	7.084616
0.024	10.777693	0.610777	8.527862
0.025	11.158447	0.589806	9.063563

the moment t_0 is a beginning of the turn-off range. Thus, the results presented in Table 7.9 and illustrated in Fig. 7.12 have been performed for $t \geq t_0 = 1.01$ s.

The functions $u_1(t)$ and $u_2(t)$ depicted in Fig. 7.12 illustrate the process of resonance discharging the filtering section LC. In the turn-off state, the Gretz bridge "cuts-off" the filtering section from the network transformer, resulting in total decay of the current flowing by the diodes of the Gretz bridge. A period of

Table 7.8 (Steady-state)

t, s	$u_1(t), V$	$i_L(t), A$	$u_2(t), V$
1.000	8.492006	0.176141	8.316112
1.002	8.498514	0.169047	7.969746
1.004	8.484778	0.158276	8.544541
1.006	8.471246	0.168583	9.015204
1.008	8.478055	0.175986	8.669466
1.010	8.482006	0.176141	8.316112
1.012	8.498514	0.169047	7.969746
1.014	8.484778	0.158276	8.544521
1.016	8.471246	0.168583	9.015204
1.018	8.478055	0.175986	8.669466
1.020	8.492006	0.176141	8.316112

a)

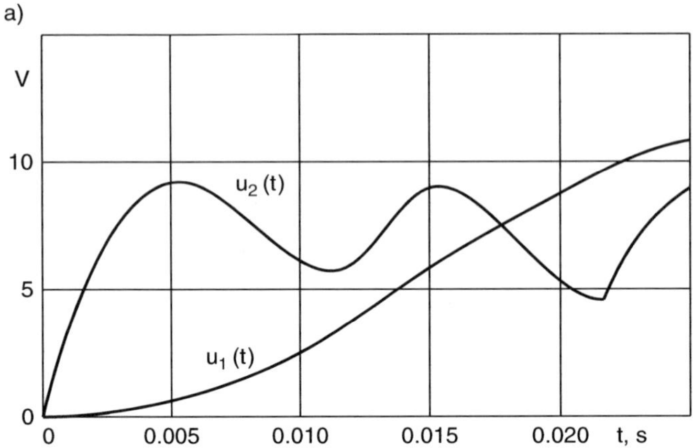

b)

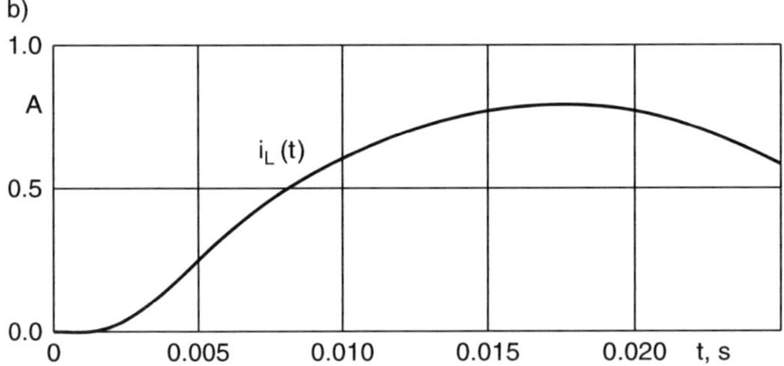

Fig. 7.9

oscillations illustrated in Fig. 7.12 is approximately equal to $T = 0.044$, s. A similar value results directly from the fundamental known formula: $T = 2\pi \sqrt{LC_{eff}}$, where $C_{eff} = C_1 C_2/(C_1 + C_2)$, $C_1 = C_2 = 1000$, µF and $L = 0.1$, H, see Fig. 7.8, [7]. The identical responses corresponding to the ones presented above have been obtained by using the specialized computer program PSpice [5].

7.4.3 The Quadruple Symmetrical Voltage Multiplier

The electrical scheme of the quadruple symmetrical voltage multiplier, analyzed in this example, is shown in Fig. 7.13.

Let us assume that the multipler under analysis is supplied with alternating voltage, which after transformation is equal to $u_s(t) = 5 \cdot \sin(2\pi \cdot 50 \cdot t)$, V. Instant values of currents flowing in the individual branches of this nonlinear circuit can be evaluated from the following differential equations.

a)

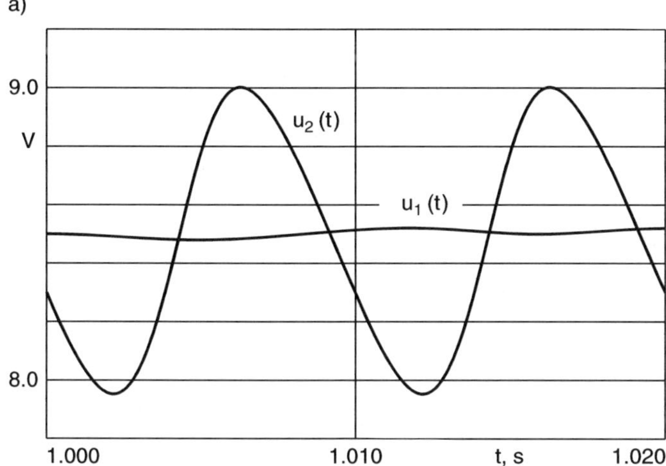

b)

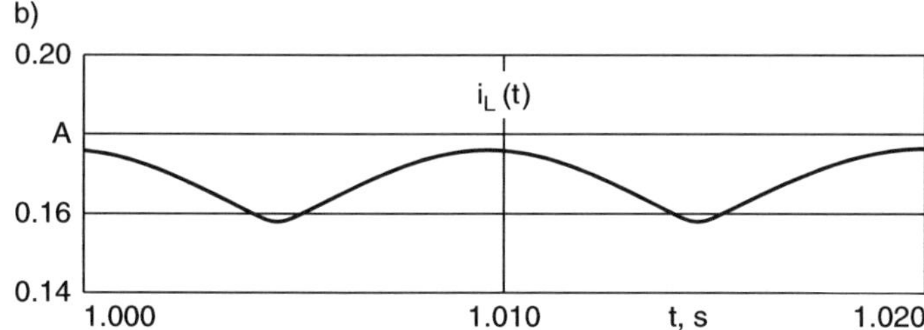

Fig. 7.10

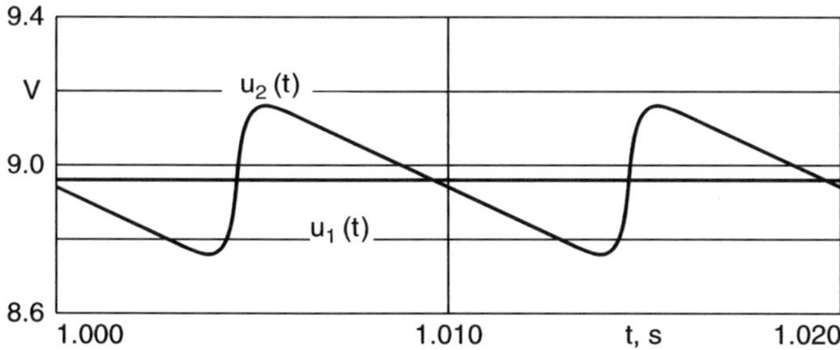

Fig. 7.11

Table 7.9

t, s	$u_1(t), V$	$i_L(t), A$	$u_2(t), V$
1.010	8.492006	0.176141	8.316112
1.020	8.218277	0.086650	6.904247
1.030	6.976727	−0.012062	6.613545
1.040	5.683332	0.027273	6.652177
1.050	5.361434	0.118177	5.886333
1.060	5.483350	0.100836	4.676457
1.070	4.952076	0.004697	4.149416
1.080	3.901983	−0.016669	4.314053
1.090	3.359306	0.056850	4.145133
1.100	3.508353	0.092291	3.314867
...	...	...	...
1.180	1.281827	0.032062	1.765368
1.190	1.476182	0.050058	1.297615
1.200	1.491962	0.008155	0.977724

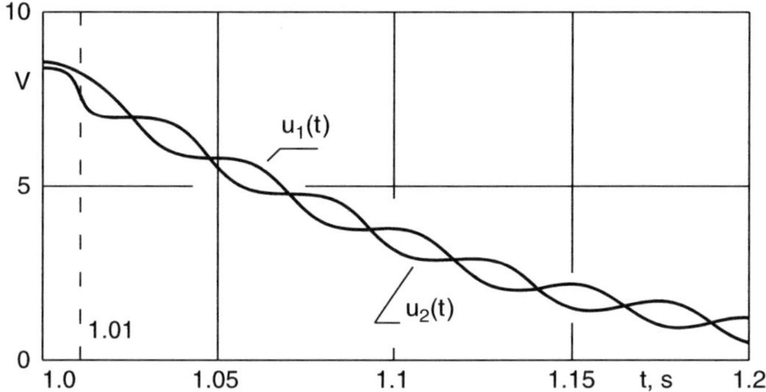

Fig. 7.12

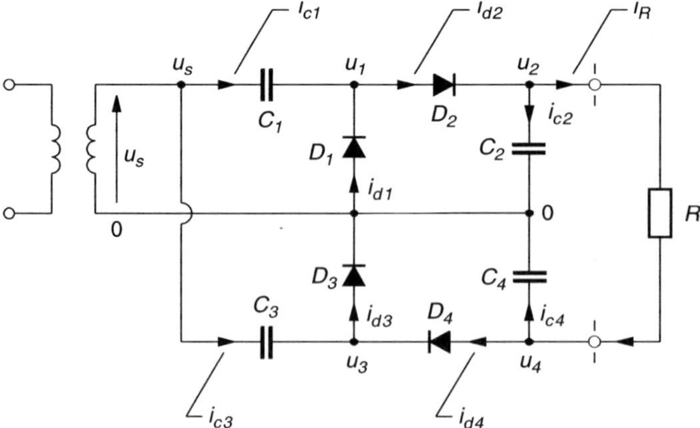

Fig. 7.13

$$i_{c1}(t) = C_1 \frac{d[u_s(t) - u_1(t)]}{dt} = C_1 \frac{du_s(t)}{dt} - C_1 \frac{du_1(t)}{dt}$$

$$i_{d1}(t) = I_s \left[\exp\left[\frac{u_1(t)}{V_T} \right] - 1 \right],$$

$$i_{d2}(t) = I_s \left[\exp\left[\frac{u_1(t) - u_2(t)}{V_T} \right] - 1 \right] \tag{7.54}$$

$$i_{c2}(t) = C_2 \frac{du_2(t)}{dt}$$

$$i_{c3}(t) = C_3 \frac{d[u_s(t) - u_3(t)]}{dt} = C_3 \frac{du_s(t)}{dt} - C_3 \frac{du_3(t)}{dt}$$

$$i_{d3}(t) = I_s \left[\exp\left[\frac{u_3(t)}{V_T} \right] - 1 \right]$$

$$i_{c4}(t) = C_2 \frac{du_4(t)}{dt}$$

$$i_r(t) = \frac{1}{R}[(u_2(t) - u_4(t)]$$

where $u_1(t)$, $u_2(t)$, $u_3(t)$ and $u_4(t)$ are the desired functions of nodal voltages. The function

$$i(t) = I_s \left[\exp\left[\frac{u_d(t)}{V_T} \right] - 1 \right], \quad I_s = 10^{-8} A, \quad V_T = 0.026 \text{ V}$$

represents the current–voltage characteristic of the individual diode [6]. According to Kirchhoff's law, sums of currents at nodes 1, 2, 3, 4, are equal to zero. In consequence, the following equations can be formulated:

$$i_{c1} + i_{d1} - i_{d2} = C_1 \frac{du_s(t)}{dt} - C_1 \frac{du_1(t)}{dt} + I_s \left[\exp\left[\frac{-u_1(t)}{V_T} \right] - 1 \right]$$

$$- I_s \left[\exp\left[\frac{u_1(t) - u_2(t)}{V_T} \right] - 1 \right] = 0$$

$$i_{d2} - i_{c2} - i_r = I_s \left[\exp\left[\frac{u_1(t) - u_2(t)}{V_T} \right] - 1 \right] - C_2 \frac{du_2(t)}{dt} - \frac{1}{R}[u_2(t) - u_4(t)] = 0$$

$$i_{c3} - i_{d3} + i_{d4} = C_3 \frac{du_s(t)}{dt} - C_3 \frac{du_3(t)}{dt} - I_s \left[\exp\left[\frac{u_3(t)}{V_T} \right] - 1 \right]$$

$$+ I_s \left[\exp\left[\frac{u_4(t) - u_3(t)}{V_T} \right] - 1 \right] = 0$$

$$i_r - i_{c4} - i_{d4} = -\frac{1}{R}[u_2(t) - u_4(t)] - C_4 \frac{du_4(t)}{dt} - I_s \left[\exp\left[\frac{u_4(t) - u_3(t)}{V_T} \right] - 1 \right] = 0$$

Naturally, the above differential equations can be written in the form of the equivalent equation system, namely:

$$\frac{du_1(t)}{dt} = \frac{du_s(t)}{dt} + \frac{I_s}{C_1}\left[\exp\left[\frac{-u_1(t)}{V_T}\right] - 1\right] - \frac{I_s}{C_1}\left[\exp\left[\frac{u_1(t) - u_2(t)}{V_T}\right] - 1\right]$$

$$\frac{du_2(t)}{dt} = \frac{I_s}{C_2}\left[\exp\left[\frac{u_1(t) - u_2(t)}{V_T}\right] - 1\right] - \frac{1}{RC_2}[u_2(t) - u_4(t)] \qquad (7.55)$$

$$\frac{du_3(t)}{dt} = \frac{du_s(t)}{dt} + \frac{I_s}{C_3}\left[\exp\left[\frac{u_3(t)}{V_T}\right] - 1\right] + \frac{I_s}{C_3}\left[\exp\left[\frac{u_4(t) - u_3(t)}{V_T}\right] - 1\right]$$

$$\frac{du_4(t)}{dt} = \frac{1}{RC_4}[u_2(t) - u_4(t)] - \frac{I_s}{C_4}\left[\exp\left[\frac{u_4(t) - u_3(t)}{V_T}\right] - 1\right]$$

In order to simplify the description of this equation system the following auxiliary notation $p(t) \equiv u_1(t), q(t) \equiv u_2(t), v(t) \equiv u_3(t), z(t) \equiv u_4(t)$ have been introduced. Consequently, the system (7.55) takes the form:

$$\frac{dp(t)}{dt} = \frac{du_s(t)}{dt} + \frac{I_s}{C_1}\left[\exp\left[\frac{-p(t)}{V_T}\right] - 1\right] - \frac{I_s}{C_1}\left[\exp\left[\frac{p(t) - q(t)}{V_T}\right] - 1\right]$$

$$= f_1[t, p(t), q(t)]$$

$$\frac{dq(t)}{dt} = \frac{I_s}{C_2}\left[\exp\left[\frac{p(t) - q(t)}{V_T}\right] - 1\right] - \frac{1}{RC_2}[q(t) - z(t)] = f_2[t, p(t), q(t), z(t)]$$

$$(7.56)$$

$$\frac{dv(t)}{dt} = \frac{du_s(t)}{dt} + \frac{I_s}{C_3}\left[\exp\left[\frac{v(t)}{V_T}\right] - 1\right] + \frac{I_s}{C_3}\left[\exp\left[\frac{z(t) - v(t)}{V_T}\right] - 1\right]$$

$$= f_3[t, q(t), v(t), z(t)]$$

$$\frac{dz(t)}{dt} = \frac{1}{RC_4}[q(t) - z(t)] - \frac{I_s}{C_4}\left[\exp\left[\frac{z(t) - v(t)}{V_T}\right] - 1\right] = f_3[t, q(t), v(t), z(t)]$$

Such formulated equation system together with the given control voltage $u_s(t)$ and initial conditions $p(t_0) = p_0, q(t_0) = q_0, v(t_0) = v_0, z(t_0) = z_0$, has a form of the typical four-dimensional initial value problem. Also, in this case the Runge–Kutta method RK 4 have been used for the numerical solving. Thus, the following formulas have been implemented in the corresponding computer program P7.13.

$$p_{n+1} = p_n + \frac{1}{6}(k_1 + 2k_2 + 2k_3 + k_4)$$

$$q_{n+1} = q_n + \frac{1}{6}(l_1 + 2l_2 + 2l_3 + l_4)$$

$$(7.57)$$

$$v_{n+1} = v_n + \frac{1}{6}(m_1 + 2m_2 + 2m_3 + m_4)$$

$$z_{n+1} = z_n + \frac{1}{6}(n_1 + 2n_2 + 2n_3 + n_4)$$

where

$$k_1 = \Delta t \cdot f_1(t_n, \; p_n, \; q_n, \; v_n, \; z_n)$$

$$l_1 = \Delta t \cdot f_2(t_n, \; p_n, \; q_n, \; v_n, \; z_n)$$

$$m_1 = \Delta t \cdot f_3(t_n, \; p_n, \; q_n, \; v_n, \; z_n)$$

$$n_1 = \Delta t \cdot f_4(t_n, \; p_n, \; q_n, \; v_n, \; z_n)$$

$$k_2 = \Delta t \cdot f_1 \left(t_n + \frac{\Delta t}{2}, \; p_n + \frac{k_1}{2}, \; q_n + \frac{l_1}{2}, \; v_n + \frac{m_1}{2}, \; z_n + \frac{n_1}{2} \right)$$

$$l_2 = \Delta t \cdot f_2 \left(t_n + \frac{\Delta t}{2}, \; p_n + \frac{k_1}{2}, \; q_n + \frac{l_1}{2}, \; v_n + \frac{m_1}{2}, \; z_n + \frac{n_1}{2} \right)$$

$$m_2 = \Delta t \cdot f_3 \left(t_n + \frac{\Delta t}{2}, \; p_n + \frac{k_1}{2}, \; q_n + \frac{l_1}{2}, \; v_n + \frac{m_1}{2}, \; z_n + \frac{n_1}{2} \right)$$

$$n_2 = \Delta t \cdot f_4 \left(t_n + \frac{\Delta t}{2}, \; p_n + \frac{k_1}{2}, \; q_n + \frac{l_1}{2}, \; v_n + \frac{m_1}{2}, \; z_n + \frac{n_1}{2} \right)$$

$$k_3 = \Delta t \cdot f_1 \left(t_n + \frac{\Delta t}{2}, \; p_n + \frac{k_2}{2}, \; q_n + \frac{l_2}{2}, \; v_n + \frac{m_2}{2}, \; z_n + \frac{n_2}{2} \right)$$

$$l_3 = \Delta t \cdot f_2 \left(t_n + \frac{\Delta t}{2}, \; p_n + \frac{k_2}{2}, \; q_n + \frac{l_2}{2}, \; v_n + \frac{m_2}{2}, \; z_n + \frac{n_2}{2} \right)$$

$$m_3 = \Delta t \cdot f_3 \left(t_n + \frac{\Delta t}{2}, \; p_n + \frac{k_2}{2}, \; q_n + \frac{l_2}{2}, \; v_n + \frac{m_2}{2}, \; z_n + \frac{n_2}{2} \right)$$

$$n_3 = \Delta t \cdot f_4 \left(t_n + \frac{\Delta t}{2}, \; p_n + \frac{k_2}{2}, \; q_n + \frac{l_2}{2}, \; v_n + \frac{m_2}{2}, \; z_n + \frac{n_2}{2} \right)$$

$$k_4 = \Delta t \cdot f_1(t_n + \Delta t, \; p_n + k_3, \; q_n + l_3, \; v_n + m_3, \; z_n + n_3)$$

$$l_4 = \Delta t \cdot f_2(t_n + \Delta t, \; p_n + k_3, \; q_n + l_3, \; v_n + m_3, \; z_n + n_3)$$

$$m_4 = \Delta t \cdot f_3(t_n + \Delta t, \; p_n + k_3, \; q_n + l_3, \; v_n + m_3, \; z_n + n_3)$$

$$n_4 = \Delta t \cdot f_4(t_n + \Delta t, \; p_n + k_3, \; q_n + l_3, \; v_n + m_3, \; z_n + n_3)$$

The above mentioned computer program P7.13 has been used to calculate discrete values of functions $u_1(t)$, $u_2(t)$, $u_3(t)$ and $u_4(t)$.

The calculations have been performed for the following input data: $u_1(t = 0) = p_0 = 0$, $u_2(t = 0) = q_0 = 0$, $u_3(t = 0) = v_0 = 0$, $u_4(t = 0) = z_0 = 0$, $C_1 = 10$, μF, $C_2 = 100$, μF $R = 50$, Ω, $\Delta t = 0.00001$, s and $u_s(t) = 5 \cdot \sin(2\pi \cdot 50 \cdot t) \cdot 1(t)$, V, where $1(t)$ is the unit step function. Most interesting results, representing the transient state and a fragment of the steady-state, are given in Tables 7.10 and 7.11 and illustrated in Fig. 7.14.

Table 7.10 (Transient state)

t, s	$u_s(t), V$	$u_1(t), V$	$u_2(t), V$	$u_3(t), V$	$u_4(t), V$
0.000	0.000000	0.000000	0.000000	0.000000	0.000000
0.001	1.545085	0.978257	0.566722	0.429552	0.000105
0.002	2.938926	1.672884	1.265473	0.425420	0.000568
0.003	4.045086	2.221521	1.822217	0.417301	0.001346
0.004	4.755283	2.568218	2.184708	0.401478	0.002356
0.005	5.000000	2.667626	2.328880	0.356556	0.003492
0.006	4.755283	2.413791	2.336832	0.093816	0.004658
0.007	4.045086	1.703594	2.335659	−0.504612	−0.105937
0.008	2.938926	0.597435	2.334309	−1.061272	−0.654085
0.009	1.545085	−0.429459	2.332646	−1.759503	−1.348033
0.010	0.000000	−0.430811	2.330615	−2.531708	−2.118882
0.011	−1.545085	−0.429552	2.328197	−3.362416	−2.890840
0.012	−2.938926	−0.425420	2.325409	−3.995884	−3.588426
0.013	−4.045086	−0.417301	2.322306	−4.543367	−4.143999
0.014	−4.755283	−0.401478	2.318975	−4.888928	−4.505304
0.015	−5.000000	−0.356556	2.315519	−4.987330	−4.648163
0.016	−4.755283	−0.093816	2.312035	−4.733279	−4.654013
0.017	−4.045086	0.616381	2.308553	−4.023081	−4.650531
0.018	−2.938926	1.722540	2.305075	−2.916922	−4.647054
0.019	−1.545085	2.914734	2.503234	−1.523081	−4.643565
0.020	0.000000	3.686072	3.273218	0.022003	−4.639802
0.021	1.545085	4.455915	4.044310	0.429552	−4.635652
0.022	2.938926	5.148521	4.741028	0.425420	−4.631134
0.023	4.045086	5.695146	5.295731	0.417301	−4.626303
0.024	4.755283	6.039861	5.656154	0.401478	−4.621244
0.025	5.000000	6.137514	5.798036	0.356555	−4.616063

Table 7.11 (Steady-state)

t, s	$u_s(t), V$	$u_1(t), V$	$u_2(t), V$	$u_3(t), V$	$u_4(t), V$
200	0.000000	4.661485	9.012270	−4.472649	−9.106814
201	1.545085	6.206570	9.003215	−2.927564	−9.097759
202	2.938926	7.600412	8.994169	−1.533723	−9.088713
203	4.045086	8.706569	8.985132	−0.427565	−9.079676
204	4.755283	9.386994	9.005873	0.282363	−9.070646
205	5.000000	9.484338	9.144166	0.356519	−9.061567
206	4.755283	9.229748	9.144936	0.093794	−9.052465
207	4.045086	8.519549	9.135842	−0.616401	−9.043370
208	2.938926	7.413389	9.126757	−1.722561	−9.034285
209	1.545085	6.019547	9.117681	−3.116403	−9.025209
210	0.000000	4.474462	9.108614	−4.661488	−9.016142
211	−1.545085	2.929377	9.099556	−6.206573	−9.007084
212	−2.938926	1.535536	9.090507	−7.600414	−8.998036
213	−4.045086	0.429379	9.081467	−8.706572	−8.988795
214	−4.755283	−0.280568	9.072435	−9.388717	−9.008014
215	−5.000000	−0.356516	9.063354	−9.486269	−9.140971
216	−4.755283	−0.093792	9.054249	−9.231679	−9.146865
217	−4.045086	0.616403	9.045153	−8.521480	−9.137769
218	−2.938926	1.722562	9.036066	−7.415320	−9.128682
219	−1.545085	3.116404	9.026989	−6.021478	−9.119604
220	0.000000	4.661489	9.017920	−4.476393	−9.110535

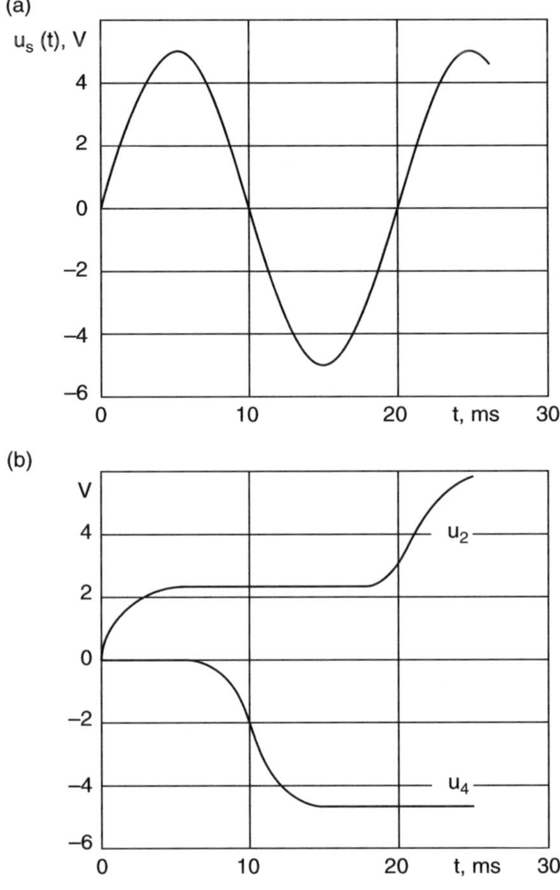

Fig. 7.14

Correctness of the presented results has been confirmed by comparison of them with the corresponding results obtained by means of the PSpice simulator. Also, in this case an excellent conformability has been achieved.

7.5 An Example of Solution of Riccati Equation Formulated for a Nonhomogenous Transmission Line Segment

Sections of nonhomogenous transmission lines are broadly used in the UHF and microwave equipment, such as the broadband impedance transformers or different kinds of filters. A description of such distributed circuits is usually made by using the reflection coefficient function $\Gamma(x, f)$, defined at any plane (cross-section), for example at the plane x as shown in Fig. 7.15.

a)

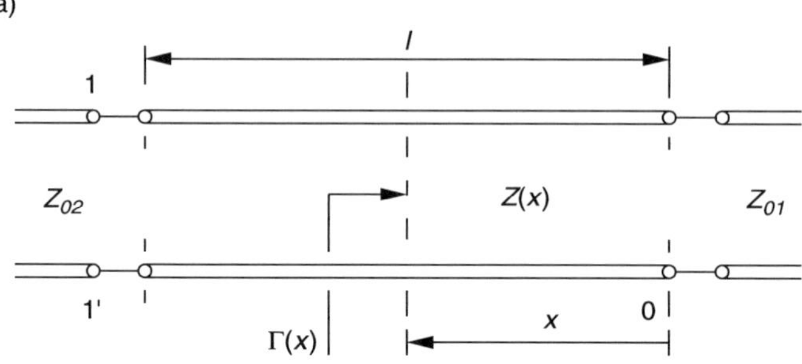

b)

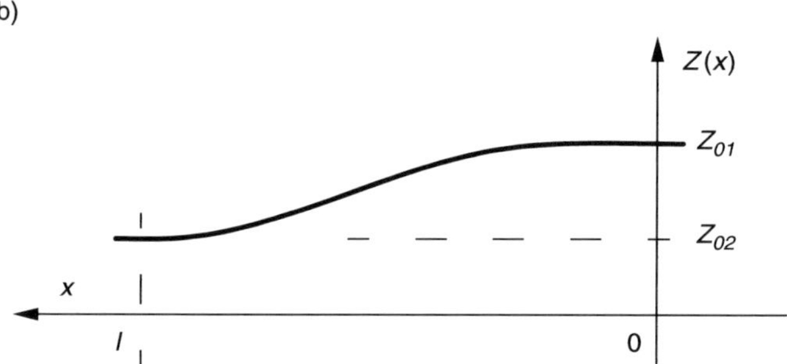

Fig. 7.15

The following differential equation of Riccati type has been derived in the literature for the reflection coefficient function $\Gamma(x)$ determined for $0 \le x \le l$ and a fixed
value of frequency f [8]:

$$\frac{d\Gamma(x)}{dx} + j2\beta\Gamma(x) + N(x)\left[1 - \Gamma^2(x)\right] = 0 \tag{7.58}$$

where

$$N(x) = \frac{1}{2}\frac{1}{Z(x)}\frac{dZ(x)}{dx} = \frac{1}{2}\frac{d}{dx}\ln\left[Z(x)\right]$$

$$\Gamma(x) = \frac{Z_{in}(x) - Z(x)}{Z_{in}(x) + Z(x)}$$

$\beta = \dfrac{2\pi}{\lambda(f)}$ is the propagation constant, $\lambda(f)$ is the wavelength and $Z(x)$ is the
function of the characteristic impedance (see Example 2.1).

In general, the reflection coefficient $\Gamma(x, f)$ is a complex quantity, $\Gamma(x) = a(x) + jb(x)$, and therefore Eq. (7.58) is equivalent to the following system of two
differential equations of the first order:

$$\frac{da(x)}{dx} = 2\beta \cdot b(x) - N(x)\left[1 - a^2(x) + b^2(x)\right]$$

$$\frac{db(x)}{dx} = -2\beta \cdot a(x) + 2N(x) \cdot a(x) \cdot b(x)$$

$$(7.59)$$

The function of local reflections $N(x)$ fully determines changes of the characteristic impedance $Z(x)$ for $0 \leq x \leq l$. In the simplest case, when $N(x) = 0$ means that the impedance $Z(x)$ is constant. Naturally, in this situation the transmission line section under analysis is homogenous and transmission line equation (known also as the Smith chart transformation) can be used for its analysis [8, 9]. Another case of interest is when

$$N(x) = 0.5 \cdot \alpha$$

$$Z(x) = Z_{01} \exp(\alpha \cdot x)$$

$$(7.60)$$

where

$$\alpha = \frac{1}{l} \ln\left(\frac{Z_{02}}{Z_{01}}\right), \; Z_{01} = Z(x = 0), \; Z_{02} = Z(x = l), \; 0 \leq x \leq l.$$

In this case, the absolute value of the reflection coefficient function, defined at the input plane $1 - 1'$ is

$$|\Gamma(x = l, \theta)| \equiv |\Gamma(\theta)| = 0.5 \cdot |\alpha \cdot l| \left|\frac{\sin(\theta)}{\theta}\right|$$

$$(7.61)$$

where

$$\theta \equiv \theta(f) = \frac{2\pi l}{\lambda(f)} = \beta l$$

is an electrical length of the line section expressed in radians. A modified version of the above line section is an exponential compensated line section, for which:

$$N(x) = 0.5 \cdot \alpha \left[1 - 0.84 \cos\left(\frac{2\pi x}{l}\right)\right], \; 0 \leq x \leq l$$

$$Z(x) = Z_{01} \exp\left[\alpha \left(x - 0.134 \cdot l \sin\left(\frac{2\pi x}{l}\right)\right)\right]$$

$$(7.62)$$

where

$$\alpha = \frac{1}{l} \ln\left(\frac{Z_{02}}{Z_{01}}\right), \; Z_{01} = Z(x = 0), \; Z_{02} = Z(x = l), \; 0 \leq x \leq l.$$

Also in this case the solution $\Gamma(x = l, \theta)$ can be found analytically. However, in the present example it has been evaluated numerically by integrating the equation system (7.59). The Runge–Kutta method RK 4 has been used for this purpose. The integration has been performed for:

$$l = 0.3\,m, \lambda = 0.3\,m, Z_{01} = 50\,\Omega, Z_{02} = 100\,\Omega,$$
$$\theta = [0, 0.25\pi, 0.5\pi, 0.75\pi, \pi, 1.25\pi, 1.5\pi, 1.75\pi, 2\pi, 2.25\pi, 2.5\pi, 3\pi]$$
$$h \equiv dx = 0.00001$$

and initial conditions

$$a(x = 0) = 0$$
$$b(x = 0) = 0$$

Some values of $a(\theta)$ and $b(\theta)$ obtained in this way are given in the second and third columns of the Table 7.12. The next column of this table includes corresponding values of $|\Gamma(\theta)| = \sqrt{a^2(\theta) + b^2(\theta)}$.

The normalized values of a function $|\Gamma(\theta)|_n = |\Gamma(\theta)|/|\Gamma(\theta = 0)|$ are given in the fifth column of Table 7.12 and illustrated in Fig. 7.16.

The values of $|\Gamma(\theta)| = \sqrt{a^2(\theta) + b^2(\theta)}$ obtained numerically are in good agreement with the corresponding exact values calculated from the following formula:

$$|\Gamma(x = l, \theta)| \equiv |\Gamma(\theta)| = 0.5 \cdot |\alpha \cdot l| \left| \frac{\sin(\theta)}{\theta} \right| \left| 1 - 0.84 \frac{\theta^2}{\theta^2 - \pi^2} \right|$$

where $\theta \equiv \theta(f) = 2\pi l/\lambda(f) = \beta l$ is the electrical length of the line section expressed in radians [8].

Table 7.12

θ, rad	$a(\theta)$	$b(\theta)$	$|\Gamma(\theta)|$	$|\Gamma(\theta)|_n$
$10^{-9}\pi$	−0.333335	0.000000	0.333335	1.000000
0.25π	−0.229729	0.220941	0.318733	0.956199
0.50π	−0.009067	0.277122	0.277271	0.831813
0.75π	0.147260	0.158500	0.216351	0.649054
1.00π	0.147989	0.004931	0.148072	0.444216
1.25π	0.062342	−0.059095	0.085900	0.257700
1.50π	0.000816	−0.039566	0.039574	0.118723
1.75π	−0.008419	−0.008699	0.012106	0.036319
2.00π	−0.000476	−0.000006	0.000476	0.001428
2.25π	0.001068	−0.001045	0.001494	0.004484
2.50π	−0.000001	−0.000010	0.000010	0.000032
2.75π	0.000625	0.000639	0.000894	0.002682
3.00π	0.000035	0.000000	0.000035	0.000106

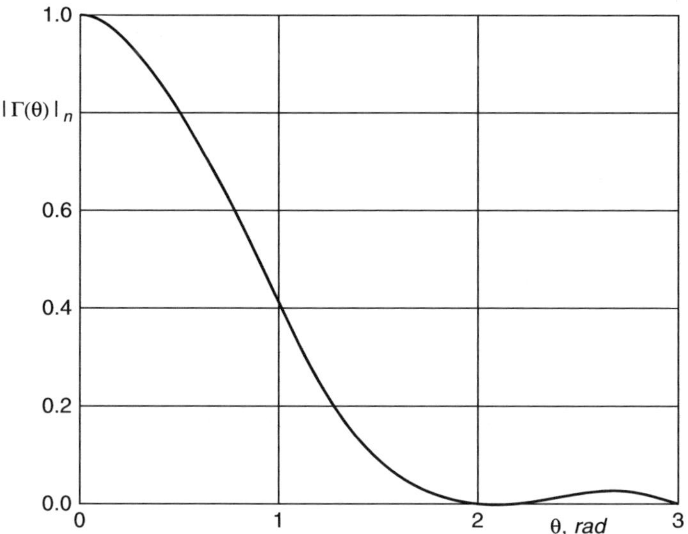

Fig. 7.16

7.6 An Example of Application of the Finite Difference Method for Solving the Linear Boundary Value Problem

Solution of the boundary problem consists in determining such function $y(x)$, which satisfies a given ordinary differential equation and at least two boundary conditions $y(x = a) = y_a$ and $y(x = b) = y_b$. An example of such problem can be given as the solution $y(x)$ satisfying the following differential equation:

$$\frac{d^2 y(x)}{dx^2} = 2 \cdot x + 3y(x) \tag{7.63}$$

and boundary conditions $y(0) = 0$, $y(1) = 1$. One of the efficient numerical methods, used for solving boundary problems with two boundary conditions, is the method of finite differences. Essential feature of this method consists in replacing the differential equation by an approximating difference equation. For this end, an integration interval $[a, b]$ should be divided into n equal subintervals (parts) determined by coordinates

$$x_i = x_0 + i \cdot h \tag{7.64}$$

where $i = 1, 2, 3, \ldots, n$, $x_0 = a$, $x_n = b$ and $h = (b - a)/n$. Let $y_i = y(x_i)$ denote a discrete value of the desired function $y(x)$. The values of $y_i = y(x_i)$ make it possible to determine approximate values of derivatives of the function, $y(x)$ using the difference expressions (6.12) and (6.18), derived in Chap. 6. Thus, we can write:

$$\frac{dy(x_i)}{dx} = y'(x_i) \approx \frac{1}{2h}(y_{i+1} - y_{i-1}) \tag{7.65}$$

$$\frac{d^2y(x_i)}{dx^2} = y''(x_i) \approx \frac{1}{h^2}(y_{i+1} - 2y_i + y_{i-1})$$

The second formula of above presented makes it possible to replace differential equation (7.63) by its difference equivalent, namely:

$$\frac{1}{h^2}(y_{i+1} - 2y_i + y_{i-1}) = 2 \cdot x_i + 3y_i \tag{7.66}$$

where $i = 1, 2, 3, \ldots, n$. Writing the difference equation (7.66) for $i = 1, 2, 3, ..,$ $n - 1$, we obtain the system of $n - 1$ algebraic equations, which are linear in this case. To focus our discussion, let us assume that $n = 10$ and correspondingly $h = (1 - 0)/10 = 0.1$. According to (7.63) and (7.66) we obtain:

$$
\begin{aligned}
y_2 - 2.03y_1 + (y_0 = 0) &= 0.002 \\
y_3 - 2.03y_2 + y_1 &= 0.004 \\
y_4 - 2.03y_3 + y_2 &= 0.006 \\
y_5 - 2.03y_4 + y_3 &= 0.008 \\
y_6 - 2.03y_5 + y_4 &= 0.010 \\
y_7 - 2.03y_6 + y_5 &= 0.012 \\
y_8 - 2.03y_7 + y_6 &= 0.014 \\
y_9 - 2.03y_8 + y_7 &= 0.016 \\
(y_{10} = 1) - 2.03y_9 + y_8 &= 0.018
\end{aligned}
\tag{7.67}
$$

The equation system (7.67) would now be presented in the matrix form, especially convenient when the Gauss elimination method is used for solving, i.e.:

$$
\begin{bmatrix}
-2.03 & 1 & 0 & 0 & 0 & 0 & 0 & 0 & 0 \\
1 & -2.03 & 1 & 0 & 0 & 0 & 0 & 0 & 0 \\
0 & 1 & -2.03 & 1 & 0 & 0 & 0 & 0 & 0 \\
0 & 0 & 1 & -2.03 & 1 & 0 & 0 & 0 & 0 \\
0 & 0 & 0 & 1 & -2.03 & 1 & 0 & 0 & 0 \\
0 & 0 & 0 & 0 & 1 & -2.03 & 1 & 0 & 0 \\
0 & 0 & 0 & 0 & 0 & 1 & -2.03 & 1 & 0 \\
0 & 0 & 0 & 0 & 0 & 0 & 1 & -2.03 & 1 \\
0 & 0 & 0 & 0 & 0 & 0 & 0 & 1 & -2.03
\end{bmatrix}
\cdot
\begin{bmatrix}
y_1 \\ y_2 \\ y_3 \\ y_4 \\ y_5 \\ y_6 \\ y_7 \\ y_8 \\ y_9
\end{bmatrix}
=
\begin{bmatrix}
0.002 \\ 0.004 \\ 0.006 \\ 0.008 \\ 0.010 \\ 0.012 \\ 0.014 \\ 0.016 \\ -0.982
\end{bmatrix}
$$

The matrix $\mathbf{A}$ of coefficients of the equation system written above is the special case of the sparse square matrix, and is called in the literature as the ribbon matrix, or more precisely three-diagonal matrix. Thus, for solving the equations formulated above the method of fast elimination is recommended, see also Example 4.4. The algorithm of this simple and efficient numerical method is described in Appendix C.

Table 7.13

x_i	y_i	$y(x_i)$	$y_i - y(x_i)$
0.0	0.000000	0.000000	0.000000
0.1	0.039417	0.039307	0.000110
0.2	0.082017	0.081820	0.000197
0.3	0.131078	0.130767	0.000311
0.4	0.190071	0.189679	0.000392
0.5	0.262766	0.262317	0.000449
0.6	0.353344	0.352868	0.000476
0.7	0.466523	0.466062	0.000461
0.8	0.607697	0.607308	0.000389
0.9	0.783102	0.782859	0.000243
1.0	1.000000	1.000000	0.000000

The values y_i given in the second column of Table 7.13 constitute the desired solution obtained in this manner.

For comparison, the third column of Table 7.13 includes the corresponding values $y(x_i)$ of the exact solution evaluated analytically [10].

$$y(x) = \frac{5}{3} \cdot \frac{\sinh(\sqrt{3} \cdot x)}{\sinh(\sqrt{3})} - \frac{2}{3} \cdot x$$

The measure of approximation of the differential equation (7.63) by the difference equation (7.66) is the set of deviations given in the fourth column.

The considered problem is an example of the linear boundary problem, for which the differential equation can be presented in the following general form:

$$y^{(n)}(x) = f_{n-1}(x)y^{(n-1)}(x) + f_{n-2}(x)y^{(n-2)}(x) + \ldots + f_1(x)y(x) + f_0(x) \quad (7.68)$$

where $y^{(k)}(x)$ denotes a derivative of order k and $f_k(x)$ is the $k-$function, bounded and continuous over a given interval $[a, b]$. In the case of problems concerning non-linear differential equations, corresponding systems of algebraic equations are also nonlinear. Consequently, the nonlinear boundary problem becomes more difficult to solve.

References

1. Lambert J.D., Computational methods in ordinary differential equations. John Wiley and Sons, New York, 1973
2. Mathews J.H., Numerical methods for mathematics, science and engineering. Prentice-Hall Intern. Inc., Englewood Cliffs, NJ, 1992
3. Mathews J.H., Numerical methods for mathematics, science and engineering. Prentice-Hall Inc., Englewood Cliffs, NJ, 1987
4. Hamming R.W., "Stable predictor–corrector methods for ordinary differential equations", J. Assoc. Comput. Mach., No. 6, 1959
5. Tront J.I., PSpice for basic circuit analysis. McGraw-Hill, New York, 2004

6. Senturia S.D., and B.D. Wedlock, Electronic circuits and applications. John Wiley and Sons, Inc., New York, 1975
7. Dorf R.C. and J.A. Svoboda, Introduction to electric circuits (4th edition). John Wiley and Sons, Inc., New York, 1999
8. Rosłoniec S., Linear microwave circuits analysis and design (in Polish). Published by: Wydawnictwa Komunikacji i Łącznosci, Warsaw, 1999
9. Rosłoniec S., Algorithms for computer-aided design of linear microwave circuits. Artech House Inc., Boston, MA, 1990
10. Shoup T.E., Applied numerical methods for the microcomputer. Prentice-Hall, Inc., Englewood Cliffs, NJ, 1984

Chapter 8
The Finite Difference Method Adopted
for Solving Laplace Boundary Value Problems

In mathematics, the boundary value problem is understood as the problem of finding a function of many variables ($n \geq 2$) satisfying a given partial differential equation and taking fixed values at prescribed points of an integration region. As an example of such problem, let us consider the following second order partial differential equation:

$$A(x, y)\frac{\partial^2 f(x, y)}{\partial x^2} + B(x, y)\frac{\partial^2 f(x, y)}{\partial x \partial y} + C(x, y)\frac{\partial^2 f(x, y)}{\partial y^2} +$$

$$D(x, y)\frac{\partial f(x, y)}{\partial x} + E(x, y)\frac{\partial f(x, y)}{\partial y} + F(x, y) \cdot f(x, y) = 0 \tag{8.1}$$

formulated for the function $f(x, y)$. The unknown function $f(x, y)$ should take at prescribed points of the two-dimensional region the fixed values, called boundary conditions. This name reflects the fact that for the majority of boundary value problems, formulated for description of various physical phenomena, values of the unknown function $f(x, y)$ are defined on the border of the given integration region. The functions $A(x, y)$, $B(x, y)$, $C(x, y)$, $D(x, y)$, $E(x, y)$ and $F(x, y)$ play the role of coefficients and for this reason they should be bounded and continuous over the given integration region. In case of many equations describing specific engineering problems, these functions take constant values as it is illustrated below by Eqs. (8.2), and (8.4). In the mathematical literature, equations of the type (8.1) are often classified according to the value of discriminant $\Delta(x, y) = B^2(x, y) - 4A(x, y) \cdot C(x, y)$ into one of the following groups: hyperbolic when $\Delta(x, y) > 0$, parabolic when $\Delta(x, y) = 0$ and elliptic when $\Delta(x, y) < 0$. The examples of partial differential equations, formulated for various physical problems are:

– Laplace equation

$$\nabla^2 U(x, y, z) = 0 \tag{8.2}$$

S. Rosłoniec, *Fundamental Numerical Methods for Electrical Engineering*,
© Springer-Verlag Berlin Heidelberg 2008

Rectangular grid

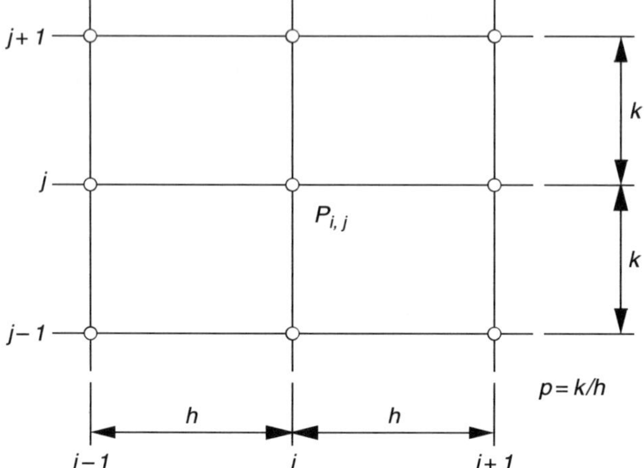

Fig. 8.1

– Poisson equation

$$\nabla^2 U(x, y, z) + f(x, y, z) = 0 \tag{8.3}$$

– Helmholtz equation

$$\nabla^2 U(x, y, z) + k^2 U(x, y, z) = 0 \tag{8.4}$$

– wave equation

$$\nabla^2 U(x, y, z) - a^2 \frac{\partial^2 U(x, y, z, t)}{\partial t^2} = 0 \tag{8.5}$$

Triangular grid

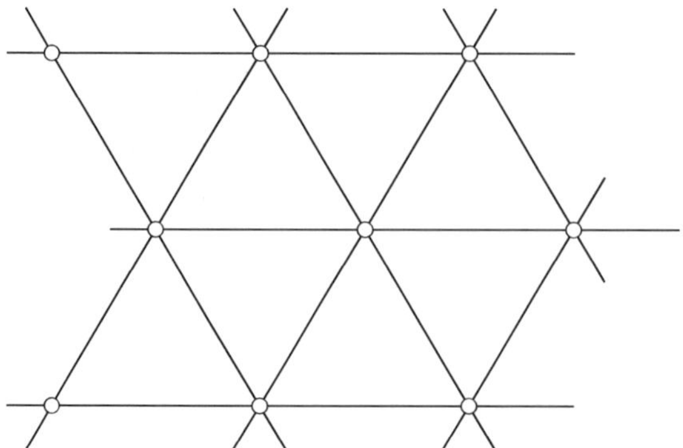

Fig. 8.2

Fig. 8.3

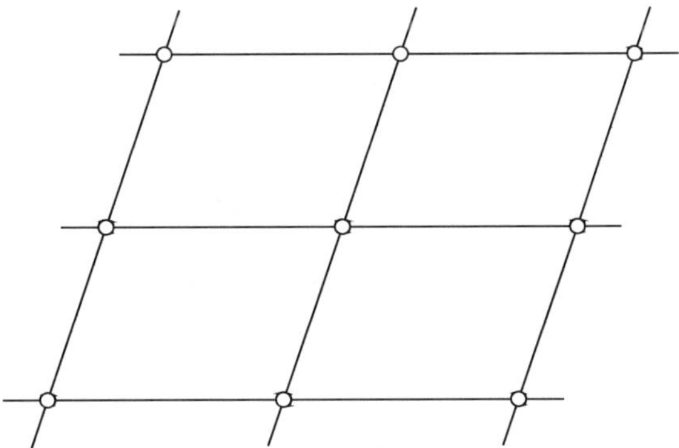

– diffusion equation

$$\nabla^2 U(x, y, z) - b^2 \frac{\partial U(x, y, z, t)}{\partial t} = 0 \qquad (8.6)$$

where the Laplace operator is defined as:

$$\nabla^2 U(x, y, z) = \frac{\partial^2 U(x, y, z)}{\partial x^2} + \frac{\partial^2 U(x, y, z)}{\partial y^2} + \frac{\partial^2 U(x, y, z)}{\partial z^2}$$

Equations (8.2)–(8.6) ennumerated above can also be written in other coordinate systems, different from cartesian one. The most common are the cylindrical

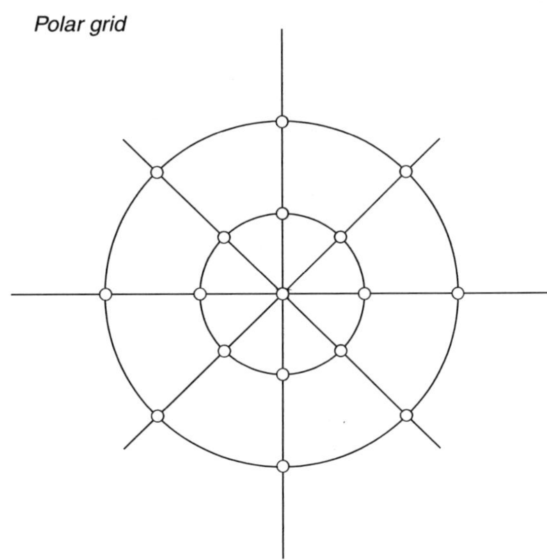

Fig. 8.4

(ρ, φ, z) and the spherical (r, φ, θ) systems. Corresponding formulas for calculating the Laplacian can be found in the literature available on this subject, for example in [1–4]. The essence of the finite difference method adopted for solving the above partial differential equations is replacement of the Laplacian of the desired function by its difference equivalent. For every specific boundary problem, we should choose such coordinate system and the discretization grid, for which the equivalent difference equation will approximate most accurately the original differential equation. Also, the boundary conditions would be satisfied to an acceptable extent. This problem is treated extensively in the literature, for example in [2]. Therefore, only a few examples of the girds commonly used are presented in Figs. 8.1, 8.2, 8.3 and 8.4.

8.1 The Interior and External Laplace Boundary Value Problems

For the Laplace equation (8.2), two boundary value problems (in the sense of Dirichlet) can be formulated. In order to explain their essential meaning, let us assume that a region V is given, for which the edge S belongs to the C^2 class (piecewise smooth). The interior boundary value problem consists in finding such harmonic function U, that satisfies Eq. (8.2) and at every point $P \rightarrow P_S \in S$ the limit of this function achieves value $U(P_S)$ that is equal to value $g(S)_{S=P_S}$ of the given boundary function $g(S)$. The external boundary value problem, in the sense of Dirichlet, consists also in finding the function U satisfying the Laplace equation (8.2) in the region V_E being the complement of the region V, i.e., $V_E = E - V$, where E denotes the Euclidean space. Moreover, for every point P_S belonging to the edge S of the region V_E, the unknown function should satisfy the following conditions:

$$U(P_S) = g(P_S)$$

$$U\left(\frac{1}{r}\right) \rightarrow 0 \quad \text{when} \quad r = \sqrt{x^2 + y^2 + z^2} \rightarrow \infty$$

For the Laplace equation, apart from the Dirichlet boundary value problems described above, we can formulate also Neumann boundary value problems, which can also be considered as internal and exterior problems. Solution of the interior Neumann problem consists in finding a function satisfying the Laplace equation (8.2) in a given region V, and the following condition:

$$\left(\frac{\partial U}{\partial n}\right)_{P=P_S} = k(P_S)$$

in which n denotes the normal direction to the surface S and $k(P)$ is a given boundary function determined over the surface S and satisfying the following condition:

$$\int_S k(P)ds = 0$$

In case of the external Neumann boundary value problem, the solution should satisfy the Laplace equation (8.2) in the complementary region V_E and the following conditions:

$$\left| \left(\frac{\partial U}{\partial n} \right)_{P=P_S} \right| = k(P_S), \quad U\left(\frac{1}{r}\right) \to 0 \quad \text{and} \quad \frac{\partial U(r)}{\partial n} \to 0 \quad \text{when}$$

$$r = \sqrt{x^2 + y^2 + z^2} \to \infty.$$

In most algorithms for numerical solving of boundary value problems, condition

$$\left| \left(\frac{\partial U}{\partial n} \right)_{P=P_S} \right| = k(P_S)$$

is usually taken into account by introducing additional fictious nodes with corresponding values of the unknown function U. For this end, the square or rectangular grid should be placed in such a way that its nodes lie on both sides of the border line S. An example of the procedure of introducing additional fictious nodes, as for example node P_F, is shown in Fig. 8.5.

The line $P_F P_Q$, perpendicular to the edge S and crossing this edge at point P_S, is drawn through the node P_F. Such new point P is then determined on this line for which $|P_S P| = |P_S P_F|$. Now the fictious potential U_F of the fictious node P_F can be calculated from the formula:

$$U_F = U_P + k(P_S) \cdot |P_S P| \tag{8.7}$$

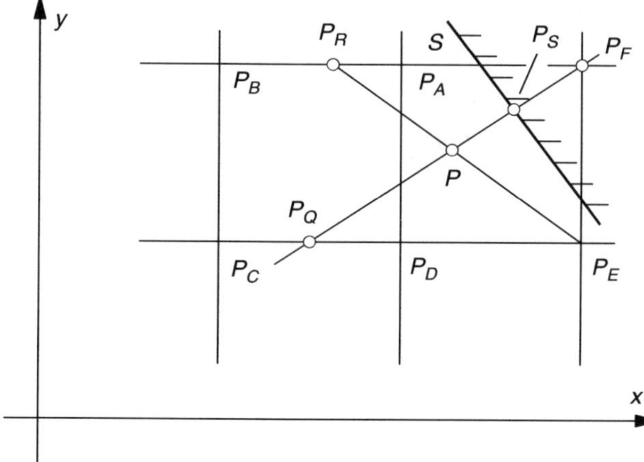

Fig. 8.5

in which $U_P \equiv U(P)$ is the coefficient calculated using the linear interpolation method, from the values of the unknown function at points P_E and P_R. In a similar way, knowing two values, $U_A = U(P_A)$ and $U_B = U(P_B)$, potential $U_R \equiv U(P_R)$ is then found. In the case when $k(P_S) = 0$, desired value of $U_F \equiv U(P_F)$ is equal to $U_P \equiv U(P)$. In particular case when $k(P_S) = 0$ and the edge S is a straight line oriented in parallel with respect to horizontal and vertical sides of the grid, determination of fictious nodes and corresponding fictious values of the desired function becomes much more easier. One simple algorithm serving to solve this particular problem will be described in Example 8.4 given below.

8.2 The Algorithm for Numerical Solving of Two-Dimensional Laplace Boundary Problems by Using the Finite Difference Method

The Laplace equation (8.2) formulated for the function $U(x, y)$, written in the system of rectangular coordinates has the form:

$$\frac{\partial^2 U(x, y)}{\partial x^2} + \frac{\partial^2 U(x, y)}{\partial y^2} = 0 \tag{8.8}$$

Numerical solution of this equation, also by means of the finite difference method, consists in replacing the second order partial derivatives of this equation by corresponding differential expressions, similar to (6.44) derived in Chap. 6. Let us assume that function $U(x, y)$ is analyzed in the close neighborhood of the point $P_{i,j} \equiv (x_i, y_j)$ at which $U(P_{i,j}) = U_{i,j}$, see Fig. 8.6.

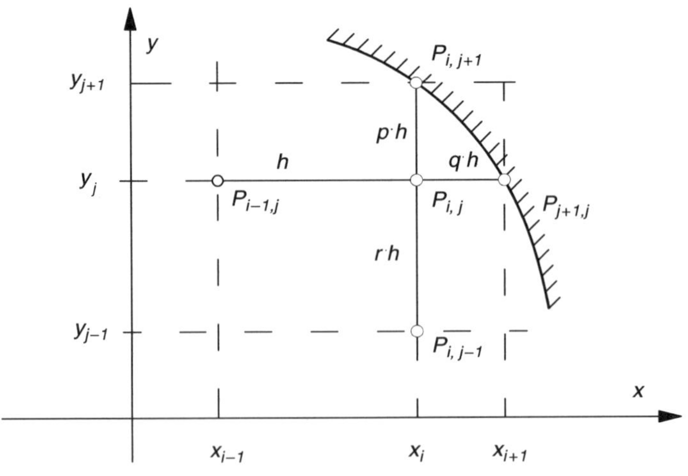

Fig. 8.6

The difference formula approximating the Laplacian (8.8) can be obtained by developing the function $U \equiv U(x, y)$ in the form of Taylor series, namely:

$$U_{i-1,j} = U_{i,j} - h \left(\frac{\partial U}{\partial x} \right)_{P_{i,j}} + \frac{1}{2!} h^2 \left(\frac{\partial^2 U}{\partial x^2} \right)_{P_{i,j}} - \frac{1}{3!} h^3 \left(\frac{\partial^3 U}{\partial x^3} \right)_{P_{i,j}} + \dots \quad (8.9)$$

$$U_{i+1,j} = U_{i,j} + qh \left(\frac{\partial U}{\partial x} \right)_{P_{i,j}} + \frac{1}{2!} q^2 h^2 \left(\frac{\partial^2 U}{\partial x^2} \right)_{P_{i,j}} + \frac{1}{3!} q^3 h^3 \left(\frac{\partial^3 U}{\partial x^3} \right)_{P_{i,j}} + \dots$$
$$(8.10)$$

$$U_{i,j+1} = U_{i,j} + ph \left(\frac{\partial U}{\partial y} \right)_{P_{i,j}} + \frac{1}{2!} p^2 h^2 \left(\frac{\partial^2 U}{\partial y^2} \right)_{P_{i,j}} + \frac{1}{3!} p^3 h^3 \left(\frac{\partial^3 U}{\partial y^3} \right)_{P_{i,j}} + \dots$$
$$(8.11)$$

$$U_{i,j-1} = U_{i,j} - rh \left(\frac{\partial U}{\partial y} \right)_{P_{i,j}} + \frac{1}{2!} r^2 h^2 \left(\frac{\partial^2 U}{\partial y^2} \right)_{P_{i,j}} - \frac{1}{3!} r^3 h^3 \left(\frac{\partial^3 U}{\partial y^3} \right)_{P_{i,j}} + \dots$$
$$(8.12)$$

where $U_{i-1,j} = U(P_{i-1,j})$, $U_{i,j+1} = U(P_{i,j+1})$, $U_{i+1,j} = U(P_{i+1,j})$, $U_{i,j-1} = U(P_{i,j-1})$ and $U_{i,j} = U(P_{i,j})$. After multiplying both sides of the series (8.9) by the coefficient q and adding them to the series (8.10) we obtain the expression:

$$qU_{i-1,j} + U_{i+1,j} = (1+q) \cdot U_{i,j} + \frac{1}{2!} h^2 q(1+q) \left(\frac{\partial^2 U}{\partial x^2} \right)_{P_{i,j}} + O(h^3)$$

Neglecting the terms including step h in the third and higher powers, it becomes:

$$\left(\frac{\partial^2 U}{\partial x^2} \right)_{P_{i,j}} \approx \frac{1}{h^2} \left[\frac{2U_{i-1,j}}{1+q} + \frac{2U_{i+1,j}}{q(1+q)} - \frac{2U_{i,j}}{q} \right] \quad (8.13)$$

In the similar way, using the series (8.11) and (8.12), multiplied by the coefficients r and p, respectively, the following difference expression approximating the second order partial derivative with respect to y can be written as:

$$\left(\frac{\partial^2 U}{\partial y^2} \right)_{P_{i,j}} \approx \frac{1}{h^2} \left[\frac{2U_{i,j+1}}{p(p+r)} + \frac{2U_{i,j-1}}{r(p+r)} - \frac{2U_{i,j}}{r \cdot p} \right] \quad (8.14)$$

Adding both sides of relations (8.13) and (8.14) we obtain the general equation:

$$\frac{2U_{i-1,j}}{1+q} + \frac{2U_{i+1,j}}{q(1+q)} + \frac{2U_{i,j+1}}{p(p+r)} + \frac{2U_{i,j-1}}{r(p+r)} - \frac{2U_{i,j}}{q} - \frac{2U_{i,j}}{r \cdot p} = 0 \quad (8.15)$$

called the difference Laplace equation of the order h^2. For the regular grid with rectangular meshes ($q = 1$, $p = r$), this equation takes a simpler form:

$$U_{i-1,j} + U_{i+1,j} + \frac{U_{i,j+1}}{p^2} + \frac{U_{i,j-1}}{p^2} - 2U_{i,j} - \frac{2U_{i,j}}{p^2} = 0 \qquad (8.16)$$

Particular case of the regular grid with rectangular meshes is the "square" grid, ($p = q = r = 1$), for which the difference equation (8.15) reduces itself to:

$$U_{i-1,j} + U_{i+1,j} + U_{i,j+1} + U_{i,j-1} - 4 \cdot U_{i,j} = 0 \qquad (8.17)$$

The difference Laplace equation in the form (8.15), (8.16) or (8.17), should be satisfied at every internal point of any given two-dimensional region G, and this property refers of course also to each point (node) of the grid defined in this region, as in Fig. 8.7.

Writing the difference Laplace equation for each internal node of the introduced grid, the system of linear equations can be obtained. The unknown variables of this system are values of the desired function $U \equiv U(x, y)$ at individual nodes. At intersection points of the grid with the edge (contour) C of the region G, values of the desired function are known, because they are equal to corresponding values of the given boundary function $g(C)$. The number of equations should be in general the same as the number of unknown values of the desired function. It is rather easy to prove that this number is the same as the number of nodes of the introduced grid. Solution of the system of linear equations obtained in this way may be obtained using several methods described in Chap. 1, such as the Jacobi, Gauss–Seidel, as well as the successive over-relaxation method (SOR) [4, 5]. For using the iterative methods mentioned above, the difference Laplace equation (8.15) should be written in the following equivalent form:

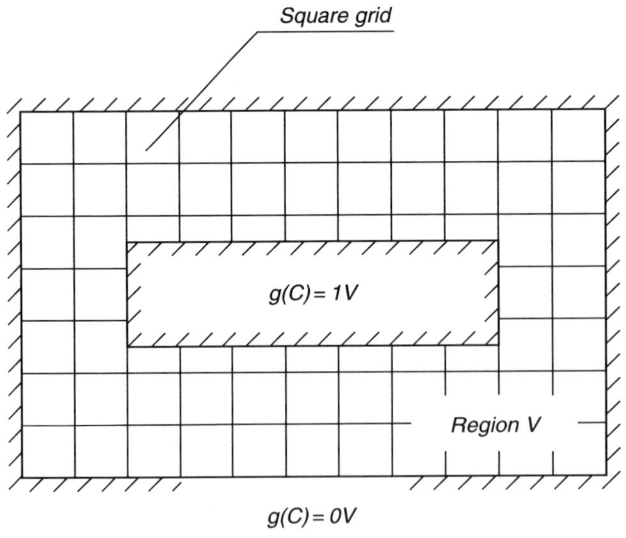

Fig. 8.7

$$U_{i,j} = \frac{pqr}{pr+q}\left[\frac{U_{i-1,j}}{1+q} + \frac{U_{i+1,j}}{q(1+q)} + \frac{U_{i,j+1}}{p(p+r)} + \frac{U_{i,j-1}}{r(p+r)}\right] \qquad (8.18)$$

Particular cases of this equation written for rectangular and square grids are, respectively:

$$U_{i,j} = \frac{p^2}{2(1+p^2)}\left[U_{i-1,j} + U_{i+1,j} + \frac{U_{i,j+1}}{p^2} + \frac{U_{i,j-1}}{p^2}\right] \qquad (8.19)$$

$$U_{i,j} = \frac{1}{4}\left[U_{i-1,j} + U_{i+1,j} + U_{i,j+1} + U_{i,j-1}\right] \qquad (8.20)$$

The method of simultaneous substitutions (Jacobi method, see Sect. 1.2.2) is the simplest iteration method, which can be used for calculating consecutive approximations of the unknown function $U_{i,j} \equiv U(x_i, y_j)$, according to formulas (8.18), (8.19) or (8.20). Let us assume therefore that k approximations of the function $U_{i,j} \equiv U(x_i, y_j)$ are known for all internal nodes of the introduced grid, namely $U_{i,j}^{(k)} \equiv U^{(k)}(x_i, y_j)$. Next, i.e. $(k+1)$ approximations of this function are calculated using the appropriate formula (8.18), (8.19) or (8.20), on the basis of the previous values of this function, determined during the previous iteration k. When the approximations $(k+1)$ for all internal nodes are known, they are substituted simultaneously in place of the previous approximate vallues (obtained during the iteration k). Thanks to that, the sequence of computations performed for individual internal nodes of the grid does not influence the values of consecutive approximations. As the criterion for terminating the calculations the following condition is used most commonly:

$$\max\ \left\{\left|U^{(k+1)}(x_i, y_j) - U^{(k)}(x_i, y_j)\right|\right\} \le \varepsilon$$

$$2 \le i \le I - 1$$

$$2 \le j \le J - 1 \qquad (8.21)$$

where ε is an arbitrarily small, positive number, determining the accuracy of the evaluated approximate solution. The method of simultaneous substitutions is seldom used for practical purposes, because convergence of its calculation process is insufficient. A more efficient version of this iterative method is the method of subsequent substitutions known also as Liebmann computional procedure.

8.2.1 The Liebmann Computational Procedure

It is well known that the method of simultaneous substitutions (Jacobi), presented in the previous subsection, does not ensure sufficiently good convergence. The reason of this disadvantege is that new more accurate values of the evaluated function are not introduced until they are calculated for all internal nodes of the grid. In the subsequent substitution method (Gauss–Seidel, see Sect. 1.2.2), each consecutive approximation is used immediately after its determination. According to this rule, value of the function $U_{i,j}^{(k)} \equiv U^{(k)}(x_i, y_j)$ calculated in k iteration for the internal node (x_i, y_j) is used immediately for calculating value of this function in the adjacent node, namely $U_{i+1,j}^{(k)} \equiv U^{(k)}(x_{i+1}, y_j)$, and so on. Such organized computational process is often called the Liebmann iteration method or Liebmann computational procedure. Figure 8.8 presents flow diagram of the computational algorithm related to this procedure adopted for solving the difference Laplace equation discussed above. The first stage of this algorithm is shown in Fig. 8.8 (b) (a). At the beginning, the data defining geometrical shape of the boundary C delimitating the interior region $G \equiv V$, parameters p, q and r of the adopted grid, as well as the boundary values $U(P_C) = g(P_C)$ for the desired function are introduced. In this case P_C denotes the point of intersection of the grid with the given contour C. Simultaneously, a corresponding "flag" is assigned to each internal node of the grid, showing which expressions, (8.18), (8.19) or (8.20), should be used to compute a consecutive approximation of the unknown function $U_{i,j} \equiv U(x_i, y_j)$. Integral part of this preparative stage is a procedure called "Initial approximation", serving to determine the initial, approximate values of the function $U_{i,j}^{(0)} \equiv U^{(0)}(x_i, y_j)$, on the basis of the known boundary function $g(C)$.

A theoretical basis for this auxiliary procedure makes the formula (8.18). The corresponding calculation process is illustrated in Example 8.1. It has been confirmed experimentally that application of this initial procedure reduces the number of iterations, necessary to obtain satisfactory, sufficiently exact approximation of the desired solution. Omission of this procedure is equivalent to assumption that at all internal nodes of the grid, initial values of the desired function are equal to zero, namely $U_{i,j}^{(0)} \equiv U^{(0)}(x_i, y_j) = 0$.

The second stage of the algorithm under discussion is illustrated in Fig. 8.8 (b). For each internal node, the calculations of $U(x_i, y_j)$ are performed iteratively according to appropriate difference formula giving consecutive, more and more accurate approximations of the desired function. The quantity used in the present algorithm to evaluate the accuracy of an approximate solution is the maximum deviation, defined for each iteration:

$$R_k = \max \left\{ \left| U_{i,j}^{(k)} - U_{i,j}^{(k-1)} \right| \right\} \tag{8.22}$$

where: $k = 1, 2, 3, 4, \ldots, 2 \leq i \leq I - 1, 2 \leq j \leq J - 1$. This deviation is determined on the basis of nodal values of the function $U_{i,j} \equiv U(x_i, y_j)$, calculated for the two consecutive iterations. In case when this deviation is not less than a predetermined positive number ε (defining accuracy of the solution), the calculation process is continued during the consecutive iteration $(k + 1)$. In the opposite case, solution obtained during k iteration can be assumed as sufficiently accurate. In the example presented below it is justified that the calculation process

Fig. 8.8 (a)

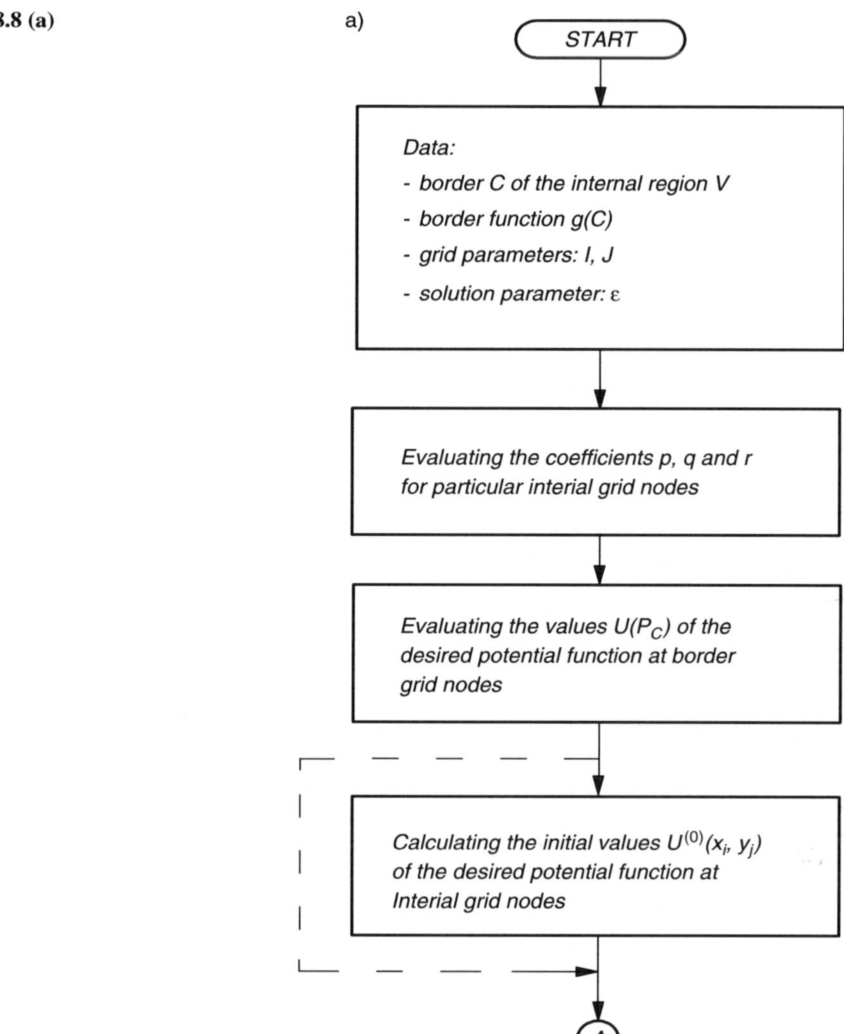

organized in such way is always convergent. It means that it is always possible to achieve good approximation of the desired solution for which $R_k \rightarrow 0$, see formula (8.22).

Example 8.1 Figure 8.9 (a) presents a transverse section of two perfect conductors with different electrical potentials. In fact, these conductors may be treated as the TEM transmission line for which the Laplace boundary value problem can be formulated. However, a transmission line of this type is not optimum in many aspects and therefore it is not recommended for telecommunication applications. It is considered here only for didactic reasons.

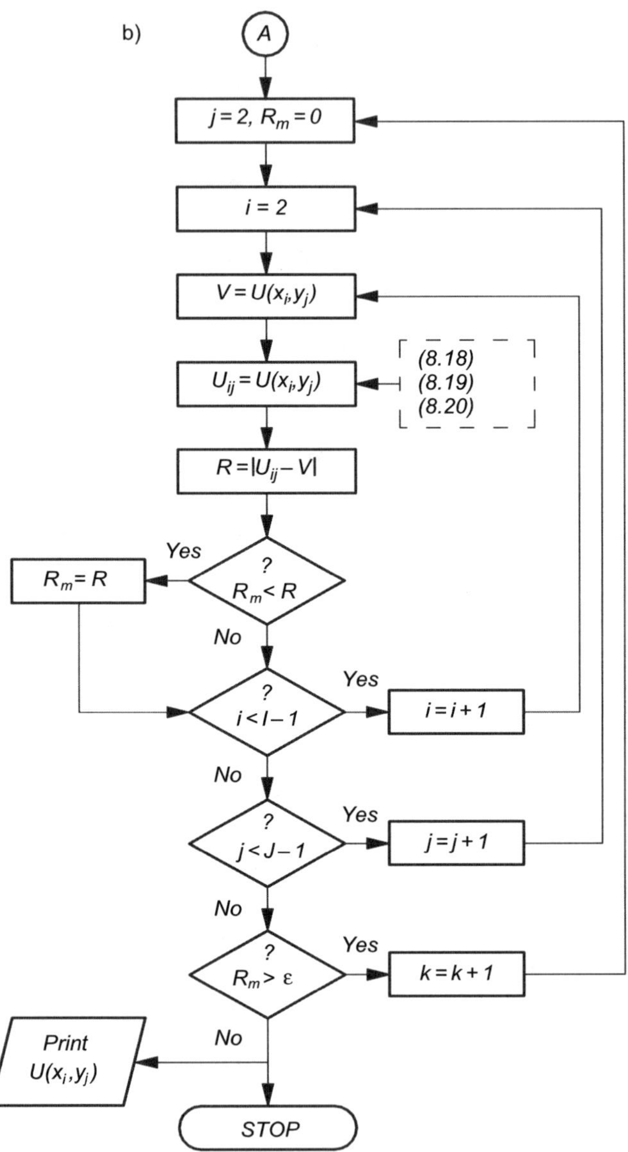

Fig. 8.8 (b)

Let us assume that electrical potentials of these conductors are equal to 1 and 0 V, respectively. Region V, for which the Laplace boundary value problem can be formulated, is the internal region limited by these conductors. Potential function $U \equiv U(x, y)$ defined over this region can be determined numerically, using rectangular grid, similar to the one shown in Fig. 8.9 (b). At all points (nodes), common to the grid and contour of the integration region, the solution takes values equal to the

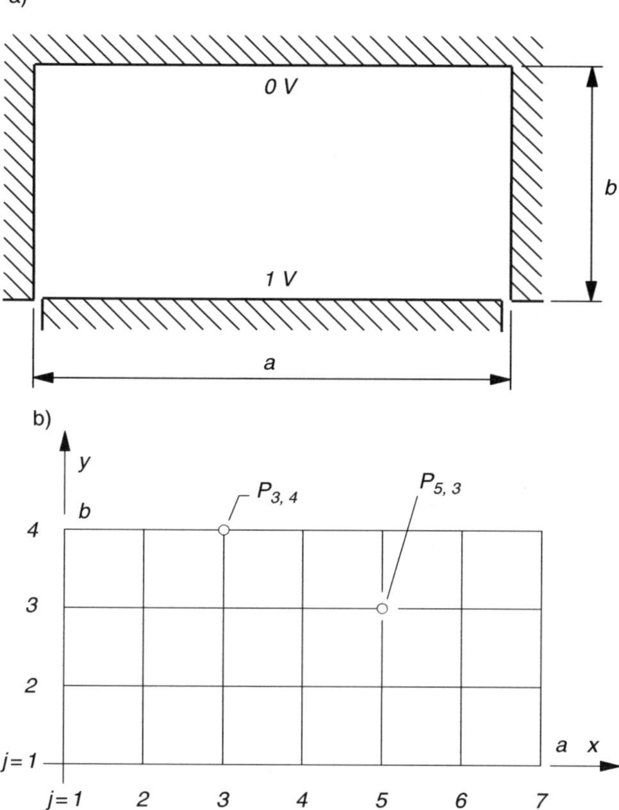

Fig. 8.9

corresponding potentials of the conductors. As the value of potential function in the slot between the two conductors, we take the arithmetical mean of potentials of both conductors. As shown in Fig. 8.9, the square grid being applied has the following parameters: $a = 12\,\text{mm}, b = 6\,\text{mm}, I = 7, J = 4, q = 1, p = r = 1$. At the boundary nodes the desired solution $U \equiv U(x, y)$ takes the following values:

$$U(x_1, y_1) = U(x_7, y_1) = 0.5,\ \text{V}$$
$$U(x_i, y_1) = 1,\ \text{V} \qquad\qquad \text{for } 2 \le i \le 6$$
$$U(x_1, y_j) = U(x_7, y_j) = 0,\ \text{V} \quad \text{for } 2 \le j \le 4$$
$$U(x_i, y_4) = 0,\ \text{V} \qquad\qquad \text{for } 1 \le i \le 7$$

$$(8.23)$$

As it is assumed above that the discretization grid adopted for this example has square meshes, and therefore relation (8.17) can be used to compute the values of function $U \equiv U(x, y)$ at $(I - 2)(J - 2) = 10$ internal nodes. Difference equations formulated in this way form the following equation system:

$$U_{1,2}^* + U_{2,3} + U_{3,2} + U_{2,1}^* - 4 \cdot U_{2,2} = 0$$
$$U_{2,2} + U_{3,3} + U_{4,2} + U_{3,1}^* - 4 \cdot U_{3,2} = 0$$
$$U_{3,2} + U_{4,3} + U_{5,2} + U_{4,1}^* - 4 \cdot U_{4,2} = 0$$
$$U_{4,2} + U_{5,3} + U_{6,2} + U_{5,1}^* - 4 \cdot U_{5,2} = 0$$
$$U_{5,2} + U_{6,3} + U_{7,2}^* + U_{6,1}^* - 4 \cdot U_{6,2} = 0$$
$$U_{1,3}^* + U_{2,4}^* + U_{3,3} + U_{2,2} - 4 \cdot U_{2,3} = 0$$
$$U_{2,3} + U_{3,4}^* + U_{4,3} + U_{3,2} - 4 \cdot U_{3,3} = 0$$
$$U_{3,3} + U_{4,4}^* + U_{5,3} + U_{4,2} - 4 \cdot U_{4,3} = 0$$
$$U_{4,3} + U_{5,4}^* + U_{6,3} + U_{5,2} - 4 \cdot U_{5,3} = 0$$
$$U_{5,3} + U_{6,4}^* + U_{7,3}^* + U_{6,2} - 4 \cdot U_{6,3} = 0$$

$$(8.24)$$

In this equation system boundary values (8.23) are additionally marked off by asterisks. Of course, the system (8.24) can be written in the following matrix form:

$$
\begin{bmatrix}
4 & -1 & 0 & 0 & 0 & -1 & 0 & 0 & 0 & 0 \\
-1 & 4 & -1 & 0 & 0 & 0 & -1 & 0 & 0 & 0 \\
0 & -1 & 4 & -1 & 0 & 0 & 0 & -1 & 0 & 0 \\
0 & 0 & -1 & 4 & -1 & 0 & 0 & 0 & -1 & 0 \\
0 & 0 & 0 & -1 & 4 & 0 & 0 & 0 & 0 & -1 \\
-1 & 0 & 0 & 0 & 0 & 4 & -1 & 0 & 0 & 0 \\
0 & -1 & 0 & 0 & 0 & -1 & 4 & -1 & 0 & 0 \\
0 & 0 & -1 & 0 & 0 & 0 & -1 & 4 & -1 & 0 \\
0 & 0 & 0 & -1 & 0 & 0 & 0 & -1 & 4 & -1 \\
0 & 0 & 0 & 0 & -1 & 0 & 0 & 0 & -1 & 4
\end{bmatrix}
\cdot
\begin{bmatrix}
U_{2,2} \\
U_{3,2} \\
U_{4,2} \\
U_{5,2} \\
U_{6,2} \\
U_{2,3} \\
U_{3,3} \\
U_{4,3} \\
U_{5,3} \\
U_{6,3}
\end{bmatrix}
=
\begin{bmatrix}
U_{1,2}^* + U_{2,1}^* \\
U_{3,1}^* \\
U_{4,1}^* \\
U_{5,1}^* \\
U_{6,1}^* + U_{7,2}^* \\
U_{1,3}^* + U_{2,4}^* \\
U_{3,4}^* \\
U_{4,4}^* \\
U_{5,4}^* \\
U_{6,4}^* + U_{7,3}^*
\end{bmatrix}
$$

$$(8.25)$$

The coefficient matrix of equation system (8.25) is a diagonally dominant matrix, and therefore convergence of the calculation process used for solving this system by means of the Gauss–Seidel iterative method is guaranteed, see Chap. 1. For this end, let us transform the equation system (8.25) into the following equivalent form:

$$U_{2,2} = \frac{1}{4}(U_{1,2}^* + U_{2,3} + U_{3,2} + U_{2,1}^*)$$

$$U_{3,2} = \frac{1}{4}(U_{2,2} + U_{3,3} + U_{4,2} + U_{3,1}^*)$$

$$U_{4,2} = \frac{1}{4}(U_{3,2} + U_{4,3} + U_{5,2} + U_{4,1}^*)$$

$$U_{5,2} = \frac{1}{4}(U_{4,2} + U_{5,3} + U_{6,2} + U_{5,1}^*)$$

$$U_{6,2} = \frac{1}{4}(U_{5,2} + U_{6,3} + U_{7,2}^* + U_{6,1}^*)$$

$$U_{2,3} = \frac{1}{4}(U^*_{1,3} + U^*_{2,4} + U_{3,3} + U_{2,2}) \qquad (8.26)$$

$$U_{3,3} = \frac{1}{4}(U_{2,3} + U^*_{3,4} + U_{4,3} + U_{3,2})$$

$$U_{4,3} = \frac{1}{4}(U_{3,3} + U^*_{4,4} + U_{5,3} + U_{4,2})$$

$$U_{5,3} = \frac{1}{4}(U_{4,3} + U^*_{5,4} + U_{6,3} + U_{5,2})$$

$$U_{6,3} = \frac{1}{4}(U_{5,3} + U^*_{6,4} + U^*_{7,3} + U_{6,2})$$

It is not difficult to verify that similar difference equations should be written for using the Liebmann computational procedure. This fact confirms the conclusion that the Liebmann method is identical to the Gauss–Seidel method, provided that equations of the system are written in appropriate order. For solving this system of linear equations by means of an arbitrary iterative method, the initial values of $U_{i,j}^{(0)} \equiv U^{(0)}(x_i, y_j)$ have to be known. Naturally, the final soultion should be independent of the adopted initial approximation. Nevertheless, the initial values of $U_{i,j}^{(0)} \equiv U^{(0)}(x_i, y_j)$ have significant influence on divergence of the calculation process. In general, that process can begin from values $U_{i,j}^{(0)} \equiv U^{(0)}(x_i, y_j) = 0$, but such approach is usually inefficient. In order to find a "more precise" initial approximation, it is possible to use a formula similar to Eq. (8.18) and given boundary conditions. As an example, let us calculate $U_{5,3}^{(0)} \equiv U^{(0)}(x_5, y_3)$, that is initial value of the solution evaluated at point $P_{5,3} \equiv P(i = 5, j = 3)$, see Figs. 8.9(b) and 8.10.

For an arbitrary internal node

$$U_{i,j}^{(0)} \equiv U^{(0)}(x_i, y_j) = \frac{pqr}{pr+q}\left[\frac{U(x_1, y_j)}{1+q} + \frac{U(x_I, y_j)}{q(1+q)} + \frac{U(x_i, y_J)}{p(p+r)} + \frac{U(x_i, y_1)}{r(p+r)}\right]$$

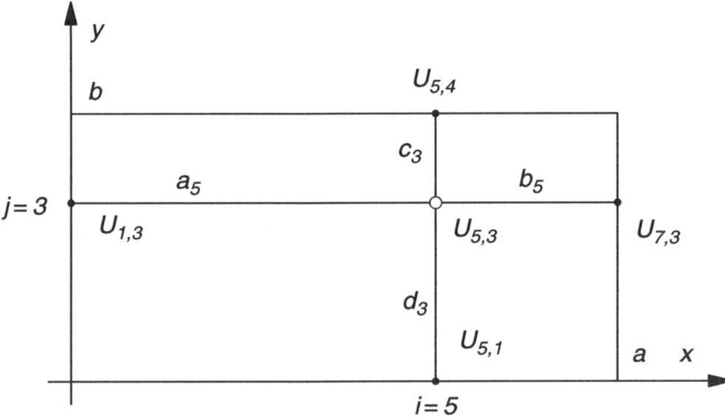

Fig. 8.10

where

$$p \equiv p(i, j) = \frac{c_j}{a_i} = \frac{b}{a} \cdot \frac{I-1}{J-1} \cdot \frac{J-j}{i-1}, q \equiv q(i) = \frac{b_i}{a_i} = \frac{I-i}{i-1},$$

$$r \equiv r(i, j) = \frac{d_j}{a_i} = \frac{b}{a} \cdot \frac{I-1}{J-1} \cdot \frac{j-1}{i-1}$$

For $a = 12\,\text{mm}$, $b = 6\,\text{mm}$, $I = 7$, $J = 4$,, $i = 5$, $j = 3$, the parameters introduced above take the following values: $p = 1/4$, $q = 1/2$ and $r = 1/2$. Thus,

$$U_{5,3}^{(0)} \equiv U^{(0)}(x_5, y_3) = \frac{pqr}{pr+q} \left[\frac{U(x_1, y_3)}{1+q} + \frac{U(x_7, y_3)}{q(1+q)} + \frac{U(x_5, y_4)}{p(p+r)} + \frac{U(x_5, y_1)}{r(p+r)} \right]$$

$$= \frac{1}{10} \left[\frac{0}{3/2} + \frac{0}{3/4} + \frac{0}{3/16} + \frac{1}{3/8} \right] = \frac{8}{30} \approx 0.2666666,\ \text{V}$$

Initial values $U_{i,j}^{(0)} \equiv U^{(0)}(x_i, y_j)$, calculated similarly for all internal nodes of the grid, are given in Table 8.1

Next approximations ($k = 1, 2, 3, \ldots$) of the desired function, $U_{i,j}^{(k)} \equiv U^{(k)}(x_i, y_j)$, are calculated according to the algorithm shown in Fig. 8.8 (b), where difference formula (8.20) is used, because meshes of the introduced discretization grid are square. Some calculation results obtained in the first, fifth, tenth and fifteenth iterations are written in Tables 8.2, 8.3, 8.4 and 8.5, respectively.

For $\varepsilon \leq 10^{-7}$, see formula (8.22), condition $R_k \leq \varepsilon$ is satisfied only by the approximate solution obtained in the 18th iteration, ($k = 18$, $R_k = 5.15 \times 10^{-8}$).

8.2.2 The Successive Over-Relaxation Method (SOR)

In order to explain the main feature of the SOR method, let us consider its algorithm adopted to solving the Laplace boundary problem. In case when the grid with square meshes, ($p = q = r = 1$) is used, the difference equation approximating the original differential equation has the form similar to one described by (8.17). The

Table 8.1 ($k = 0$)

$j/i \rightarrow$	2	3	4	5	6
2	0.47619047	0.53333333	0.54545454	0.53333333	0.47619047
3	0.23809523	0.26666666	0.27272727	0.26666666	0.23809523

Table 8.2 ($k = 1$, $R_k = 6.07 \times 10^{-2}$)

$j/i \rightarrow$	2	3	4	5	6
2	0.44285714	0.56374458	0.59245129	0.58382711	0.45548058
3	0.17738095	0.25346320	0.27814529	0.27501691	0.18262437

Table 8.3 ($k = 5$, $R_k = 7.98 \times 10^{-4}$)

$j/i \rightarrow$	2	3	4	5	6
2	0.43754156	0.57635557	0.60879599	0.57625900	0.43757975
3	0.17395714	0.25804123	0.28129719	0.25790405	0.17387095

Table 8.4 ($k = 10$, $R_k = 2.20 \times 10^{-5}$)

$j/i \rightarrow$	2	3	4	5	6
2	0.43739010	0.57577734	0.60809664	0.57576701	0.43737746
3	0.17374866	0.25758931	0.28081889	0.25758219	0.17373991

Table 8.5 ($k = 15$, $R_k = 5.36 \times 10^{-7}$)

$j/i \rightarrow$	2	3	4	5	6
2	0.43737411	0.57575802	0.60808116	0.57575782	0.43737382
3	0.17373763	0.25757606	0.28080832	0.25757589	0.17373743

notion of residuum of this equation formulated for node (x_i, y_j) is now introduced, namely

$$\text{Res}_{i,j} = U_{i-1,j} + U_{i+1,j} + U_{i,j+1} + U_{i,j-1} - 4 \cdot U_{i,j} \qquad (8.27)$$

In general, the value of residuum (8.27) can be negative, zero or positive. Of course, the Eq. (8.27) is exactly satisfied at node (x_i, y_j), if $\text{Res}_{i,j} = 0$. According to (8.27), residuum $\text{Res}_{i,j}$ will change by -4, if the function $U_{i,j} \equiv U(x_i, y_j)$ will be incremented by 1. Simultaneously, residua evaluated at four adjacent nodes will increase by 1. Consequently, if we intend to reduce the residua $\text{Res}_{i,j}$ up to zero it is necessary to add $(\text{Res}_{i,j})/4$ to the function value $U_{i,j} \equiv U(x_i, y_j)$ computed for this node, that is (x_i, y_j). This operation will of course result in unproportional changes of residua in the adjacent nodes. Reducing in this way residua $\text{Res}_{i,j}$ successively in each internal node, we obtain more precise approximations of the desired solutions. After terminating calculations for all internal nodes, the process should be repeated from the beginning in the next iteration. The computational process organized in this way (relaxation method) is identical to the method of successive substitutions (Liebmann computational procedure), which is in turn a particular version of the Gauss–Seidel method. It has been confirmed by numerous numerical experiments that significant acceleration of convergence of the calculational process can be achieved by changing the mesh point value $U_{i,j} \equiv U(x_i, y_j)$ by an increment, greater than $(\text{Res}_{i,j})/4$. A method extrapolating this unique property is called the SOR method [5, 6]. Seeking possibly precise description of the relevant algorithm, let us assume that the value of the function $U_{i,j} \equiv U(x_i, y_j)$ obtained using this method in the $(k-1)$ iteration is equal to $U_{i,j}^{(k-1)} \equiv U^{(k-1)}(x_i, y_j)$. Let us assume also that $U_{i,j}^{(L,k)} \equiv U^{(L,k)}(x_i, y_j)$ denotes the value of this function obtained during the $(k-1)$ iteration by means of the Liebmann successive substitution method. When the function $U_{i,j} \equiv U(x_i, y_j)$ is determined using the SOR method, its discrete value in the k iteration is calculated from the following, extrapolating formula:

Table 8.6 ($k = 1$, $R_k = 7.17 \times 10^{-2}$)

$j/i \rightarrow$	2	3	4	5	6
2	0.43766388	0.56698200	0.60070873	0.59407990	0.45521657
3	0.16642121	0.24917477	0.28013625	0.27985568	0.17530401

$$U_{i,j}^{(k)} = U_{i,j}^{(k-1)} + \omega \cdot [U_{i,j}^{(L,k)} - U_{i,j}^{(k-1)}] \tag{8.28}$$

where $1 \leq \omega < 2$ is the relaxation coefficient. Naturally, the convergence speed of the iterative computational process depends on the value of coefficient ω. In the extreme case, when $\omega = 1$, this speed attains its minimum, and the SOR method transforms itself to the Liebmann successive substitution method. Evaluating the optimum value of ω, for which the most rapid convergence can be achieved is a rather complex issue, remaining beyond the scope of this book. In practice, it is evaluated most frequently from the following formula:

$$\omega = \frac{4}{2 + \sqrt{4 - \left[\cos(\frac{\pi}{I-1}) + \cos(\frac{\pi}{J-1})\right]^2}} \tag{8.29}$$

where I and J are maximum indexes of the applied rectangular grid [4].

Example 8.2 Some consecutive approximate solutions of the boundary problem presented in previous example, calculated by means of the SOR method, are given in Tables 8.6, 8.7 and 8.8. All calculations have been performed for $\omega = 1.16$ evaluated according to formula (8.29). Initial approximations ($k = 0$), for these solutions are given in Table 8.1.

For $\varepsilon \leq 10^{-7}$, see formula (8.22), the condition $R_k \leq \varepsilon$ is satisfied by the approximate solution obtained in the 11th iteration ($k = 11$, $R_k = 8.88 \times 10^{-8}$). According to the results obtained in Example 8.1, the method of successive substitutions ($\omega = 1$) makes it possible obtaining a good approximate solution (the same order of accuracy, defined by $R_k \leq \varepsilon = 10^{-7}$), only after 18 iterations ($k = 18$, $R_k = 5.15 \times 10^{-8}$). An influence of the over-relaxation coefficient ω on the convergence obtained in the present example (expressed by k_{min}) is illustrated by the data written in Table 8.9.

Minimum numbers k_{min} of necessary iterations, given in Table 8.9 has been determined for $R_k \leq \varepsilon = 10^{-7}$. These results fully confirm usefulness of the relation (8.29) for computing the over-relaxation coefficient ω, taking the value close to optimum.

Table 8.7 ($k = 5$, $R_k = 1.79 \times 10^{-3}$)

$j/i \rightarrow$	2	3	4	5	6
2	0.43818196	0.57607508	0.60825387	0.57582314	0.43739248
3	0.17408658	0.25770889	0.28088501	0.25760877	0.17374662

Table 8.8 $(k = 10, R_k = 4.64 \times 10^{-7})$

$j/i \rightarrow$	2	3	4	5	6
2	0.43737384	0.57575764	0.60808084	0.57575759	0.43737374
3	0.17373741	0.25757581	0.28080809	0.25757576	0.17373737

Another, equally important problem consists in choosing the mesh size h of the grid, which undoubtedly influences on accuracy of the calculational process. It has been confirmed in the literature that the approximate solution is a function of even powers in h and is related to the accurate solution U_R of the differential equation (but not of the difference equation) by the following formula $U_{i,j}(h) = U_R + a_2 h^2 + a_4 h^4 + \ldots$, in which $a_2, a_4, \ldots$ are constant coefficients [2]. This formula constitutes a theoretical basis of the Richardson extrapolation procedure. An essence of this procedure consists in multiple solution of the boundary problem for different values of the step size h and on subsequent extrapolation of these results for the case $h = 0$. This problem is illustrated below by computational results given in Example 8.3.

Example 8.3 In present example, the problem considered in Examples 8.1 and 8.2 has been solved again (five times) by using the grids with different square meshes. The corresponding grid parameters and mesh sizes h (in millimetres) are given in the three first rows of the Table 8.10.

In the last row of this table, the values of the approximate solution $U_{i,j} \equiv U(x_i, y_j)$ evaluated at the point P_A are given, see Fig. 8.11. Coordinates of the point P_A are: $x = 2$ mm and $y = 2$ mm. The following condition $R_k \leq \varepsilon = 10^{-9}$ has been adopted as the stop criterion, see formula (8.22). The presented values of U_A have been used to evaluate the extrapolating polynomial $U_A(h) = U_R + a_2 h^2 + a_4 h^4$ where $U_R = 0.439283788$, $a_2 = -4.766666666 \times 10^{-4}$ and $a_4 = -1.433600001 \times 10^{-6}$. Thus, the value $U_A(h \rightarrow 0) = U_R = 0.439283788$ can be treated as the exact value of U_A at the point P_A. In this case, the absolute value of difference $|U_R - U(h = 0.125)|$ is less than 7.5×10^{-6}.

Here, it should be pointed out that point P_A is identical to all internal nodes (grid points) $((I - 1)/6 + 1, (J - 1)/3 + 1)$ of each discretization grids being used.

Table 8.9

ω	1.00	1.05	1.10	1.15	1.20	1.25	1.30	1.35	1.40
k_{min}	18	16	14	12	11	12	13	16	19

Table 8.10

I	7	13	25	49	97
J	4	7	13	25	49
h	2	1	0.5	0.25	0.125
ω_{opt}	1.155	1.427	1.659	1.812	1.901
U_A	0.437373737	0.438805713	0.439164532	0.439253991	0.439276340

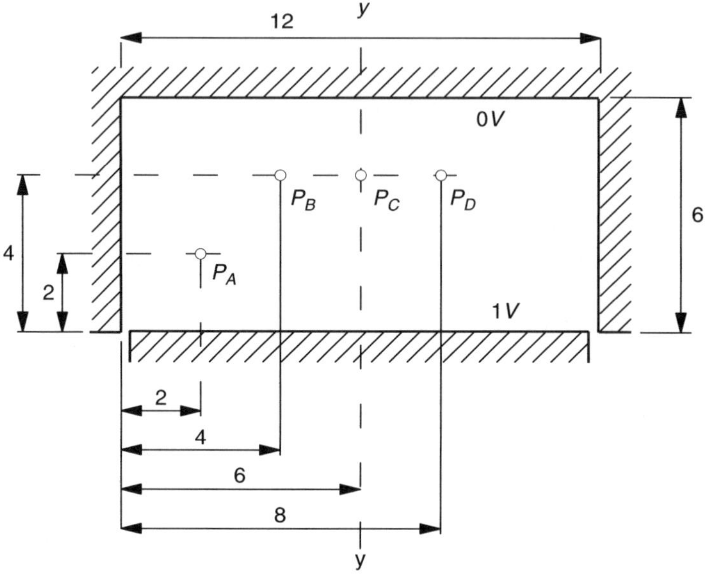

Fig. 8.11

8.3 Difference Formulas for Numerical Calculation of a Normal Component of an Electric Field Vector at Good Conducting Planes

In the previous sections, it has been assumed that function $U_{i,j} \equiv U(x_i, y_j)$, where $1 \leq i \leq I, 1 \leq j \leq J$, is the function of scalar potential of the electric field. According to general principles of electrodynamics, the function of the scalar potential makes it possible evaluating the vector electric field $\mathbf{E}_{i,j} \equiv \mathbf{E}(x_i, y_j)$ over the same internal region, i.e., for $1 \leq i \leq I, 1 \leq j \leq J$. For this purpose, the following fundamental formula can be used:

$$\mathbf{E}(P_{i,j}) = -\nabla U(x_i, y_j) = -\mathbf{i}_x \frac{\partial U(x_i, y_j)}{\partial x} - \mathbf{i}_y \frac{\partial U(x_i, y_j)}{\partial y} \qquad (8.30)$$

where: $\mathbf{i}_x$ and $\mathbf{i}_y$ are unity vectors (versors) of the utlilized cartesian coordinate system. The partial derivatives appearing in the formula (8.30) can be calculated numerically by using the appropriate difference formulas, discussed in Chap. 6. To this end, the second order central difference formulas are used most frequently. Thus,

$$\frac{\partial U(x_i, y_j)}{\partial x} \approx \frac{U(x_{i+1}, y_j) - U(x_{i-1}, y_j)}{2h} = \frac{U_{i+1,j} - U_{i-1,j}}{2h}$$
$$\frac{\partial U(x_i, y_j)}{\partial y} \approx \frac{U(x_i, y_{j+1}) - U(x_i, y_{j-1})}{2k} = \frac{U_{i,j+1} - U_{i,j-1}}{2k} \qquad (8.31)$$

where h and k are sufficiently small "steps" referring to the variables x and y, respectively, see Fig. 8.1. For calculating the partial derivatives at external nodes lying on the contour C of the given inernal region, the appropriate one-side approximation should be used, see relation (6.11).

Example 8.4 Tables 8.11, 8.12 and 8.13 present some values of the function $U_{i,j} \equiv U(x_i, y_j)$, evaluated in the previous, namely 8.3 example for $1 \leq i \leq 97$ and $1 \leq j \leq 49$.

The above values of function $U_{i,j} \equiv U(x_i, y_j)$ were subsequently used to calculate the components E_x and E_y of electric field vectors $\mathbf{E}$ at points $P_B \equiv P_{33,33}$, $P_C \equiv P_{49,33}$ and $P_D \equiv P_{65,33}$, which are also indicated in Fig. 8.11. The electric field vectors, calculated according to (8.30) and (8.31), are equal to:

$$\mathbf{E}_B = \mathbf{E}(P_B) = -\mathbf{i}_x 26.857048 + \mathbf{i}_y 141.680308, \ V/m$$
$$\mathbf{E}_C = \mathbf{E}(P_C) = \mathbf{i}_x 0.000000 + \mathbf{i}_y 151.716468, \ V/m$$
$$\mathbf{E}_D = \mathbf{E}(P_D) = \mathbf{i}_x 26.857048 + \mathbf{i}_y 141.680308, \ V/m$$

Vector $\mathbf{E}_C$ is directed parallel to the y-axis ($E_{Cx} = 0$). It confirms the fact that numerical values of the function $U_{i,j} \equiv U(x_i, y_j)$ we have found are symmetrically distributed (mirror reflection symmetry) with respect to the symmetry line $y - y$, at which the point $P_C \equiv P_{49,33}$ lies. The points $P_B \equiv P_{33,33}$ and $P_D \equiv P_{65,33}$, see Fig. 8.11, lie symmetrically with respect to line $y - y$, and therefore components E_y of the electric field vectors, determined at these points should be equal. For the same reason, components E_x of these vectors should have equal absolute values and opposite signs. These requirements are satisfied by vectors $\mathbf{E}_B$ and $\mathbf{E}_D$ evaluated above. It proves the fact that all calculations are correct and accuracy is

Table 8.11 ($h = 0.125\,mm$)

$j/i \rightarrow$	32	33	34
34	0.240537830	0.243837103	0.246896645
33	0.257934984	0.261419628	0.264649246
32	0.275595163	0.279257180	0.282649139

Table 8.12 ($h = 0.125\,mm$)

$j/i \rightarrow$	48	49	50
34	0.267812180	0.267900714	0.267812180
33	0.286679136	0.286772204	0.286679136
32	0.305732516	0.305829831	0.3057322516

Table 8.13 ($h = 0.125\,mm$)

$j/i \rightarrow$	64	65	66
34	0.246896643	0.243837101	0.240537828
33	0.264649244	0.261419626	0.257934982
32	0.282649137	0.279257178	0.275595161

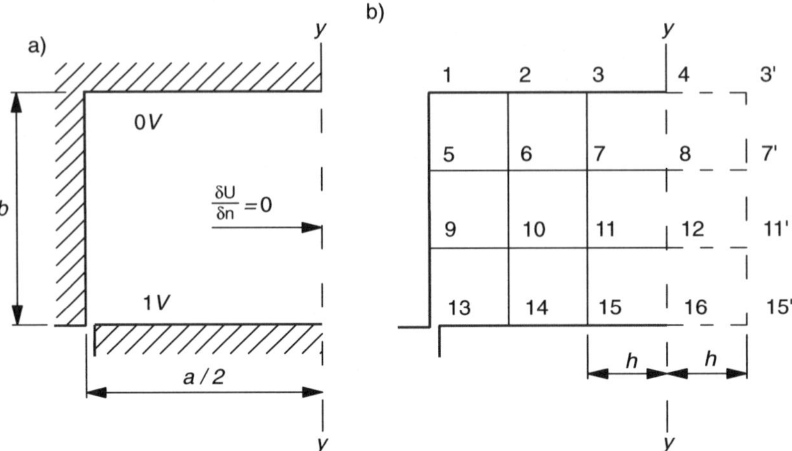

Fig. 8.12

sufficient, $R_k \le \varepsilon = 10^{-9}$. The conductor configuration, shown in Figs. 8.9 and 8.11, has mirror reflection symmetry with respect to the line $y - y$, and therefore the region in which the functions $U_{i,j} \equiv U(x_i, y_j)$ and $\mathbf{E}_{i,j} \equiv \mathbf{E}(x_i, y_j)$ are calculated, can be limited to the subregion shown in Fig. 8.12. The function $U_{i,j} \equiv U(x_i, y_j)$ being calculated over this subregion should satisfy the following condition:

$$\frac{\partial U(x, y)}{\partial n} = 0 \tag{8.32}$$

at every point belonging to the division line $y - y$. The directional derivative (8.32) is calculated in the direction n that is normal to line $y - y$ at point $P(x, y)$. In the computational procedure the condition (8.32) can be easily taken into account in the manner described in Sect. 8.1.

In the case of conductors shown in Fig. 8.12(a) it is necessary to introduce fictious nodes, which are mirror reflections of the nodes lying closest to the division line $y - y$. Assuming that $I = 7$, $J = 4$, see Fig. 8.9(b), we obtain an equivalent region, as in Fig. 8.12(b). Fictious nodes $3'$, $7'$, $11'$ and $15'$ are mirror reflections of the real nodes $3, 7, 11$ and 15. Assume further that k approximations of the desired function $U_{i,j}^{(k)} \equiv U^{(k)}(x_i, y_j)$ at all internal nodes, (6, 7, 8, 10, 11 and 12) are known. In the next $(k + 1)$ iteration we first calculate values of $U_{i,j}^{(k+1)} \equiv U^{(k+1)}(x_i, y_j)$ at nodes $6, 7, 10$ and 11, using the SOR. When the values of $U_{i,j}^{(k+1)} \equiv U^{(k+1)}(x_i, y_j)$ evaluated at real nodes 7 and 11 (internal nodes lying closest to the division line $y - y$) are known, then we assign these values also to the associated fictious nodes, $7'$ and $11'$, respectively. Next, in the final stage of the $(k + 1)$ iteration, based on the values $U_{i,j}^{(k+1)} \equiv U^{(k+1)}(x_i, y_j)$ calculated at the real nodes $4, 7, 11, 16$ and fictious nodes $7'$, $11'$, we calculate the approximations $U_{i,j}^{(k+1)} \equiv U^{(k+1)}(x_i, y_j)$ of the desired function at internal nodes 8 and 12 lying on the division line $y - y$.

Table 8.14 ($h = 0.125\,mm$)

$j/i \rightarrow$	32	33	34
34	0.240537839	0.243837112	0.246896654
33	0.257934993	0.261419637	0.264649255
32	0.275595173	0.279257190	0.282649149

The computational process described above is now repeated iteratively, until maximum difference R_k, see formula (8.22), would be less than a given small number ε defining accuracy of calculations. Some values of $U_{i,j} \equiv U(x_i, y_j)$, obtained in this manner for $a = 12\,mm$, $b = 6\,mm$, $I = 49$, $J = 49$, $h = 0.125\,mm$ and $\varepsilon = 10^{-9}$ are given in Table 8.14

It should be pointed out that these values are very close to the corresponding values written in Table 8.11. Undoubtedly, this fact confirms correctness of the approach being employed.

8.4 Examples of Computation of the Characteristic Impedance and Attenuation Coefficient for Some TEM Transmission Lines

The fundamental parameter of the TEM transmission line is its characteristic impedance Z_0. Physical meaning of this circuit parameter is explained in Example 2.1. According to [7, 8], the characteristic impedance of an arbitrary air (ε_0, μ_0) TEM transmission line can be calculated from the following general formula:

$$Z_0 = \sqrt{\frac{\mu_0}{\varepsilon_0}} \cdot \frac{U}{\oint_{S_2} E_n \cdot ds} = \sqrt{\frac{\mu_0}{\varepsilon_0}} \cdot \frac{U}{Q/\varepsilon_0} = \sqrt{\mu_0 \varepsilon_0} \cdot \frac{1}{C} = \frac{1}{\upsilon} \cdot \frac{1}{C}, \Omega \qquad (8.33)$$

where $\eta_0 = \sqrt{\mu_0/\varepsilon_0} = 120\,\pi \approx 377, \Omega$ is the wave impedance of the open free space, U denotes the difference of potentials (voltage) between two conductors of the line (inner and outer), E_n is the normal component of electric field vector defined on the border S_1 of the external conductor, and ds denotes an infinitesimally short section of the integration contour. In this case, $\upsilon = 1/\sqrt{\varepsilon_0 \mu_0} \approx 2.997925 \times 10^8$ m/s is the velocity of light in free space and C denotes a line capacity per unit length. The way to compute the characteristic impedance Z_0 is open if a distribution $E_n(S_1)$ of the component E_n at the boundary line S_1 is known. Similar distribution $E_n(S_2)$ should be evaluated on the border line S_2 of the inner conductor. These distributions make it possible calculating the attenuation coefficient $\alpha = \text{Re}[\gamma = \alpha + j\beta]$. To this end, the following relation can be used:

$$\alpha = \frac{1}{2 \cdot \sigma \cdot \delta} \cdot \sqrt{\frac{\varepsilon_0}{\mu_0}} \cdot \frac{\oint_{S_1+S_2} |E_n|^2 \cdot ds}{\oint_{S_1} E_n \cdot ds} \qquad (8.34)$$

where σ is conductivity of the material (metal) used to manufacture the conductors and δ denotes skin depth [8]. It follows from formulas (8.33) and (8.34) that basic problem, which should be solved in order to find Z_0 and α, is determination of the distributions of normal component of electric field vector on the surfaces of two conductors creating the transmission line. These distributions should be evaluated as accurately as possible in the manner discussed in Sects. 8.2 and 8.3. Unfortunately, we cannot use the central difference formulas for numerical calculation of partial derivatives of the function $U_{i,j} \equiv U(x_i, y_j)$ at points lying on the boundaries of the conductors. Only one-side approximations of the derivatives can be evaluated at these external points (grid nodes). It is obvious that these approximations are less accurate than corresponding approximations performed using the central difference formulas. One simple and efficient method of increasing approximation accuracy of derivatives is the Runge interpolation procedure, described in Example 6.1. In the case $k = 2$, see relation (6.29), it is based on two one-side difference approximations calculated for various step sizes, for example h and $2h$. An effective accuracy of this two-step procedure is close to that obtained when using the central difference formulas (8.31).

8.4.1 The Shielded Triplate Stripline

The transverse section of a shielded triplate stripline is shown in Fig. 8.13(a). An internal region V of this TEM transmission line is limited to the space closed between inner and outer conductors. Usually, this region is fulfiled by dry air that is homogenous lossless medium with relative permittivity $\varepsilon_r = 1$ and relative permeability $\mu_r = 1$ [8, 9].

The presented transverse section has mirror reflection symmetry with respect to lines $x - x$ and $y - y$. Similar symmetry characterizes also distribution of the potential function $U_{i,j} \equiv U(x_i, y_j)$. Thanks to this double symmetry, the problem of finding distribution of the function $U_{i,j} \equiv U(x_i, y_j)$ in the region V can be reduced to the similar problem, solved for the four times smaller subregion $(V/4)$, shown in Fig. 8.13(b). Such reduced problem can be solved by means of the method similar to that used in Example 8.4. Consequently, values of the function $U_{i,j} \equiv U(x_i, y_j)$ at nodes lying on the division lines $x - x$ and $y - y$ are calculated by means of the fictious nodes being mirror reflections of the nodes situated very near to these division lines. Grids, most appropriate to the analysis of this type of the line, Fig. 8.13(a), are the ones having square meshes and the size h chosen in such a way that the distances between all adjacent nodes (internal and external) are equal. Unfortunately, it is not always possible to satisfy this condition. In such cases, some boundary lines of the conductors do not coincide with the lines of the grids. For example, in case of nodes, situated most closely to the side edge of the inner conductor of the line shown in Fig. 8.13(b), the distance from this edge is less than h. Values of potential function $U_{i,j} \equiv U(x_i, y_j)$ at these nodes should be calculated using the appropriate formula, (8.18) or (8.19), respectively.

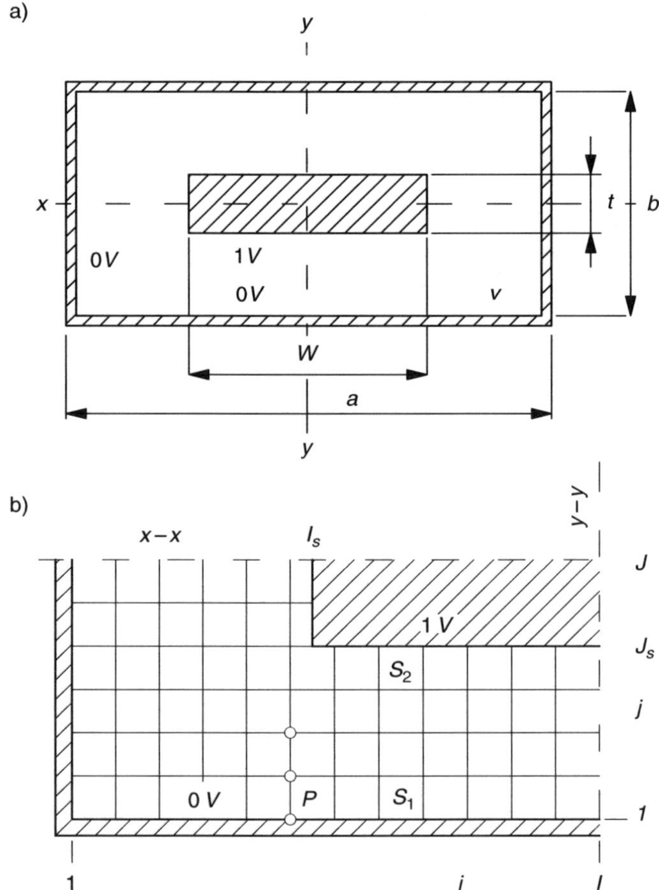

Fig. 8.13

The normal component of electric field vector E_n at the border of the outer conductor is determined by means of two one-side approximations of partial derivatives of the function $U_{i,j} \equiv U(x_i, y_j)$, calculated for two different step size. For example, computation of E_n at point $P \equiv P(i, 1)$, see Figs. 8.13(b) and 8.14, is performed as follows.

First, the initial approximation $E_n(i, 1)$ is calculated from

$$E_n^{(1)}(i, 1) \approx \frac{U(i, 2) - U(i, 1)}{h}$$

The second approximation of the component $E_n(i, 1)$ is calculated using the step $2 \cdot h$, according to similar difference formula. Hence,

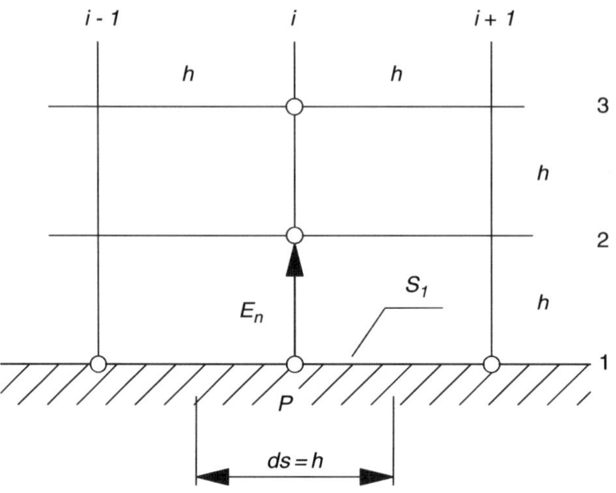

Fig. 8.14

$$E_n^{(2)}(i, 1) \approx \frac{U(i, 3) - U(i, 1)}{2 \cdot h}$$

Finally, the normal component $E_n(i, 1)$, evaluated according to the Runge procedure, is:

$$E_n(i, 1) \approx E_n^{(1)}(i, 1) + \frac{E_n^{(1)}(i, 1) - E_n^{(2)}(i, 1)}{(2h/h)^1 - 1} = 2 \cdot E_n^{(1)}(i, 1) - E_n^{(2)}(i, 1)$$

The dimensions of the subregion $(V/4)$, under analysis, see Fig. 8.13(b), are: $a/2 = 10\,\text{mm}, b/2 = 2.5\,\text{mm}, t/2 = 0.5\,\text{mm}$ and $W/2 = 2\,\text{mm}$. Into this subregion the grid with square meshes determined by $I = 201, I_S = 161, J = 51$ and $J_S = 41$ have been introduced. Thus, the distances between all adjacent nodes (including the distances between nodes situated inside and on the edge of the subregion $(V/4)$) are the same and equal to $h = 0.05\,\text{mm}$. The computation of discrete values of the potential function $U_{i,j} \equiv U(x_i, y_j)$ have been performed with accuracy $R_k \leq 1.34 \cdot 10^{-6}$ for the following internal grid points (nodes):

$P(i, j)$, where $2 \leq i \leq 201$ when $2 \leq j \leq 40$,

$P(i, j)$, where $2 \leq i \leq 160$ when $41 \leq j \leq 51$

Some final values of the function $U_{i,j} \equiv U(x_i, y_j)$, V are given in Table 8.15.

Normal components E_n of the electric field vector have been calculated for the following nodes: $P(1, j)$ when $2 \leq j \leq 51$ and $P(i, 1)$ when $2 \leq i \leq 201$. All these nodes lie on the border line S_1 of the outer conductor. The corresponding values of

Table 8.15

$j/i \rightarrow$	98	99	100	101	102
51	0.1221401	0.1260761	0.1301396	0.1343347	0.1386661
50	0.1220789	0.1260128	0.1300741	0.1342670	0.1385960
49	0.1218944	0.1258222	0.1298769	0.1340631	0.1383852
...	...	...	...	...	...
21	0.0712702	0.0735313	0.0758620	0.0782645	0.0807408
20	0.0681315	0.0702914	0.0725179	0.0748126	0.0771777
19	0.0649281	0.0669851	0.0691053	0.0712904	0.0735422
...	...	...	...	...	...
3	0.0075850	0.0078237	0.0080696	0.0083228	0.0085836
2	0.0037943	0.0039137	0.0040367	0.0041633	0.0042938

Table 8.16

$i \rightarrow$	98	99	100	101	102
$E_n(i, 1), V/m$	75.92194	78.30988	80.77040	83.30561	85.91492

Table 8.17

$j \rightarrow$	38	39	40	41	42
$E_n(1, j), V/m$	6.72368	6.81143	6.89245	6.96666	7.03396

$U_{i,j} \equiv U(x_i, y_j)$, V, given in Table 8.15, have been used to this end. Several instant values of E_n, chosen from the set $\{E_n(1, j), E_n(i, 1)\}$ defined above, are given in Tables 8.16 and 8.17.

The values of normal component E_n evaluated above make it possible to calculate the characteristic impedance Z_0 of the transmission line under analysis. However, it should be pointed out that subregion $(V/4)$ is only one from the four symmetrical parts of the internal region V. In other words, the distributions of E_n over three remaining subregions are also symmetrical. Naturally, the distribution $E_n(S_1)$ evaluated over the whole boundary S_1 of the outer conductor has to be used to calculate the characteristic impedance Z_0 according to formula (8.33). The characteristic impedance Z_0 of the shielded triplate stripline under consideration calculated in this way is equal to $Z_0 = 55.65\ \Omega$. This approximate value differs only by $0.03\ \Omega$ from the corresponding more accurate value of $Z_0 = 55.62\ \Omega$ given in the literature [10]. Increasing the width W of the inner conductor, see Fig. 8.13(a), to 6 mm results in decreasing the characteristic impedance to a level of $42.96\ \Omega$. In opposite case of decreasing the width W to 2 mm, the characteristic impedance Z_0 attains the value of $79.07\ \Omega$. These two impedances differ from the corresponding more accurate values published in [10, 11] by less than $0.04\ \Omega$.

8.4.2 The Square Coaxial Line

The transverse section of an air square coaxial line is shown in Fig. 8.15. This TEM transmission line can be treated as the special case of shielded triplate stripline discussed earlier, see Fig. 8.13(a). Of course, in this case $a = b$ and $W = t$.

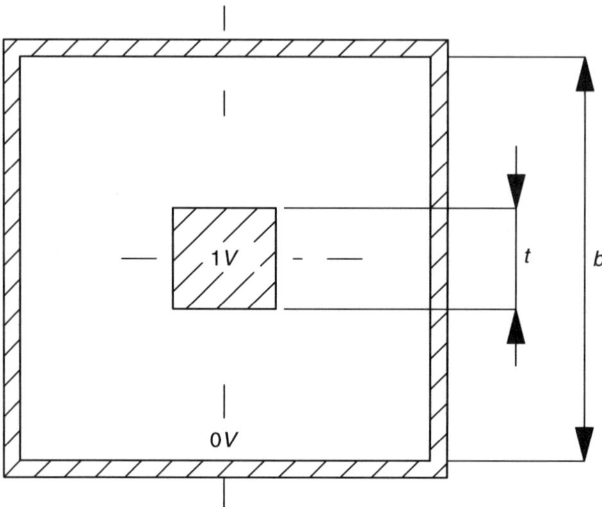

Fig. 8.15

A procedure for numerical simulation of that transmission line includes:

– evaluation of the distribution of potential function $U_{i,j} \equiv U(x_i, y_j)$ over the whole internal region V (limited space between the inner and outer conductors),
– evaluation of the distribution of electric field vector $\mathbf{E}_{i,j} \equiv \mathbf{E}(x_i, y_j)$ over the whole internal region V and
– calculating the characteristic impedance Z_0.

At the first stage, the distribution of potential function $U_{i,j} \equiv U(x_i, y_j)$ over the whole internal region V is evaluated similarly as in the case of shielded triplate stripline analyzed in Sect. 8.4.1. For clarity of further considerations, let us assume that geometrical dimensions of the transmission line under discussion are: $b = 20\,\text{mm}$ and $t = 8\,\text{mm}$. It means that dimensions of the subregion $(V/4)$ are: 10 mm and 4 mm, respectively. Into this subregion the grid with square meshes characterized by $I = J = 51, I_S = J_S = 31$ and mesh size $h = 0.05\,\text{mm}$ have been introduced. The calculations of discrete values of the potential function $U_{i,j} \equiv U(x_i, y_j)$ over the subregion $(V/4)$ have been carried out with accuracy $R_k \leq 8.94 \cdot 10^{-7}$, see relation (8.22). Some instance results of these iterative calculations are given in Table 8.18.

As it has been explained earlier the distribution of $U_{i,j} \equiv U(x_i, y_j)$ constitutes a basis for evaluating the related distribution of electric field vector $\mathbf{E}_{i,j} \equiv \mathbf{E}(x_i, y_j)$. It should be pointed out once again that distribution of $\mathbf{E}_{i,j} \equiv \mathbf{E}(x_i, y_j)$ has to be evaluated as accurate as possible, especially at the border line S_1 of the outer conductor. In the next stage, the distribution of $\mathbf{E} \equiv \mathbf{E}(S_1) = E_n(S_1)$ is used to calculate characteristic impedance Z_0 according to formula (8.33). The above approach implemented

Table 8.18

$j/i \rightarrow$	18	19	20	21	22
51	0.5345663	0.5685376	0.6028721	0.6375675	0.6726155
50	0.5343854	0.5683555	0.6026919	0.6373909	0.6724456
49	0.5338403	0.5678081	0.6021483	0.6368591	0.6719331
...	...	...	...	...	...
21	0.3003669	0.3189443	0.3375928	0.3562825	0.3749717
20	0.2851210	0.3025906	0.3200916	0.3375929	0.3550536
19	0.2698196	0.2862051	0.3025904	0.3189443	0.3352270
...	...	...	...	...	...
3	0.0293639	0.0310278	0.0326733	0.0342972	0.0358962
2	0.0146779	0.0155092	0.0163311	0.0171423	0.0179411
1	0.0000000	0.0000000	0.0000000	0.0000000	0.0000000

Table 8.19

$t/b \rightarrow$	0.1	0.3	0.5	0.7	0.9
$Z_0(t/b)$, Ω	132.15	66.81	36.78	18.02	5.07
Z_0, Ω	132.65	66.87	36.81	18.02	5.07

in analysis of the square coaxial line under consideration yields $Z_0 = 49.78 \, \Omega$. This approximate value differs from the value $49.82 \, \Omega$, obtained analytically by less than $0.04 \, \Omega$, [11, 12]. Discrete values of the function $Z_0(t/b)$ obtained in much the same way for different values of ratio t/b, see Fig. 8.15, are written in the second row of Table 8.19.

In the third row of the table the appropriate exact values of impedance Z_0, Ω (evaluated analytically) are given for comparison [11].

8.4.3 The Triplate Stripline

Figure 8.16 presents a transverse section of the unshielded triplate stripline fulfiled by dry air that is homogenous lossless medium characterized by relative permittivity $\varepsilon_r = 1$ and relative permeability $\mu_r = 1$.

A width of its external, equipotential conducting planes (of the outer conductor) should be sufficiently large for the electric field intensity in the regions $1 - 1'$ and $2 - 2'$. Of course, this intensity should be sufficiently small in comparison with the electric field intensity in the closest neighborhood of the inner conductor. The triplate stripline, in which this condition is satisfied, can be analyzed similarly as the shielded stripline discussed earlier, in the case, when the side walls (of height b) are sufficiently distant from the inner conductor. The condition for vanishing of the electric field vector at side walls should be of course satisfied for this distance. A correctness of that approach can be justified by the calculation results given in Table 8.20. Presented values of the normal component E_n of the electric vector

a)

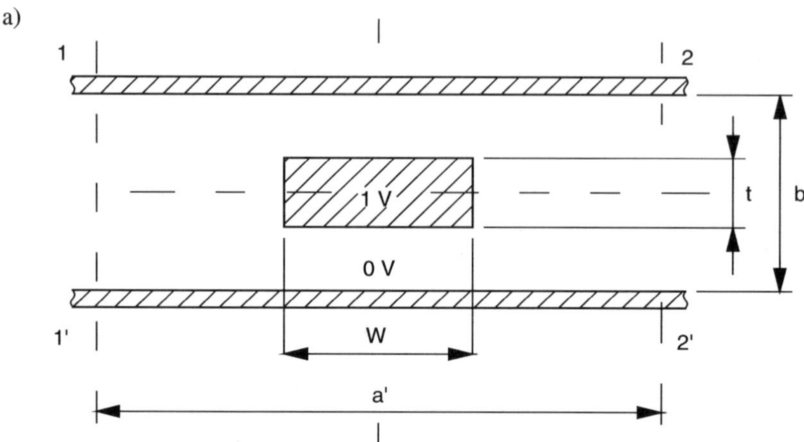

b)

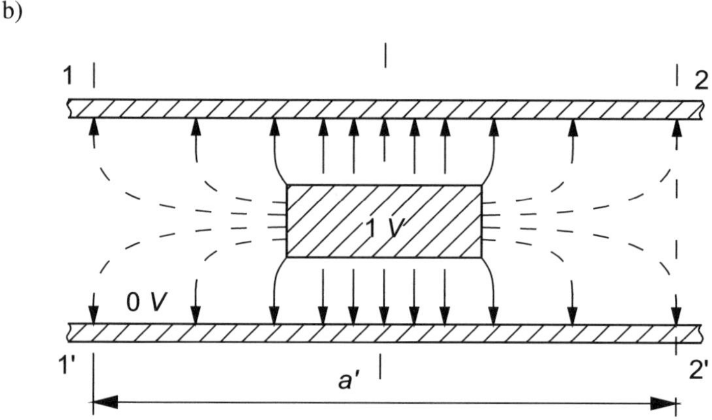

Fig. 8.16

$\mathbf{E}_n$, have been evaluated at the side walls for: $b/2 = 2.5\,\text{mm}$, $t/2 = 0.5\,\text{mm}$, $W/2 = 2\,\text{mm}$, $h = 0.05\,\text{mm}$ and some values of $a/2$. The appropriate values of potential function $U_{i,j} \equiv U(x_i, y_j)$ have been calculated with an accuracy of $R_k \leq 1.18 \times 10^{-6}$, using the square grid with mesh size of $h = 0.05\,\text{mm}$.

Table 8.20

$a/2$, mm	4.0	6.0	8.0	10	12
$E_n(1, 51)$, V/m	350.1233	90.6190	25.6384	7.3217	2.1164
$E_n(1, 41)$, V/m	328.0744	86.0941	24.3859	6.9666	2.0160
$E_n(1, 31)$, V/m	268.5441	73.0280	20.7430	5.9292	1.7178
$E_n(1, 21)$, V/m	186.2990	52.8722	15.0698	4.3101	1.2504
$E_n(1, 11)$, V/m	94.4789	27.7189	7.9228	2.2673	0.6586
$Z_0(a)$, Ω	52.3963	55.3969	55.6321	55.6492	55.6495

A minimum width a of the shielded triplate stripline, see Fig. 8.13(a), for which it can be treated as equivalent to unshielded one under discussion, Fig. 8.16, is determined most frequently according to the criterion of the minimum change of the characteristic impedance $Z_0(a)$. It has been confirmed numerically that function $Z_0(a)$ calculated for the shielded triplate stripline has a one-side "saturation region". It is therefore not recommended to increase the width a beyond some threshold value of a'. This conclusion, significant for practical applications, is well illustrated by values of characteristic impedance written in the seventh row of Table 8.20. They show that characteristic impedance of the air unshielded triplate stripline, for which: $b = 5\,\text{mm}$, $t = 1\,\text{mm}$ and $W = 4\,\text{mm}$, attains maximum value of $Z_0 = 55.65\,\Omega$ when the width a' of its semi-opened outer conductor is greater than 20 mm.

8.4.4 The Shielded Inverted Microstrip Line

Another version of the shielded stripline widely used in the microwave technology is the shielded inverted microstrip line. The transverse section of this transmission line is shown in Fig. 8.17(a).

The thin ($t << b$) lossless dielectric layer with small permittivity ε_r plays mainly the role of mechanical support, holding the inner conductor (strip) of width W in proper position with respect to the surrounding outer conductor. The transverse section presented here has mirror reflection symmetry with respect to vertical plane $y - y$. Due to this symmetry, the problem of determining the potential function $U_{i,j} \equiv U(x_i, y_j)$ over the internal region V can be reduced to a similar problem solved for two times smaller subregion ($V/2$), shown in Fig. 8.17(b). This region is

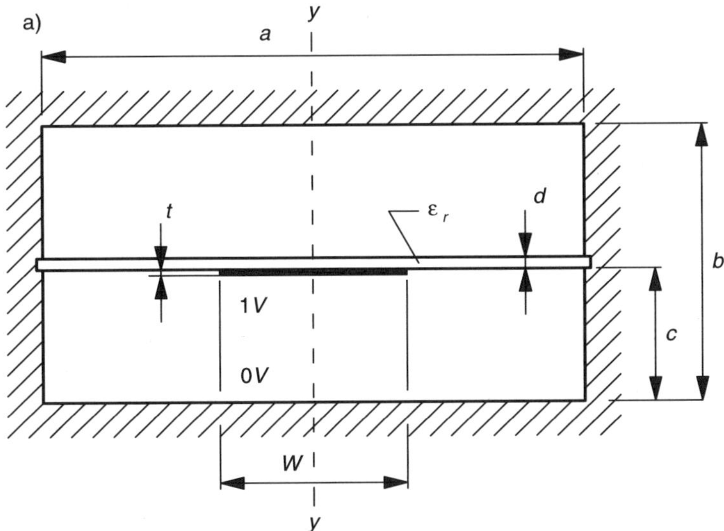

Fig. 8.17 (Continued)

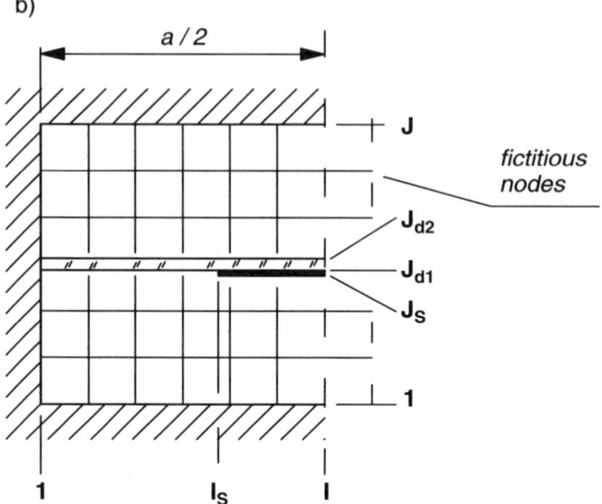

Fig. 8.17

electrically inhomogenous, because permittivity ε_r of the dielectrical layer differs from 1, i.e., $\varepsilon_r > 1$. In order to guarantee the clarity of further considerations, let us assume that border line of the inner and outer conductors coincides with the lines of the used discretization grid with square meshes. This assumption holds also for the border line of the dielectric layer. For the problem defined in this manner, values of the potential function $U_{i,j} \equiv U(x_i, y_j)$ at internal nodes which do not belong to the border air–dielectric are calculated using formulas (8.20) and (8.28). In case of nodes situated on this border, see Fig. 8.18(a), the Laplace equation (8.2) is not satisfied and consequently, formula (8.20) resulting from this equation cannot be used [13]. For these nodes, the following more general formula is suitable:

$$U_{i,j} = \frac{U_{i,j+1} + \varepsilon_r \cdot U_{i,j-1}}{2(1 + \varepsilon_r)} + \frac{U_{i-1,j} + U_{i+1,j}}{4} \tag{8.35}$$

The difference formula (8.35), see Appendix E, results from the following equation of electrodynamics

$$\nabla \cdot \mathbf{D} = \nabla \cdot (\varepsilon_r \varepsilon_0 \nabla U) = 0 \tag{8.36}$$

telling that at a surface of dielectric substrate no storage of electrical charge occurs. For methodological reasons, it is recommended to divide the inhomogenous region $(V/2)$ into four smaller, electrically homogenous similar to those shown in Fig. 8.19.

Values of potential function $U_{i,j} \equiv U(x_i, y_j)$, at points belonging to the symmetry plane $y - y$, that is at side edges of the subregions $(V/2)_1$, $(V/2)_3$, $(V/2)_4$, can be found identically as these are described in the previous examples. Fictious nodes used for this purpose are placed at nodes of the grid, which are closest to the line $y - y$. For calculating potential function $U_{i,j} \equiv U(x_i, y_j)$ at nodes situated on the borders between two subregions with different permittivity ε_r, the relation (8.35)

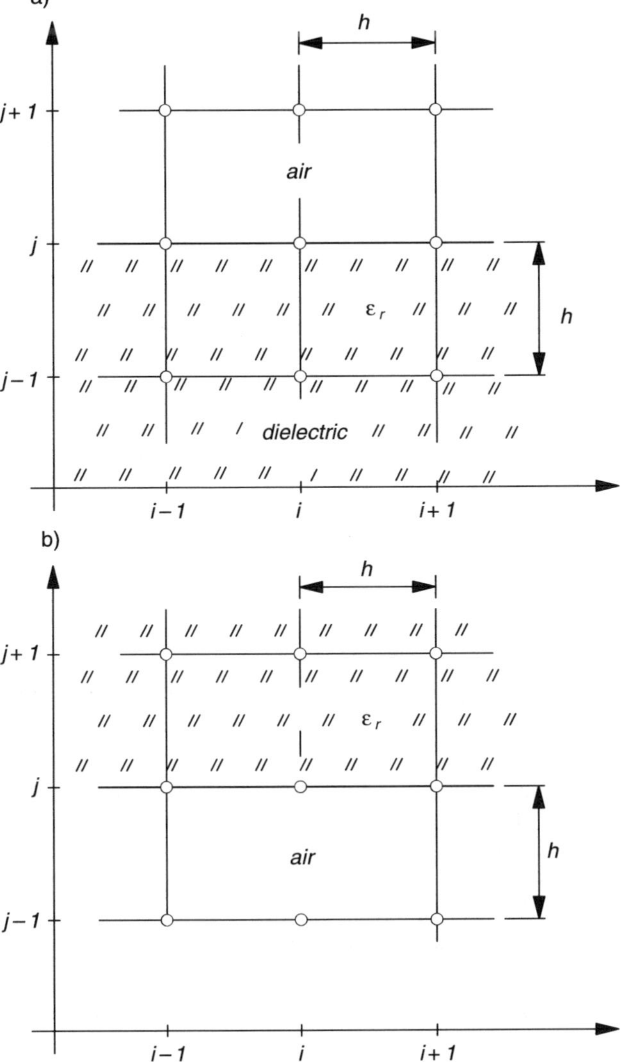

Fig. 8.18

should be used, remembering about proper interpretation of the adopted notation. In other words, relation (8.35) is satisfied at points situated on the border between two media shown in Fig. 8.18(a). In case of dielectric media in reverse configuration, see Fig. 8.18(b), the appropriate formula corresponding to (8.35) is:

$$U_{i,j} = \frac{U_{i,j-1} + \varepsilon_r \cdot U_{i,j+1}}{2(1 + \varepsilon_r)} + \frac{U_{i-1,j} + U_{i+1,j}}{4} \tag{8.37}$$

The methodical recommendations given above have been employed in the analysis of the transmission line shown in Fig. 8.17 and characterized by: $a = 310 \cdot h$,

Fig. 8.19

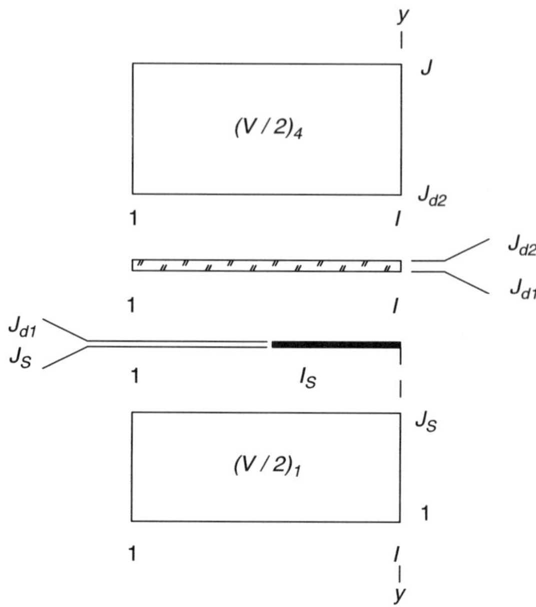

$b = 60 \cdot h$, $c = 24 \cdot h$, $d = 12 \cdot h$, $t = 1 \cdot h$ and $W = 60 \cdot h$, where $h = 0.0508$ mm (2 mils) is the size of the grid mesh. The permittivity of the dielectric layer is equal to $\varepsilon_r = 2.65$. For these parameters, grid lines defined by $i = 1$, $i = I = 156$, $I_S = 126$, $j = 1$, $j = J = 61$, $J_S = 24$, $J_{d1} = 25$, $J_{d2} = 37$ agree with corresponding borders of the inner and outer conductors, as well as with the border of the dielectric layer. Table 8.21 contains some discrete values of the function $U_{i,j} \equiv U(x_i, y_j)$, evaluated with accuracy of $R_k \leq 1.132 \times 10^{-6}$. As it was repeated many times the potential function $U_{i,j} \equiv U(x_i, y_j)$ constitutes a basis for evaluating distribution of the electric field vector $\mathbf{E}(x_i, y_j)$. In order to calculate this distribution at the surface of the outer conductor the extrapolating procedure described in Sect. 8.4.1 is recommended. Some values of E_n, calculated in this way in the region close to the surface of the outer conductor, are written in Tables 8.22 and 8.23.

According to the laws of electrodynamics, the surface charge density at an arbitrary point P of a conducting surface is $q(P) = \varepsilon_0 \cdot \varepsilon_r(P) \cdot E_n(P)$. Hence, an electric charge stored on the unit length of the outer line surface S_1 is equal to:

$$Q = \varepsilon_0 \cdot \oint_{S_1} \varepsilon_r(P) \cdot E_n(P) ds \qquad (8.38)$$

where $\varepsilon_0 = 8.854184 \cdot 10^{-12}$, F/m is permittivity of the free space. Dividing this unit charge by voltage $U = 1$, V between inner and outer conductors yields the unit line capacity $C = Q/U$. If the line medium is inhomogenous two stages are necessary

Table 8.21

$j/i \rightarrow$	123	124	125	126	127
23	0.7194310	0.7702185	0.8291498	0.8911311	0.9166453
22	0.6901418	0.7325579	0.7771443	0.8187293	0.8462229
21	0.6572688	0.6927266	0.7281418	0.7604189	0.7856923
...	...	...	...	...	...
11	0.3178634	0.3277425	0.3372938	0.3464353	0.3550958
10	0.2854351	0.2941333	0.3025497	0.3106194	0.3182848
9	0.2532179	0.2608053	0.2681526	0.2752080	0.2819243
...	...	...	...	...	...
3	0.0628770	0.0646617	0.0663948	0.0680658	0.0696672
2	0.0314283	0.0323180	0.0331820	0.0340156	0.0348145
1	0.0000000	0.0000000	0.0000000	0.0000000	0.0000000

Table 8.22

$i \rightarrow$	123	124	125	126	127
$E_n(i, 1), V/m$	618.4661	635.9290	652.8858	669.2596	684.9518

Table 8.23

$j \rightarrow$	2...	24	25	26...	59
$E_n(1, j), V/m$	0.355955	7.018490	7.221045	7.288838	0.709235

to evaluate characteristic impedance Z_0 and effective line permittivity ε_{eff}, [8, 10]. In the first stage, the unit line capacity $C_0 = C(\varepsilon_r = 1)$ is evaluated when the dielectric layer is removed. Naturally, in the second stage the unit line capacity C_r is evaluated in the same manner when the dielectric layer is present. When the unit capacities C_0 and C_r are known, it is possible to find the characteristic impedance Z_0 and the phase velocity v from the following formulas:

$$Z_0 = \frac{1}{v_0} \cdot \frac{1}{\sqrt{C_0 \cdot C_r}}, \Omega$$
$$v = v_0 \sqrt{\frac{C_0}{C_r}} = \frac{v_0}{\sqrt{\varepsilon_{eff}}}$$

(8.39)

where $v_0 \equiv c = 2.997925 \cdot 10^8$, m/s is the light velocity in free space. An effective line permittivity ε_{eff} occurring in the above formulas is defined as $\varepsilon_{eff} = C_r/C_0$. In case of the shielded inverted microstrip line under discussion, see Fig. 8.17, $C_0 = 5.508917 \cdot 10^{-11}$, F/m, $C_r = C(\varepsilon_r = 2.65) = 6.810233 \cdot 10^{-11}$, F/m, $Z_0 = 54.458$, Ω and $v = v_0 \cdot 0.899398$. An influence of permittivity ε_r of the dielectric layer on C_r, Z_0 and v is illustrated by the data given in Table 8.24.

Table 8.24

ε_r	C_r, pF/m	Z_0, Ω	v/v_0
1.00	55.08917	60.55	1.000000
2.65	68.10233	54.46	0.899398
3.74	73.81895	52.31	0.863870

8.4.5 *The Shielded Slab Line*

The transverse section of a shielded slab line is shown in Fig. 8.20(a). This kind of TEM transmission line is widely used in the microwave technology because their use offers several manufacturing advantages and excellent electrical properties. As most of air–dielectric lines the shielded slab line is particularly recommended for operation at high peak and average powers. Therefore, the computation of its characteristic impedance Z_0 and attenuation coefficient α is a problem of considerable importance for practice. In the present section, it is shown how this problem can be effectively solved by means of the finite difference method.

Also in this case the three-stage approach specified at the beginning of Sect. 8.4.2 will be adopted. It follows from Fig. 8.20 that a distribution of the potential function $U_{i,j} \equiv U(x_i, y_j)$ over the internal region V (limited by inner and outer conductors) is mirror reflection symmetrical with respect to the horizontal line $x - x$ and vertical line $y - y$. Due to this double symmetry, the problem of finding distribution

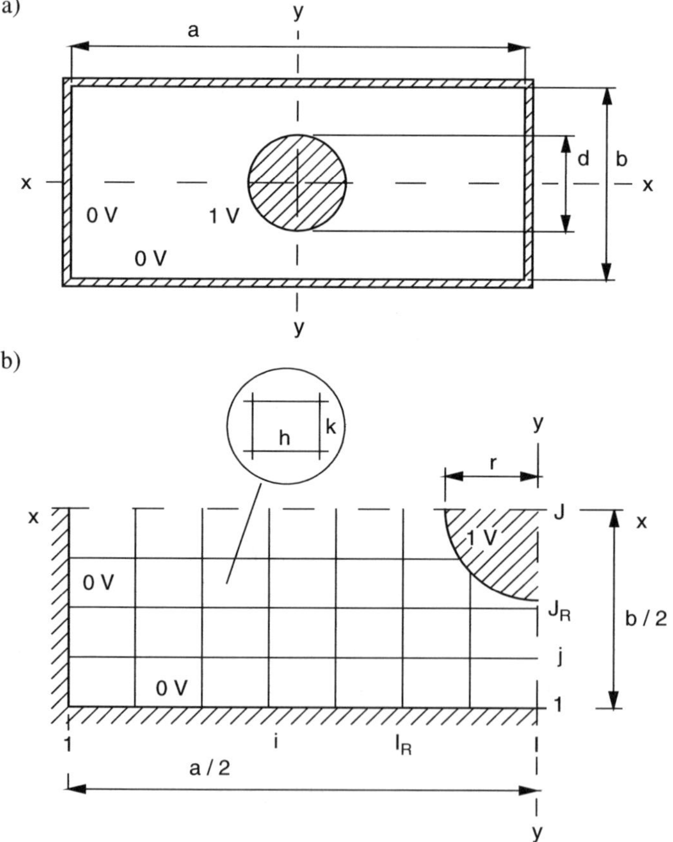

Fig. 8.20

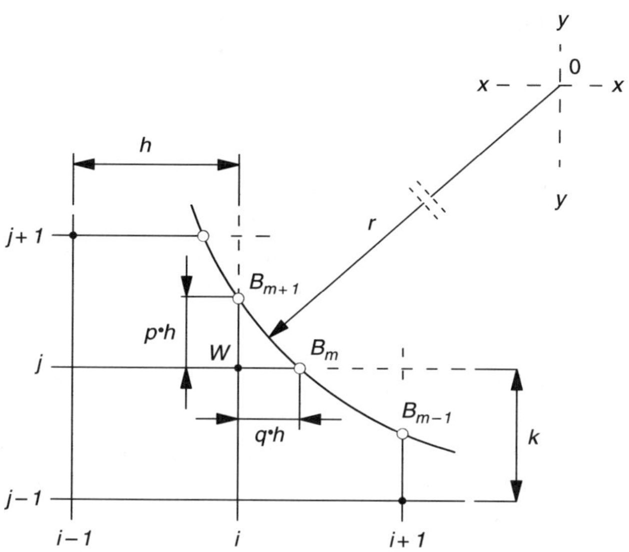

Fig. 8.21

of $U_{i,j} \equiv U(x_i, y_j)$ can be reduced to the solution of similar problem over the four times smaller subregion $(V/4)$, shown in Fig. 8.20(b). Mesh sizes h and k of the rectangular discretization grid should be chosen in a such manner that its lines coincide with the border of outer conductor and the symmetry lines $x - x$ and $y - y$. This requirement is satisfied when $h = a/[2(I - 1)]$ and $k = b/[2(J - 1)]$, where I and J are maximum values of indexes i and j, respectively. Unfortunately, due to circular shape of the inner conductor the adopted rectangular grid cross its border at points situated at unequal distances with respect to the closest grid points (nodes). In Fig. 8.21 border nodes determined in this way are denoted by small circles.

To each border node one of the parameters p or q defining the distance to the nearest grid node (coastal) lying on the same vertical or horizontal line, should be assigned. Similarly as in the previous examples, it is assumed that potential function $U_{i,j} \equiv U(x_i, y_j)$ takes at the border nodes of the inner conductor the value of $U = 1V$, and potential of the outer conductor is equal to $U = 0V$. Values of the function $U_{i,j} \equiv U(x_i, y_j)$ at coastal nodes, marked in Fig. 8.21 by dots, should be computed using the general formula (8.18). In case of remaining, internal nodes of the subregion $(V/4)$, consecutive approximated values of the potential function $U_{i,j} \equiv U(x_i, y_j)$ are next calculated according to formula (8.19), where $p = k/h$. As it has been mentioned above the distribution of the potential function $U_{i,j} \equiv U(x_i, y_j)$ has mirror reflection symmetry with respect to lines $x - x$ and $y - y$. Thus, values of $U_{i,j} \equiv U(x_i, y_j)$ at nodes belonging to these symmetry lines can be calculated using the manner, in which auxiliary fictious nodes are introduced. Essential features of this simple computation technique have been explained and illustrated in the previous examples. According to the recommendations given above, in the first stage of computation process, the coordinates of nodes lying on the inner conductor, see Figs. 8.20 and 8.21, should be determined. For example, the coordinates of border node B_m belonging to the inner conductor and grid horizontal line denoted by index j are

$$y_m = (j-1)k, \; x_m = a/2 - \sqrt{r^2 - (b/2 - y_m)^2} \qquad (8.40)$$

where a, b and $r = d/2$ are geometrical dimensions of the transverse section shown in Fig. 8.20(b). The coordinates (8.40) can be represented by the indexes j and $i_B(j)$ defined as:

$$j = y_m/k + 1, \; i_B(j) = \text{int}(x_m/h) + 2 \qquad (8.41)$$

where the function $\text{int}(a)$ assigns to the argument a the greatest integer not exceeding a. Similarly, coordinates of border node B_{m+1}, belonging to the vertical line denoted by index i, can be calculated using the formulas:

$$x_{m+1} = (i-1)h, \; y_{m+1} = b/2 - \sqrt{r^2 - (a/2 - x_{m+1})^2} \qquad (8.42)$$

The indexes corresponding to coordinates (8.8) are:

$$i = x_{m+1}/h + 1, \; j_B(i) = \text{int}(y_{m+1}/k) + 2 \qquad (8.43)$$

The formulas (8.40, 8.41, 8.43) make it possible to calculate the coordinates and related to them indexes of all border nodes lying on the inner conductor and belonging to the subregion $(V/4)$. When coordinates of these border nodes are known, it is possible to determine the indexes i and j of the corresponding coastal nodes. For instance, with the border node B_m, see Fig. 8.21, the coastal node W is connected, which lies on the same line with an index $i = \text{int}(x_m/h) + 1$. Coefficient q, determining the distance between these nodes, is equal to $q = \text{frac}(x_m/h)$, where the function $\text{frac}(a)$ assigns the fractional part to the argument a. Similarly, we can find indexes of coastal nodes with respect to the border nodes lying on the vertical lines of the grid. The node W, shown in Fig. 8.21, is also a coastal node with respect to border node B_{m+1}, on the basis of which relations (8.42) have been derived. Hence, the index j of node W can be evaluated from the formula $j = \text{int}(y_{m+1}/k) + 1$, in which coordinate y_{m+1} is described by relation (8.42). Value of the parameter p, defining the distance between the nodes B_{m+1} and W under consideration, is equal to $p = \text{frac}(y_{m+1}/k)$.

All computations, performed during the first stage described above, have the character of preparatory single-time calculations. The computations of the second stage are also single-time. They serve to determine initial approximation of the desired solution $U_{i,j} \equiv U(x_i, y_j)$. They are performed similarly, as in Example 8.1, based on the values of $U_{i,j} \equiv U(x_i, y_j)$ given for the border points (nodes) of the subregion $(V/4)$, as in Fig. 8.22.

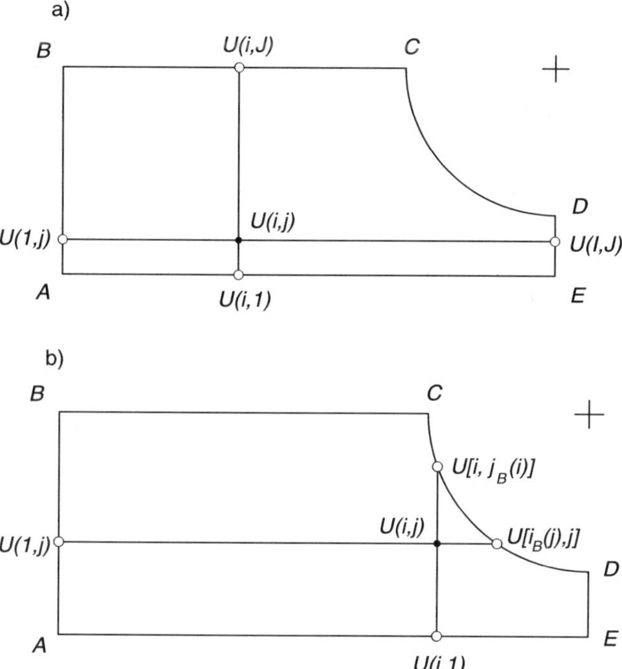

Fig. 8.22

In the computer program SSL elaborated for this purpose the following initial
(border) values of the potential function $U_{i,j} \equiv U(x_i, y_j)$ have been used:

- edge $AB \rightarrow U_{i,j} \equiv U(x_i, y_j) = 0$, V,
- edge $AE \rightarrow U_{i,j} \equiv U(x_i, y_j) = 0$, V,
- edge $CD \rightarrow U_{i,j} \equiv U(x_i, y_j) = 1$, V,
- edge $U_{i,j} \equiv U(x_i, y_j)$ varies linearly along the edge BC from 0, V at point B to 1, V at point C,
- function $U_{i,j} \equiv U(x_i, y_j)$ varies linearly along the edge ED from 0, V at point E to 1, V at point D.

During the third computation stage the consecutive approximations of the function $U_{i,j} \equiv U(x_i, y_j)$ have been evaluated by means of the SOR method. As the stop condition, the inequality $R_k \leq 9.53 \times 10^{-7}$ has been used. Some discrete values of the potential function $U_{i,j} \equiv U(x_i, y_j)$ obtained in this way are given in Table 8.25. The calculations have been carried out for: $a/2 = 12$ mm, $b/2 = 2.5$ mm, $r = d/2 = 1.3712$ mm and $h = k = 0.05$ mm In this case, indexes of the introduced discretization grid satisfy the inequalities $1 \leq i \leq I = 241$ and $1 \leq j \leq J = 51$. Some values of normal component E_n of the electric field vector calculated on the basis of the $U_{i,j} \equiv U(x_i, y_j)$ distribution are given in Tables 8.26 and 8.27.

Table 8.25

$j/i \rightarrow$	198	199	200	201	202
51	0.5996876	0.6193628	0.6397175	0.6607763	0.6825693
50	0.5993603	0.6190230	0.6393644	0.6604101	0.6821886
49	0.5983796	0.6180045	0.6383060	0.6593112	0.6810468
...	...	...	...	...	...
20	0.3244321	0.3342438	0.3443144	0.3546471	0.3652439
...	...	...	...	...	...
3	0.0355999	0.0366335	0.0376900	0.0387693	0.0398711
2	0.0178057	0.0183224	0.0188508	0.0193903	0.0199410
1	0.0000000	0.0000000	0.0000000	0.0000000	0.0000000

Table 8.26

$i \rightarrow$	198	199	200	201	202
$E_n(i, 1)$, V/m	356.2306	366.5601	377.1306	387.9214	398.9308

Table 8.27

$j \rightarrow$	2...	24	25	26...	51
$E_n(1, j)$, V/m	0.0495018	1.0373171	1.0735520	1.1087151	1.5610012

Finally, the characteristic impedance Z_0 of the shielded slab line under analysis have been calculated according to formula (8.33). The result of the calculations is $Z_0 = 50.057$ Ω. This value differs from 50 Ω, calculated by means of formula (2.34), by less than 0.06 Ω. In this case, the influence of side walls on the distribution of $U_{i,j} \equiv U(x_i, y_j)$ and indirectly also on the characteristic impedance Z_0 is negligibly small. This conclusion confirms also the computational results given in Tables 8.26 and 8.27. The intensity of the electric field at side walls is relatively small in comparison with its value evaluated at the central line plane, ($i = 202$). When side walls of the outer conductor are sufficiently distant from the inner round conductor, the electrical parameters of the shielded slab line are very close to corresponding parameters of the unshielded slab line with the same geometrical dimensions b and d, discussed already in Example 2.2.

In a special case when $a = b = 5$ mm, see Fig. 8.20, the line under consideration becomes a coaxial line with the square outer conductor. Tables 8.28, 8.29 and 8.30 present some values of $U_{i,j} \equiv U(x_i, y_j)$, $E_n(i, 1)$, V/m and $E_n(1, j)$, V/m calculated for the line in question assuming that $h = k = 0.05$ mm. In this case, indexes of the grid satisfy the inequalities $1 \le i \le I = 51$ and $1 \le j \le J = 51$. Computations of the potential function $U_{i,j} \equiv U(x_i, y_j)$ over the suitable subregion $(V/4)$ have been performed with an accuracy determined by condition $R_k \le 8.94 \times 10^{-7}$, see expression (8.22).

The characteristic impedance of this special line version is equal to $Z_0 = 40.602$ Ω and differs by less than 0.06 Ω from the value of $Z_0 = 40.547$, Ω given in Table 4.1 of the handbook [11]. Performing similar computations using the grid with smaller meshes ($h = k = 0.025$ mm) the characteristic impedance

Table 8.28

$j/i \rightarrow$	18	19	20	21	22
51	0.7091850	0.7575211	0.8070557	0.8577519	0.9097104
50	0.7086132	0.7569216	0.8064758	0.8573708	0.9092808
49	0.7068861	0.7550766	0.8045543	0.8554748	0.9080939
...	...	...	...	...	...
3	0.0343164	0.0363016	0.0382766	0.0402396	0.0421886
2	0.0171561	0.0181429	0.0191353	0.0201164	0.0210903
1	0.0000000	0.0000000	0.0000000	0.0000000	0.0000000

Table 8.29

$i \rightarrow$	2...	24	25	26...	51
$E_n(i, 1)$, V/m	20.2578	460.1502	479.0576	497.7260	777.4435

Table 8.30

$j \rightarrow$	2...	24	25	26...	51
$E_n(1, j)$, V/m	20.2579	460.1514	479.0587	497.7277	777.4435

of value $Z_0 = 40.579$ Ω is achieved. This final result differs from the value of $Z_0 = 40.547$ Ω by less than 0.04 Ω. In this situation further reduction of meshes of the discretization grid is unnecessary.

8.4.6 Shielded Edge Coupled Triplate Striplines

The finite difference method presented in this chapter can be easily adopted to analysis of coupled TEM transmission lines. One of them is the air–dielectric edge coupled triplate stripline whose transverse section is shown in Fig. 8.23(a).

The conducting surfaces of this transmission line are distributed symmetrically with respect to the plane $y - y$. Due to this symmetry, it can be analyzed by means of the method of even mode $(++)$ and odd mode $(+-)$ excitations, explained already in Example 3.3. The circuit representing the coupled lines for even mode excitation is a single transmission line with a transverse section as shown in Fig. 8.23(b). The characteristic impedance of this transmission line is denoted by $Z_0^{++} \equiv Z_{0e}$. Similarly, the circuit representing the coupled lines for odd mode excitation is also a single transmission line with a transverse section as shown in Fig. 8.23(c). Its characteristic impedance is denoted by $Z_0^{+-} \equiv Z_{0o}$. Both transverse sections shown in Fig. 8.23(b) and 8.23(c) are symmetrical with respect to the plane $x - x$. Thus, the problem of evaluation of the potential function $U_{i,j}^{(++)} \equiv U^{(++)}(x_i, y_j)$ over the region $(V/2)^{++}$ can be reduced to the similar problem over the subregion $(V/4)^{++}$ shown in Fig. 8.24(a). In the similar way, the function $U_{i,j}^{(+-)} \equiv U^{(+-)}(x_i, y_j)$ can be evaluated on the basis of the subregion $(V/4)^{+-}$ shown in Fig. 8.24(b).

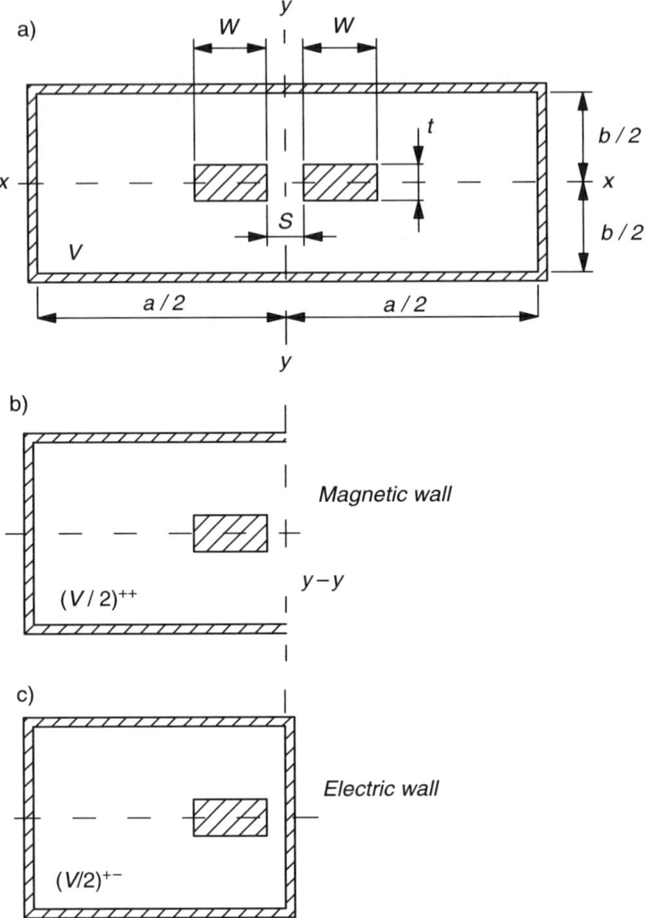

Fig. 8.23

The distribution of the potential function $U_{i,j}^{++} \equiv U^{++}(x_i, y_j)$ over the subregion $(V/4)^{++}$, see Fig. 8.24(a), can be evaluated in the manner similar to that employed in Sect. 8.4.1 for analysis of the shielded triplate stripline. Of course, the same approach is suitable for evaluating the distribution of the function $U_{i,j}^{(+-)} \equiv U^{(+-)}(x_i, y_j)$ over the subregion $(V/4)^{+-}$ shown in Fig. 8.24(b). These distributions make a basis for evaluating the corresponding distributions of the electric field vector, i.e., $\mathbf{E}_{i,j}^{++} \equiv \mathbf{E}^{++}(x_i, y_j)$ and $\mathbf{E}_{i,j}^{+-} \equiv \mathbf{E}^{+-}(x_i, y_j)$, respectively. The appropriate difference formulas described in detail in previous sections (for instance in Sect. 8.4.1) can be used for this purpose. Integral parts of these electric field vector distributions are distributions of the normal component of the electric field vector evaluated on the surfaces S_e and S_o of outer conductors of the transmission lines shown in Figs. 8.23(b) and 8.23(c), respectively. These distributions, namely

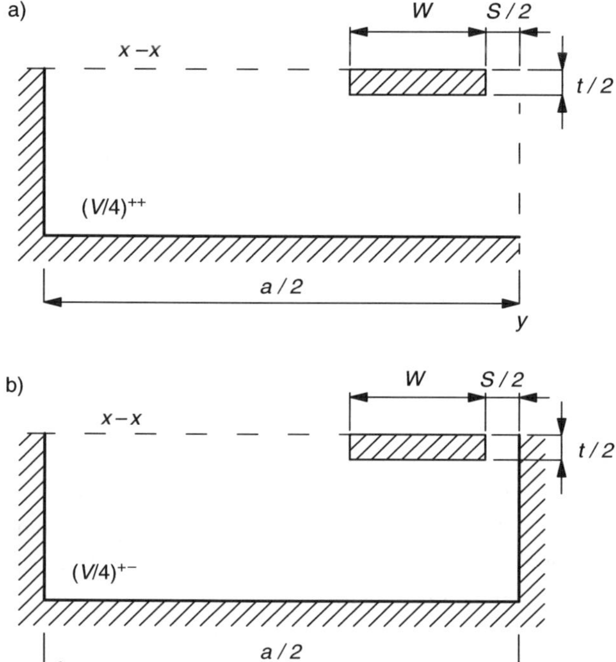

Fig. 8.24

$E_n^{++} \equiv E_n^{++}(S_e)$ and $E_n^{+-} \equiv E_n^{+-}(S_o)$ make it possible to calculate the characteristic impedances $Z_0^{++} \equiv Z_{0e}$ and $Z_0^{+-} \equiv Z_{0o}$ according to general formula (8.33). The computation procedure outlined above has been employed for analysis of the shielded air–dielectric edge coupled striplines, see Fig. 8.23(a), with the following parameters: $a = 40.00$ mm, $b = 5.00$ mm, $S = 2.40$ mm, $t = 1.00$ mm and $W = 4.00$ mm. Equal rectangular grids have been used for covering the subregions $(V/4)^{++}$ and $(V/4)^{+-}$, see Fig. 8.24. Node positions of these grids are defined by indexes, (i, j), where $1 \leq i \leq I = 251$ and $1 \leq j \leq J = 51$. The dimensions: $a = 40.00$ mm, $b = 5.00$ mm and indexes $I = 251$, $J = 51$ define sizes of the grid meshes univocally, that are $h = (a/2)/(I - 1) = 0.08$ mm and $k = p \cdot h = (b/2)/(J - 1) = 0.05$ mm. In order to find initial approximation $U_{i,j}^{++(0)} \equiv U^{++(0)}(x_i, y_j)$ of the function $U_{i,j}^{++} \equiv U^{++}(x_i, y_j)$ it was assumed that this function takes on the borders of the subregion $(V/4)^{++}$ the following values:

$$U_{i,j}^{++(0)} = 0, \text{V} \qquad\qquad \text{for } i = 1, \ 1 \leq j \leq J$$

$$U_{i,j}^{++(0)} = 0, \text{V} \qquad\qquad \text{for } j = 1, \ 1 \leq i \leq I$$

$$U_{i,j}^{++(0)} = 1 \cdot \frac{i - 1}{I_W - 1}, \text{V} \qquad \text{for } j = J, \ 1 \leq i \leq I_W$$

$$U_{i,j}^{++(0)} = 1, \text{V} \qquad\qquad\qquad \text{for } j = J, \; I_W \leq i \leq I_S$$

$$U_{i,j}^{++(0)} = 1 - [1 - U_{i,j}^{++(0)}(I, J)] \cdot \frac{i - I_S}{I - I_S}, \text{V} \qquad \text{for } j = J, \; I_S \leq i \leq I$$

$$U_{i,j}^{++(0)} = U_{i,j}^{++(0)}(I, J) \cdot \frac{j-1}{J-1}, \text{V} \qquad\qquad \text{for } i = I, \; 1 \leq j \leq J$$

where: $I_s = 236$, $I_W = 186$, $J_s = 41$ and

$$U_{i,j}^{++(0)}(I, J) = \frac{p_{I,J} q_{I,J} r_{I,J}}{p_{I,J} r_{I,J} + q_{I,J}} \left[\frac{1}{1 + q_{I,J}} + \frac{1}{q_{I,J}(1 + q_{I,J})} + \frac{0}{p_{I,J}(p_{I,J} + r_{I,J})} \right.$$
$$\left. + \frac{0}{r_{I,J}(p_{I,J} + r_{I,J})} \right], \text{V}$$

is an initial value of the determined potential function at the terminal node (I, J). Thanks to symmetry of the coupled lines under consideration (see Figs. 8.6 and 8.23), $q_{I,J} = 1$ and $p_{I,J} = r_{I,J} = b/S = 2.5$. Hence, initial value of the potential function at node (I, J) is equal to $U_{i,j}^{++(0)}(I, J) = b^2/(b^2 + S^2) = 25/29$, V. Table 8.31 presents some values of potential function $U_{i,j}^{++} \equiv U^{++}(x_i, y_j)$ determined by means of the SOR method. As the stop criterion the condition $R_k \leq 1.013 \times 10^{-6}$ has been used, see relation (8.22).

Some values of the distribution $E_n^{++} \equiv E_n^{++}(S_e)$ evaluated on the basis of the potential function $U_{i,j}^{++} \equiv U^{++}(x_i, y_j)$ are given in Table 8.32.

Finally, the characteristic impedance calculated on the basis of the distribution $E_n^{++} \equiv E_n^{++}(S_e)$ is equal to $Z_0^{++} \equiv Z_{0e} = 59.3999 \; \Omega$. As it has been mentioned above the function $U_{i,j}^{+-} \equiv U^{+-}(x_i, y_j)$ is evaluated in the similar manner over the subregion $(V/4)^{+-}$, see Fig. 8.24(b). Some values of this function calculated with accuracy defined by $R_k \leq 1.014 \times 10^{-6}$, are given in Table 8.33.

Table 8.31

$j/i \rightarrow$	234	235	236	237	228
40	0.9648619	0.9595679	0.9465962	0.9012761	0.8623688
39	0.9307016	0.9221351	0.9058289	0.8736286	0.8413264
38	0.8979273	0.8877244	0.8712706	0.8460212	0.8189873
...	...	...	...	...	...
20	0.4338699	0.4297129	0.4253570	0.4208516	0.4162554
...	...	...	...	...	...
3	0.0453387	0.0449792	0.0446116	0.0442398	0.0438680
2	0.0226683	0.0224887	0.0223054	0.0221199	0.0219343
1	0.0000000	0.0000000	0.0000000	0.0000000	0.0000000

Table 8.32

$i \rightarrow$	2...	235	236	237...	251
$E_n(i, 1)$, V/m	0.0065168	449.7598	446.1025	442.3996	408.8472

Table 8.33

$j/i \rightarrow$	234	235	236	237	228
40	0.9501752	0.9389871	0.9128662	0.8249545	0.7437695
39	0.9023052	0.8838077	0.8498692	0.7844080	0.7150580
38	0.8572562	0.8346599	0.7991868	0.7453806	0.6853173
...	...	...	...	...	...
20	0.3457720	0.3313840	0.3158930	0.2993351	0.2817674
...	...	...	...	...	...
3	0.0341457	0.0326655	0.0310969	0.0294421	0.0277045
2	0.0170641	0.0163241	0.0155399	0.0147129	0.0138445
1	0.0000000	0.0000000	0.0000000	0.0000000	0.0000000

The characteristic impedance calculated on the basis of this distribution is equal to $Z_0^{+-} \equiv Z_{0o} = 50.6935 \ \Omega$.

The analysis procedure presented above has been repeated many times for various thickness t of the internal strips, see Fig. 8.23(a). Calculation results, illustrating an influence of this thickness on characteristic impedances Z_{Oe}, Z_{0o} and coupling coefficient $k = (Z_{Oe} - Z_{Oo})/(Z_{Oe} + Z_{Oo})$ are given Table 8.34.

Values of impedances Z_{Oe}, Z_{0o} calculated for $t = 0$, mm and $a \rightarrow \infty$ differ by less than 0.89, Ω from the exact values $Z_{Oe} = 79.9899$, Ω and $Z_{Oo} = 71.3354$, Ω. These reference values have been found from the following formulas:

$$Z_{0e} = 29.976 \cdot \pi \sqrt{\frac{\mu_r}{\varepsilon_r}} \cdot \frac{K'(k_e)}{K(k_e)}, \quad \Omega$$

$$Z_{0o} = 29.976 \cdot \pi \sqrt{\frac{\mu_r}{\varepsilon_r}} \cdot \frac{K'(k_o)}{K(k_o)}, \quad \Omega$$

(8.44)

where:

$$k_e = \text{th}\left(\frac{\pi W}{2b}\right) \text{th}\left(\frac{\pi}{2} \cdot \frac{W+S}{b}\right), \quad k_o = \text{th}\left(\frac{\pi W}{2b}\right) \text{cth}\left(\frac{\pi}{2} \cdot \frac{W+S}{b}\right),$$

$\mu_r = 1$ and $\varepsilon_r = 1$. The term $K(k)$ denotes the complete elliptic integral of the first kind, and $K'(k)$ is the same integral associated (complementary) with $K(k)$, [7][14]. The calculation of integral $K(k)$, when its modulus k is known, is not complicated, and can be done by using the effective algorithm described in Appendix F. It should also be pointed out that the same algorithm can be applied to calculate the complementary integral $K'(k)$.

Table 8.34 ($a = 40.0\,\text{mm}$, $b = 5.0\,\text{mm}$, $S = 2.4\,\text{mm}$, $W = 4.0$, mm)

t, mm	Z_{Oe}, Ω	Z_{Oo}, Ω	k	R_k
1.000	59.3999	50.6935	0.079082	1.014×10^{-6}
0.500	67.9508	59.0282	0.070268	1.098×10^{-6}
0.000	79.1041	70.7488	0.055756	1.013×10^{-6}

Table 8.35 ($a = 40.0\,\text{mm}$, $b = 5.0\,\text{mm}$, $S = 2.4\,\text{mm}$, W = 4.0, mm)

t, mm	Z_{Oe}, Ω	Z_{Oo}, Ω	k	R_k
1.000	59.6877	50.7736	0.080699	$< 10^{-9}$
0.500	68.2943	59.1456	0.071788	$< 10^{-9}$
0.000	79.7297	71.1520	0.056850	$< 10^{-9}$

The difference 0.89, Ω of impedances evaluated above shows that the performed numerical caclulations are not accurate enough. Therefore, numerical analysis of the coupled lines under consideration has been repeated for the new grid with twice reduced mesh sizes now equal to $h = k = 0.025$, mm. The results of these improved calculations are given in Table 8.35.

In this case impedances Z_{Oe}, Z_{0o} evaluated numerically for the coupled lines with infinitely thin inner conductors ($t \approx 0$) differ from their exact values $Z_{Oe} = 79.9899$, Ω, $Z_{Oo} = 71.3354$, Ω by less than 0.27, Ω. This difference seems to be acceptable for the most applications.

References

1. Forsythe G.E. and W.R. Wasow, Finite-difference methods for partial differential equations. John Wiley and Sons, New York, 1960
2. Moon P., Spencer D.E., Field theory for engineers. Van Nostrandt Comp, New York, 1971
3. Kong J.A., Electromagnetic wave theory. John Wiley and Sons, New York, 1983
4. Mathews J.H., Numerical methods for mathematics, science and engineering. Prentice-Hall Intern. Inc., Englewood Cliffs, NJ, 1992
5. Shoup T.E., A practical guide to computer methods for engineers. Prentice-Hall, Englewood Cliffs, NJ, 1979
6. Mathews J.H., Numerical methods for mathematics, science and engineering. Prentice-Hall, Inc, Englewood Cliffs, NJ, 1987
7. Matthaei G.L., L.Young and E.M.T. Jones, Microwave filters, impedance matching networks and coupling structures. Artech House Inc. Boston, MA, 1980
8. Schneider M.V., "Computation of impedance and attenuation of TEM-lines by finite difference method". IEEE Trans., MTT-13, Microwave Theory and Techniques, November 1965
9. Wadell B.C., Transmission line design handbook. Artech House Inc., Boston, MA, 1991
10. Green H.E., "The numerical solution of some important transmission-line problems". IEEE Trans., MTT-13, Microwave Theory and Techniques, September 1965
11. Gunston M.A.R., Microwave transmission line impedance data. Van Nostrandt Reinhold Comp., New York, 1972
12. Conning S.W., "The characteristic impedance of square coaxial line". IEEE Trans., MTT-12, Microwave Theory and Techniques, July 1964
13. Brenner H.E., "Numerical solution of TEM-line problems involving inhomogeneous media". IEEE Trans., MTT-15, Microwave Theory and Techniques, August 1967
14. Rosłoniec S., Algorithms for computer-aided design of linear microwave circuits. Artech House Inc., Boston, MA, 1990

Appendix A
Equation of a Plane in Three-Dimensional Space

Let us assume that in three-dimensional space ($x \equiv x_1, y \equiv x_2, z \equiv x_3$) the vector $\mathbf{R} \equiv [x_0 - 0, y_0 - 0, z_0 - 0]$ is perpendicular (normal) to the plane S at point $R \equiv (x_0, y_0, z_0)$, as it is shown in Fig. A.1.

The plane S can be treated as an infinite set of points $Q \equiv (x, y, z)$, for which the vector $\mathbf{Q} \equiv [x - x_0, y - y_0, z - z_0]$ is perpendicular to the vector $\mathbf{R}$. Vectors $\mathbf{R}$ and $\mathbf{Q}$ are perpendicular, if and only if their dot product is equal to zero. This condition can be written in the form of equation as:

$$\mathbf{R} \cdot \mathbf{Q} \equiv (x_0 - 0)(x - x_0) + (y_0 - 0)(y - y_0) + (z_0 - 0)(z - z_0) = 0 \qquad (A.1)$$

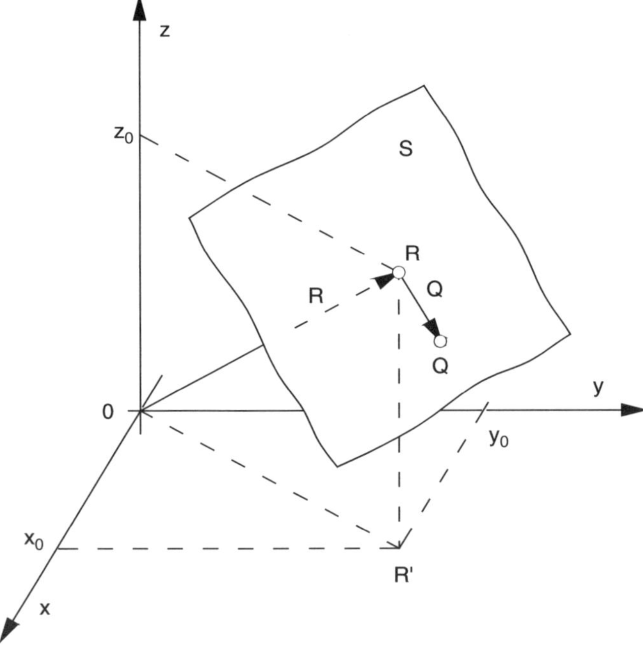

Fig. A.1

Equation (A.1) can be easily transformed to the following equivalent form:

$$x_0 \cdot x + y_0 \cdot y + z_0 \cdot z = (x_0^2 + y_0^2 + z_0^2) \tag{A.2}$$

similar to linear equations occurring in the system (1.8).

Appendix B
The Inverse of the Given Nonsingular Square Matrix

Let $\mathbf{A}$ be a square nonsingular matrix:

$$\mathbf{A} = \begin{bmatrix} a_{11} & a_{12} & \ldots & a_{1n} \\ a_{21} & a_{22} & \ldots & a_{2n} \\ \vdots & \vdots & \ldots & \vdots \\ a_{n1} & a_{n2} & \ldots & a_{nn} \end{bmatrix} \tag{B.1}$$

whose determinant $D \neq 0$. The matrix $\mathbf{A}^{-1}$ is the inverse of the matrix $\mathbf{A}$, when their product is equal to the unitary matrix, i.e., $\mathbf{A} \cdot \mathbf{A}^{-1} = \mathbf{A}^{-1} \cdot \mathbf{A} = \mathbf{E}$. Determinant of the inverse matrix $\mathbf{A}^{-1}$ equals $1/D$. In order to determine the inverse $\mathbf{A}^{-1}$, we introduce the notion of the minor of an element a_{ij} and the cofactor A_{ij} of this element. Minor of element a_{ij} of the square matrix $\mathbf{A}$ is defined as the determinant of the matrix of the rank $(n - 1)$ obtained by crossing out the row i and column j from the original matrix $\mathbf{A}$. The cofactor A_{ij} of the element a_{ij} of the matrix $\mathbf{A}$ is the product of the minor of this element and the multiplier $(-1)^{(i+j)}$. When the cofactors of all elements of the matrix $\mathbf{A}$ and its determinant D are known, the inverse of this matrix can be written as:

$$\mathbf{A}^{-1} = \frac{1}{D} \begin{bmatrix} A_{11} & A_{21} & \ldots & A_{n1} \\ A_{12} & A_{22} & \ldots & A_{n2} \\ \vdots & \vdots & \ldots & \vdots \\ A_{1n} & A_{2n} & \ldots & A_{nn} \end{bmatrix} \tag{B.2}$$

Example B.1 Let us consider the nonsingular matrix:

$$\mathbf{A} = \begin{bmatrix} 2 & -1 \\ 3 & 1 \end{bmatrix}, \quad \det \mathbf{A} = 5$$

The cofactors evaluated for this matrix are: $A_{11} = (-1)^2 \cdot 1 = 1$, $A_{12} = (-1)^3 \cdot 3 = -3$, $A_{21} = (-1)^3 \cdot (-1) = 1$ and $A_{22} = (-1)^4 \cdot 2 = 2$. Thus, the inverse of the matrix $\mathbf{A}$ is:

$$\mathbf{A}^{-1} = \frac{1}{\det \mathbf{A}} \begin{bmatrix} A_{11} & A_{21} \\ A_{12} & A_{22} \end{bmatrix} = \frac{1}{5} \begin{bmatrix} 1 & 1 \\ -3 & 2 \end{bmatrix}, \quad \det \mathbf{A}^{-1} = \frac{1}{5} \cdot \frac{2}{5} - \frac{-3}{5} \cdot \frac{1}{5} = \frac{5}{25} = \frac{1}{5}$$

The product $\mathbf{A} \cdot \mathbf{A}^{-1}$ is equal to the unitary matrix $\mathbf{E}$, namely:

$$\mathbf{A} \cdot \mathbf{A}^{-1} = \frac{1}{5} \begin{bmatrix} 2 & -1 \\ 3 & 1 \end{bmatrix} \cdot \begin{bmatrix} 1 & 1 \\ -3 & 2 \end{bmatrix} = \begin{bmatrix} 1 & 0 \\ 0 & 1 \end{bmatrix}$$

In this simple manner, the correctness of the performed calculations has been confirmed.

Appendix C
The Fast Elimination Method

The fast elimination method discussed below can be treated as a simplified version of the Gauss elimination method adopted for solving large linear equation systems with tridiagonal matrices of coefficients. The term "fast" emphasizes a fact that the computational process is relatively faster because only tridiagonal coefficients are taken into account. In its first stage, pairs of auxiliary coefficients are recursively determined for each of $n - 1$ equations creating the system. During the second stage, these coefficients are then used to find the values of the unknowns. In order to explain an essence of this method, let us consider a tridiagonal equation system written in the following general form:

$$
\begin{aligned}
b_1 x_1 + c_1 x_2 &= d_1 \\
a_2 x_1 + b_2 x_2 + c_2 x_3 &= d_2 \\
a_3 x_2 + b_3 x_3 + c_3 x_4 &= d_3 \qquad\qquad \text{(C.1)}
\end{aligned}
$$

$$
\cdots\cdots\cdots\cdots\cdots\cdots\cdots\cdots\cdots\cdots\cdots\cdots\cdots
$$

$$
\begin{aligned}
a_{n-1} x_{n-2} + b_{n-1} x_{n-1} + c_{n-1} x_n &= d_{n-1} \\
a_n x_{n-1} + b_n x_n &= d_n
\end{aligned}
$$

For further considerations it is assumed that all the coefficients of the main diagonal b_i, where $i = 1, 2, 3, \ldots, n$, are different from zero. It follows from the literature that this condition is satisfied by majority of equation systems describing the real engineering problems. In case of the method under discussion, similarly as in the Gauss elimination method, two stages are distinguished, namely the upward and backward movement. During the first stage, each unknown variable x_i, where $i = 1, 2, 3, \ldots, n - 1$, is expressed in the form of a linear function:

$$
x_i = A_i x_{i+1} + B_i \qquad\qquad \text{(C.2)}
$$

where A_i and B_i are recursive coefficients. Following this rule, from the first equation of the system (C.1) it follows that:

$$
x_1 = -\frac{c_1}{b_1} x_2 + \frac{d_1}{b_1} = A_1 x_2 + B_1 \qquad\qquad \text{(C.3)}
$$

where $A_1 = -c_1/b_1$ and $B_1 = d_1/b_1$. The relation (C.3) introduced into the second equation of the system (C.1) makes it possible to write the following relation $a_2(A_1x_2 + B_1) + b_2x_2 + c_2x_3 = d_2$ as:

$$x_2 = -\frac{c_2}{a_2A_1 + b_2}x_3 + \frac{d_2 - a_2B_1}{a_2A_1 + b_2} = A_2x_3 + B_2 \tag{C.4}$$

where $A_2 = -c_2/(a_2A_1 + b_2)$, $B_2 = (d_2 - a_2B_1)/(a_2A_1 + b_2)$. Similarly, for $1 \le i \le n - 1$ the remaining coefficients are:

$$A_i = -\frac{c_i}{a_iA_{i-1} + b_i}, \quad B_i = \frac{d_i - a_iB_{i-1}}{a_iA_{i-1} + b_i} \tag{C.5}$$

The values of all coefficients A_1, B_1, A_2, B_2, ..., A_{n-1}, B_{n-1}, evaluated in this manner have to be stored in the computer memory. In the second stage (backward movement) values of the unknowns x_i are consecutively calculated, starting from x_n. The unknown x_n is computed by solving the following equation system:

$$x_{n-1} = A_{n-1}x_n + B_{n-1} \tag{C.6}$$
$$a_nx_{n-1} + b_nx_n = d_n$$

formulated from (C.2) defined for $i = n - 1$ and the last equation of the original system (C.1). A solution of the equation system (C.6) is:

$$x_n = \frac{d_n - a_nB_{n-1}}{a_nA_{n-1} + b_n} \tag{C.7}$$

The remaining unknowns x_i are next calculated using the relation (C.2) and the coefficients A_i and B_i evaluated in the first stage. Theoretically, in the computational process described above an operation of division by zero, or infinitesimal number may occur, see relations (C.5) and (C.7). It has been proved in the literature that such "danger" is absent, if for $1 \le i \le n$ the following inequalities are satisfied:

$$|b_i| \ge |a_i| + |c_i| \tag{C.8}$$

and at least one of the inequalities (C.8) is acute. In other words, matrix of coefficients $\mathbf{A}$ should be diagonally dominant.

Appendix D
The Doolittle Formulas Making Possible Presentation of a Nonsingular Square Matrix in the form of the Product of Two Triangular Matrices

The matrix equation $[a_{ij}]_{nn} \equiv \mathbf{A} = \mathbf{L} \cdot \mathbf{U}$ discussed in Sect. 1.1.3 is equivalent to the system of n^2 linear equations:

$$a_{ij} = \sum_{q=1}^{r} l_{iq} \cdot u_{qj}, \quad \text{where } r = \min(i, j) \tag{D.1}$$

In the Crout method it is assumed that all the diagonal elements of the matrix $\mathbf{U}$ are equal 1, i.e., $u_{ii} = 1$ for $i = 1, 2, 3, \ldots, n$. Due to this assumption, the sum of elements of the triangular matrices $\mathbf{L}$ and $\mathbf{U}$, determined in the process of LU decomposition reduces to n^2. In order to find the calculation formulas, appropriate for this particular decomposition, consider the general Eq. (D.1), written for the element a_{ik} of the row i and column j, assuming that $i \geq k$.

$$a_{ik} = \sum_{q=1}^{k} l_{iq} \cdot u_{qk} = \sum_{q=1}^{k-1} l_{iq} \cdot u_{qk} + l_{ik} \cdot u_{kk} \tag{D.2}$$

From equation (D.2), assuming that $u_{kk} = 1$, we obtain the following recursive formula:

$$l_{ik} = a_{ik} - \sum_{q=1}^{k-1} l_{iq} \cdot u_{qk}, \quad \text{where } i = k, k+1, k+2, \ldots, n \tag{D.3}$$

Let us consider Eq. (D.1) again, written for an element a_{kj} of the row k and column j, assuming that $j > k$.

$$a_{kj} = \sum_{q=1}^{k} l_{kq} \cdot u_{qj} = \sum_{q=1}^{k-1} l_{kq} \cdot u_{qj} + l_{kk} \cdot u_{kj} \tag{D.4}$$

After some elementary transformations, Eq. (A.4) takes the form:

$$u_{kj} = \frac{1}{l_{kk}} \left[a_{kj} - \sum_{q=1}^{k-1} l_{kq} \cdot u_{qj} \right], \quad \text{where } j = k+1, k+2, \ldots, n \qquad (D.5)$$

Equations (D.3) and (D.5) derived above are particular cases of the Doolittle formulas. They constitute a theoretical basis for the computer program CROUT, which has been used to decomposite the matrix $\mathbf{A} = \mathbf{L} \cdot \mathbf{U}$ given below.

$$\mathbf{A}$$

$$\begin{bmatrix}
1 & 2 & 3 & 4 & 5 & 6 \\
2 & 7 & 12 & 17 & 22 & 27 \\
4 & 13 & 28 & 43 & 58 & 73 \\
7 & 22 & 46 & 80 & 114 & 148 \\
11 & 34 & 70 & 120 & 185 & 250 \\
16 & 49 & 100 & 170 & 260 & 371
\end{bmatrix}$$

$$\mathbf{L} \qquad\qquad\qquad \mathbf{U}$$

$$= \begin{bmatrix}
1 & 0 & 0 & 0 & 0 & 0 \\
2 & 3 & 0 & 0 & 0 & 0 \\
4 & 5 & 6 & 0 & 0 & 0 \\
7 & 8 & 9 & 10 & 0 & 0 \\
11 & 12 & 13 & 14 & 15 & 0 \\
16 & 17 & 18 & 19 & 20 & 21
\end{bmatrix} \cdot \begin{bmatrix}
1 & 2 & 3 & 4 & 5 & 6 \\
0 & 1 & 2 & 3 & 4 & 5 \\
0 & 0 & 1 & 2 & 3 & 4 \\
0 & 0 & 0 & 1 & 2 & 3 \\
0 & 0 & 0 & 0 & 1 & 2 \\
0 & 0 & 0 & 0 & 0 & 1
\end{bmatrix}$$

Appendix E
Difference Formula for Calculation of the Electric Potential at Points Lying on the Border Between two Looseless Dielectric Media Without Electrical Charges

Let us consider the vector $\mathbf{D}$ of the electric induction determined at points $E \equiv (x_0 + h, y_0)$, $W \equiv (x_0 - h, y_0)$, $N \equiv (x_0, y_0 + h)$ and $S \equiv (x_0, y_0 - h)$ of the inhomogeneous dielectric area shown in Fig. E.1.

The components of this electric induction vector $\mathbf{D}$ at these particular points can be approximated by the following differences:

$$D_{xE} \approx 1 \cdot \varepsilon_0 \cdot E_{xE} = \varepsilon_0 \frac{U_E - U_0}{h}$$

$$D_{xW} \approx \varepsilon_r \cdot \varepsilon_0 \cdot E_{xW} = \varepsilon_r \cdot \varepsilon_0 \frac{U_0 - U_W}{h}$$

$$D_{yN} \approx \varepsilon_0 \frac{1 + \varepsilon_r}{2} E_{yN} = \varepsilon_0 \frac{1 + \varepsilon_r}{2} \cdot \frac{U_N - U_0}{h} \tag{E.1}$$

$$D_{yS} \approx \varepsilon_0 \frac{1 + \varepsilon_r}{2} E_{yS} = \varepsilon_0 \frac{1 + \varepsilon_r}{2} \cdot \frac{U_0 - U_S}{h}$$

The differences (E.1) make it possible replacing the differential equation $\nabla \cdot \mathbf{D} = \nabla \cdot (\varepsilon_r \varepsilon_0 \nabla U) = 0$, defined at point $O \equiv (x_0, y_0)$, by the following difference equivalent:

$$\nabla \cdot \mathbf{D}(x_0, y_0) = \frac{\partial D_x}{\partial x} + \frac{\partial D_y}{\partial y} \approx \frac{D_{xE} - D_{xW}}{2h} + \frac{D_{yN} - D_{yS}}{2h}$$

$$= \varepsilon_0 \frac{U_E - U_0}{2h^2} - \varepsilon_r \varepsilon_0 \frac{U_0 - U_W}{2h^2} + \varepsilon_0 \frac{(1 + \varepsilon_r)}{2} \cdot \frac{U_N - U_0}{2h^2} \tag{E.2}$$

$$- \varepsilon_0 \frac{(1 + \varepsilon_r)}{2} \cdot \frac{U_0 - U_S}{2h^2} = 0$$

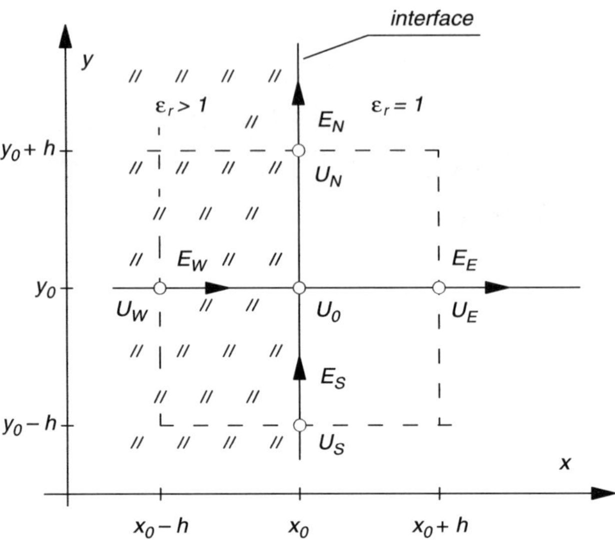

Fig. E.1

Performing some elementary transformations, Eq. (E.2) simplifies to the form:

$$U_0 = \frac{U_E + \varepsilon_r \cdot U_W}{2(1 + \varepsilon_r)} + \frac{U_N + U_S}{4} \qquad \text{(E.3)}$$

identical to relation (8.35).

Appendix F
Complete Elliptic Integrals of the First Kind

The complete elliptic integral of the first kind $K(k)$ is defined as:

$$K(k) = \int_0^1 \frac{\mathrm{d}t}{\sqrt{(1 - t^2)(1 - k^2 t^2)}} \tag{F.1}$$

where k ($0 \leq k \leq 1$) is the modulus of $K(k)$. The associated (complementary) integral $K'(k)$ is defined as:

$$K'(k) = K(k') \tag{F.2}$$

where $k' = \sqrt{1 - k^2}$ is the complementary modulus.

The calculation of integral $K(k)$, when its modulus k is known, is not complicated, and can be done by using the algorithm presented below. Hence, let us consider two infinite mathematical series (a_n) and (b_n) defined as:

$$
\begin{aligned}
a_0 &= 1 + k & b_0 &= 1 - k \\
a_1 &= \frac{a_0 + b_0}{2} & b_1 &= \sqrt{a_0 b_0} \\
&\cdots\cdots\cdots & &\cdots\cdots\cdots \\
a_{n+1} &= \frac{a_n + b_n}{2} & b_{n+1} &= \sqrt{a_n b_n}
\end{aligned}
\tag{F.3}
$$

Series defined in this manner converge to a common limit, usually denoted as $\mu(a_0, b_0) = \mu(k)$. Then:

$$\lim_{n \to \infty} (a_n) = \lim_{n \to \infty} (b_n) = \mu(k) \tag{F.4}$$

Finally, the value of integral $K(k)$ is related to limit $\mu(k)$ as follows:

$$K(k) = \frac{\pi}{2\mu(k)} \tag{F.5}$$

where $\pi = 3.141592653589\ldots$ The integral $K'(k)$ can be calculated in a similar way. Of course, in this case the limit $\mu(k')$ has to be calculated instead of $\mu(k)$. The series (a_n) and (b_n) are rapidly convergent, and in the most cases only a few iterations, for instance $n = 5$, must be taken into account. Some calculation results for the complete elliptic integrals of the first kind are given in Table F.1.

Table F.1

k^2	$K(k)$	$K'(k)$
0.00	1.570 796 326 794	$\rightarrow \infty$
0.01	1.574 745 561 317	3.695 637 362 989
0.10	1.612 441 348 720	2.578 092 113 348
0.50	1.854 074 677 301	1.854 074 677 301
0.90	2.578 092 113 348	1.612 441 348 720
0.99	3.695 637 362 989	1.574 745 561317
1.00	$\rightarrow \infty$	1.570 796 326 794

Subject Index